Lecture Notes
in Control and Information Sciences   353

Editors: M. Thoma, M. Morari

Claudio Bonivento, Alberto Isidori,
Lorenzo Marconi, Carlo Rossi (Eds.)

Advances in Control Theory and Applications

 Springer

Editors

Claudio Bonivento
Lorenzo Marconi
Carlo Rossi

DEIS-CASY
University of Bologna
Viale Risorgimento, 2
40136 Bologna - Italy
Email: cbonivento@deis.unibo.it
lmarconi@deis.unibo.it
crossi@deis.unibo.it

Alberto Isidori

Dipartimento di Informatica e Sistemistica
Sapienza - Università di Roma
Via Eudossiana 18
00184 Rome Italy

Library of Congress Control Number: 2007920184

ISSN print edition: 0170-8643
ISSN electronic edition: 1610-7411
ISBN-10   3-540-70700-X Springer Berlin Heidelberg New York
ISBN-13   978-3-540-70700-4 Springer Berlin Heidelberg New York

Springer is a part of Springer Science+Business Media
springer.com
© Springer-Verlag Berlin Heidelberg 2007

Typesetting: by the authors and SPS using a Springer LaTeX macro package

Printed on acid-free paper   SPIN: 11930259   89/SPS   5 4 3 2 1 0

Preface

This volume can be considered a direct outcome of the special scientific "meeting-in-the-fortress" on "Advances in Control Theory and Applications" organized in Bertinoro, Italy, by the Centre of research on Complex Automated Systems (CASY), Department of Electronics Computer and Systems of the University of Bologna, during the week May 22–26, 2006. The inspiring idea of that workshop was to provide a forum for exchange of ideas between theory-oriented and application-oriented researchers working on various systems and control problems. The meeting offered an opportunity for formal presentations of research results as well as for informal discussions about ideas and problems, case-studies, limitations and potentials of existing and emerging theories. The main goal of the meeting was to facilitate cross-fertilization between different theoretical and applicative areas. Emphasis was put on identification of new theoretical developments and research directions, as needed by recent progresses in applications and problems which are still looking for a theoretical support and effective rigorous solutions. The technical programme consisted of twenty-five main lectures delivered by distinguished scholars and was complemented by a number of poster presentations prepared by post doctoral fellows and PhD students currently working at CASY. Out of the twenty five lectures given in Bertinoro, fifteen are reported here in written form. They are organized as separate contributions and listed according to the alphabetic order of the first author, as follows.

Modeling and Control of Autonomous Helicopters by Manuel Béjar, Anibal Ollero, Federico Cuesta, presents an overview on the modeling and model-based control of autonomous helicopters.

Efficient Quantization in the Average Consensus Problem by Ruggero Carli, Sandro Zampieri deals with the average consensus problem where a set of linear systems has to be driven to the same final state which corresponds to the average of their initial states.

Human-Robot Interaction Control Using Force and Vision by Agostino De Santis, Vincenzo Lippiello, Bruno Siciliano, Luigi Villani focuses on techniques for augmenting safety by means of control systems, starting from the idea of mimicking sensing and actuation of humans.

A Dissipation Inequality for the Minimum Phase Property of Nonlinear Control Systems by Christian Ebenbauer, Frank Allgöwer discusses a new characterization of the minimum phase property of nonlinear control systems in terms of a dissipation inequality.

Input disturbance suppression for port-Hamiltonian systems: an internal model approach by Luca Gentili, Andrea Paoli, Claudio Bonivento presents a comprehensive port-Hamiltonian systems framework to deal with input disturbance suppression problems.

A Systems Theory View of Petri nets by Alessandro Giua, Carla Seatzu focuses on Petri nets as a family of powerful discrete event models whose interest has grown in parallel with the development of the theory of discrete event systems.

Wireless Sensing with Power Constraints by Orhan C. Imer, Tamer Başar introduces two conceptual models for wireless sensing and control with power-limited sensors and controllers.

The Important State Coordinates of a Nonlinear System by Arthur J. Krener offers an alternative way of evaluating the relative importance of the state coordinates of a nonlinear control system.

On Decentralized and Distributed Control of Partially-Observed Discrete Event Systems by Stéphane Lafortune surveys recent work of the author with several collaborators on decentralized control of discrete event systems.

A Unifying Approach to the Design of Nonlinear Output Regulators by Lorenzo Marconi, Alberto Isidori aims to propose a unique vision able to frame a number of results recently proposed in literature to tackle problems of output regulation for nonlinear systems.

Controller Design through Random Sampling: an Example by Maria Prandini, Marco C. Campi, Simone Garatti presents the 'scenario approach', an innovative technology for solving convex optimization problems with an infinite number of constraints.

Digital Control of High Performance Power Supplies for a Synchrotron Light Source by Carlo Rossi, Andrea Tilli, Manuel Toniato discusses some aspects of an advanced control strategy for a class of quadrupole magnet power supply, where variable output current has to be imposed.

Distributed PCHD-Systems, from the Lumped to the Distributed Parameter Case by Kurt Schlacher extends the Hamiltonian approach to a class of distributed parameter Hamiltonian systems, which preserves some useful properties of the well known class of Port Controlled Hamiltonian systems with dissipation.

Observability and the Design of Fault Tolerant Estimation Using Structural Analysis by Marcel Staroswiecki presents a structural analysis approach for the design of fault tolerant estimation algorithms.

Robust hybrid control systems: an overview of some recent results by Andrew R. Teel gives an overview of a new framework for analyzing hybrid dynamical systems.

We are grateful to all the outstanding colleagues and friends who accepted to participate to the Bertinoro workshop and to contribute to the success of that initiative with inspiring presentations, fruitful interactions and technical discussions, namely Frank Allgöwer, Karl Åström, Tamer Başar, Marco Campi, Tryphon Georgiou, Alessandro Giua, Lino Guzzella, Arthur Krener, Stéphane Lafortune, Manfred Morari, Steve Morse, Anibal Ollero, Laurent Praly, Anders Rantzer, Giorgio Rizzoni, Kurt Schlacher, Bruno Siciliano, Marcel Staroswiecki, Andrew Teel, Roberto Tempo, Arijan van der Schaft, Yutaka Yamamoto, Sandro Zampieri. We warmly thank in particular those of them who spend time in addition in order to prepare their revised written texts collected in this volume. We are sure that this effort will be useful for many young scientists and skilled professionals operating in different technical areas around the world.

We are indebted with many individuals and institutions for their support and help. In particular, we thank Manfred Morari who promptly accepted our idea of publishing this book in the LNCIS series, Thomas Ditzinger and Heather King, Engineering Editorial of Springer-Verlag, for the precious assistance, Roberto Naldi for the accurate editing work. Finally, the funding supports given by the Institue of Advanced Studies and the Department DEIS both of the University of Bologna, and the hospitality offered by the Bertinoro Residential Centre are gratefully acknowledged.

Bologna,
28 November, 2006

Claudio Bonivento
Alberto Isidori
Lorenzo Marconi
Carlo Rossi

Contents

Modeling and Control of Autonomous Helicopters

Manuel Béjar[1], Anibal Ollero[2], and Federico Cuesta[2]

[1] Universidad Pablo de Olavide
   mbejdom@upo.es
[2] Universidad de Sevilla
   {aollero,fede}@cartuja.us.es

Summary. This chapter presents an overview on the modeling and model-based control of autonomous helicopters. Firstly it introduces some of the platforms and control architectures that has been developed in the last 15 years. Later, the Chapter considers the modeling of the helicopter and the identification techniques. Then, it overviews different linear and non-linear model-based control approaches. This section also includes experiments on the control of the helicopter vertical motion that illustrate the presented techniques and point out the interest of nonlinear analysis methods to study the dynamic behavior of the helicopter. Finally, the Chapter presents open research lines coming from two challenging applications: the autonomous landing in oscillating platforms and the lifting and transporting of a single load with several helicopters.

Keywords: Autonomous Helicopter, Helicopter Modeling and Identification, Autonomous Helicopter Control, Autonomous Landing.

1 Introduction

In the last decade Unmanned Aerial Vehicles (UAVs) have attracted a significant interest. UAVs avoid the human risk inherent to human-piloted aerial vehicles, particularly in missions in hostile environments, and they can be smaller and more maneuverable. The exploitation costs can be also lower that the in manned aircrafts.

UAVs have been widely used for military applications. Recently, the evolution of UAV technologies, the miniaturization of the sensors and cameras and the new advances in communication and control systems, point to a wide range of civilian applications such as natural disasters, inspection, search and rescue, traffic surveillance and law enforcement.

Remotely piloted and autonomous helicopters have been extensively used for applications involving aerial and lateral views including aerial photography, cinematography, inspection and other aerial robotic applications. The maneuverability and hovering ability of helicopters and other VTOL design are main requirements in many of these applications. However, helicopters are more difficult to control than fixed wing aircrafts. In fact, they require critically stabilization loops which are coupled to the displacement behaviors.

C. Bonivento et al. (Eds.): Adv. in Control Theory and Applications, LNCIS 353, pp. 1–29, 2007.
springerlink.com   © Springer-Verlag Berlin Heidelberg 2007

Fig. 1 shows the reference systems used in helicopter control. The position and orientation of a helicopter is usually controlled by means of 5 control inputs: the main rotor collective pitch which has a direct effect on the helicopter height (z axis in the X-Y-Z system); the longitudinal cyclic which modifies the helicopter pitch angle (rotation about the y^b axis in the $x^b - y^b - z^b$ system) and the longitudinal translation; the lateral cyclic, which affects the helicopter roll angle (rotation about the x^b axis in $x^b - y^b - z^b$ system) and the lateral translation; the tail rotor which controls the heading (yaw motion) of the helicopter (rotation about the z^b axis in $x^b - y^b - z^b$ system) and compensates the anti-torque generated by the main rotor; and the throttle control. It is a multivariable non-linear system with strong coupling in some control loops.

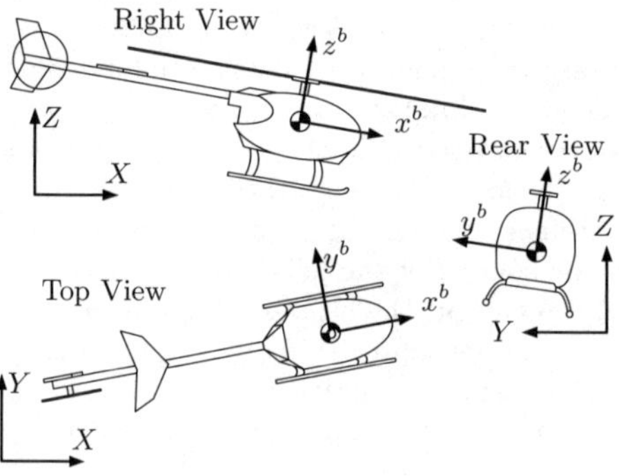

Fig. 1. Inertial (XYZ) and body ($x^b y^b z^b$) coordinate systems of the helicopter

Autonomous helicopter control has been a well known control benchmark. Different approaches can be used including model based control and other approaches based on learning from human operators and rule-based techniques. Thus, fuzzy logic with rules generated by the observation of a human pilot and consultation with helicopter experts is the approach used in [5]. In [30] PD control loops with gains tuned by trial and error are implemented. In [29], the controller is generated by using training data gathered while a human teacher controls the helicopter. In [4] learning is based on the direct mapping of sensor inputs to actuator control via an artificial neural network. Then, the neural network controller was used for the helicopter hovering. The analysis of the pilot's execution of aggressive manoeuvres from flight test data is the base of the method presented in [12] to develop a full-non-linear dynamic model of a helicopter. In this Chapter only model-based analysis and control techniques are considered.

Section 2 of this Chapter introduces some of the platforms that have been used for the experimentation of control techniques and also the control architectures developed for aerial robotics. Section 3 is devoted to modeling including model development and identification. Section 4 deals with model based control techniques and Section 5 points to open research lines. Finally, sections 6 and 7 are devoted to the Conclusions and References.

2　Platforms

Autonomous helicopters are very valuable platforms for aerial robotic. Thus, many different Universities and Research Centers have developed experimentation platforms since the beginning of the nineties. The usual approach has been the adaptation of remotely piloted helicopters that are available in the hobby and aerial photography market. Furthermore, commercial platforms designed and implemented by companies have been also used. The most well known and widely used autonomous helicopters are the Yamaha R50 and Rmax platforms (see Fig. 2). They have been commercially used in Japan for crop spraying. Other commercial platforms developed mainly for military applications are the Fire Scout from Northrop Grumman and the Camcopter from Schiebel.

Fig. 2. Yamaha Rmax platform

The Robotics Institute at Carnegie Mellon University (CMU) conducted since the early nineties an autonomous helicopter project. They have developed different prototypes from small electrical radio controlled vehicles to autonomous helicopters using the Yamaha R50 platform. The autonomous CMU helicopter won the AUVSI aerial robotic competition in 1997.

The University of Southern California (USC) carried out several autonomous helicopter projects since 1991, developing prototypes, such as the AVATAR (Autonomous Vehicle Aerial Tracking and Retrieval/Reconnaissance) prototypes presented in 1994 and 1997. The AVATAR helicopter won the AUVSI Aerial Robotics competition in 1994.

The University of Berkeley also developed autonomous helicopters in the Berkeley AeRobot project, BEAR, in which the autonomous aerial robot is a test bed for an integrated approach to intelligent systems.

The Georgia Institute of Technology (GIT) has the Unmanned Aerial Vehicle Research facility and developed several platforms and aerial autonomous systems during the last decade. GIT also won the AUVSI aerial robotics competition.

In Europe the University of Linköping leaded the WITAS project which was a long term basic research project involving cooperation with other Universities and private companies [8]. The Yamaha Rmax helicopter was used for demonstration in the WITAS project. Moreover, several Universities such as the

Technical University of Berlin, ETH Zurich [9], and Universidad Politécnica de Madrid [1] are using the adaptation of conventional radio controlled helicopters with different autonomous capabilities.

Fig. 3 shows MARVIN developed by the Technical University of Berlin [33], which won the AUVSI Aerial Robotics Competition in 2000. Later, they developed Marvin II in the framework of the COMETS European IST project on the coordination and control of multiple heterogeneous vehicles (http://grvc.us.es/comets). These helicopters are shown in Fig. 3.

Marvin I

Marvin II

Fig. 3. The Marvin autonomous helicopters flying in experiments of the COMETS project

The GRVC group of the University of Seville (http://grvc.us.es) developed several teleoperated and autonomous platforms in the framework of the above mentioned COMETS project, which was coordinated by the researchers of this University, and the Spanish CROMAT project on the coordination of aerial and ground robots (http://grvc.us.es/cromat). Fig. 4 shows some of these platforms.

Control Architectures

On board control architectures for UAV integrate a variety of sensor information including GPS, 3-axis rate gyro, 3-axis accelerometer, aircraft attitude reference sensor, compass, altitude sensors among others. Furthermore, low level motion servo-controllers are implemented to control the vehicle typically in different control modes. Intelligent control architectures also include environment perception, object tracking, and local reactive (obstacle avoidance) and planning capabilities. The on-board control hardware is linked to an operator ground controller which is used to send commands and GPS corrections to the on-board controller and to visualize information transmitted from the UAV. In many projects these controllers are now implemented by means of laptops.

The University of Southern California (USC) developed a behavior-based architecture for the control of the AVATAR autonomous helicopter [10]. The low-level behaviors correspond to the generation of the four input commands of the helicopter (collective throttle, tail rotor, longitudinal and lateral cyclic). The second level implements short-term goal behaviors: transition to altitude and

HELIV jointly developed by the University of Seville and Helivision in the COMETS project.

HERO 1 developed by the University of Seville (2004)

HERO 3 developed by the University of Seville in the CROMAT project

Fig. 4. Helicopters developed at the University of Seville

lateral velocity. The highest-level behavior, navigation control, is responsible for long-term goals such as moving to a particular position and heading.

Intelligent control architectures for unmanned air vehicles (helicopters) are also researched at Berkeley. The hierarchical architecture segments the control tasks into different layer of abstraction in which planning, interaction with the environment and control activities are involved. The hierarchical flight management system [19] has a stabilization/tracking layer, a trajectory generation layer, responsible for generating a desired trajectory or a sequence of flight modes, and a layer which switches between several strategy planners. Both continuous and discrete event systems are considered. In order to model these control systems, hybrid system theory has been proposed (see for example [21]).

GIT also developed autonomous helicopter control systems and research in flight controls, avionics and software systems.

The control architecture developed at the University of Seville includes a low level control system based on the DSP TMI2812 and a PC104 to implement complex control strategies eventually involving environment perception functions. Several control strategies can be implemented including manual guidance with automatic stabilization and hovering, and fully autonomous flight.

To conclude this section it should be noted that a practical difficulty in most autonomous helicopter projects is the need of experienced pilots for their development and application. Other relevant issues are the following:

- Strong need of mechanical maintenance and testing of platforms, particularly in the low cost platforms built by adapting small radio control helicopters.
- Relevance of the weight and power consumption particularly in small helicopters. This issue imposes strict requirements on the hardware to be used on-board. Thus, on-board UAV control hardware is an ideal application for new embedded control systems involving microcontrollers, DSPs, and embedded PCs with real-time Operating Systems.
- Relevance of the mechatronic design involving mechanical, sensing and control joint design.

- Strict safety and reliability constraints imposing extensive testing before implementation which may involve the application of hardware in the loop techniques.

3 Modeling

Modeling UAV dynamics is a challenging research area. The full model of a helicopter, including flexibility of the rotors and fuselage, dynamics of the actuators and combustion engine, is very complex.

In most cases, the helicopter is considered as a rigid body, whose inputs are forces and torques applied to the center of mass and whose outputs are the linear position and velocity of the center of mass, as well as the rotation angles and angular velocities.

Furthermore, the relations between the actual control inputs of the helicopter and the above mentioned forces and torques should be considered in the model. In general, these relations involve the aerodynamics of the fuselage and the effect of stabilizers. However, it has been pointed out that these stabilizers effects can be ignored at low speeds [22].

In [20] a mathematical model and its experimental identification for a model helicopter are presented. The model of the interactions between the stabilizer flybar and the main rotor blade is also included, showing its effects on the stability of the model helicopter. The identification from input-output data, collected when a human pilot is controlling the vehicle, is difficult because it is not possible to study the individual effect of each control input (the pilot has to apply more than one input to maintain the stability). To overcome this, the identification of the parameters is performed on a SISO basis, using four specially-built stands to restrict the motion of the helicopter to one degree of freedom. For example, one of these stands only allowed vertical motion. Thus only the main rotor collective input was excited and only the vertical displacement was measured. In these conditions, the simplified linear transfer SISO function from collective input to the vertical motion was identified using standard identification techniques.

In [28] a parameterized model of the Yamaha R-50 autonomous helicopter is identified using frequency domain methods. The stabilizer bar is also taken into account. The model was validated with special flight experiments using doublet-like control inputs in hover and forward flight, showing its ability to predict the time domain response of the helicopter to control inputs. At CMU, a high-order linear model of the R-50 Yamaha helicopter is used for control. This model was extracted by using the MOSCA (Modeling for Flight Simulation and Control Analysis) with a non-linear simulation model of the helicopter [25].

The work in [7] also presents a complete nonlinear helicopter model. Modeling is based on the Blade Element Momentum Theory which is a combination of the Blade Element Theory and Momentum Theory. Aerodynamic effects such as effective translational lift, traverse flow and ground effect, are considered. In contrast to previous works, the author highlights the presence of gyroscopic effect in pitching and rolling movements of rotors. The influence of wind gust is also studied. It shows the strong effect of wind disturbances on lateral position and how the altitude is also affected due to the Effective Translational lift. The complete model has been validated by means of several experiments with the MARVIN I autonomous helicopter, shown in Fig. 3.

3.1 Model Development

Generally, the model of a helicopter can be divided into five different subsystems: Servo Dynamics, Engine Dynamics, Aerodynamics (Main Rotor, Tail Rotor and Fuselage), Force and Moment Generation, and Rigid Body Dynamics. Connections between these subsystems are shown in Fig. 5. In this figure, Pc (main rotor collective pitch), Pt (tail rotor collective pitch), Px (longitudinal cyclic pitch), Py (lateral cyclic pitch), and Pth (Throttle control), are the input signals mentioned in Section 1 of this Chapter. $Pcmd$ comprises the five corresponding commanded signals before actuator dynamics.

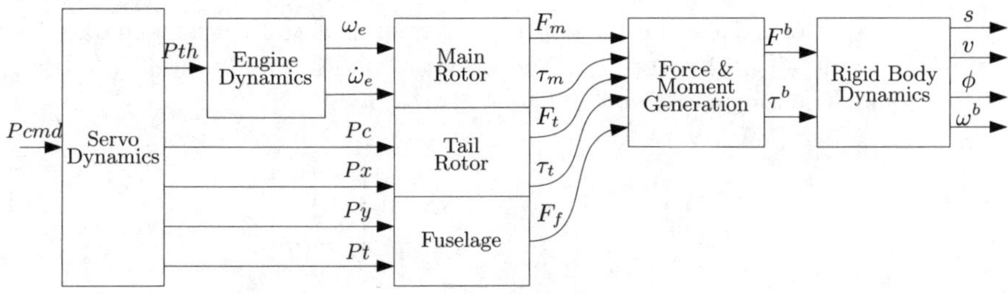

Fig. 5. Helicopter Dynamics

Likewise, $s = \begin{bmatrix} x\ y\ z \end{bmatrix}^T$ and $v = \begin{bmatrix} vx\ vy\ vz \end{bmatrix}^T$ are, respectively, the linear position and velocity of the helicopter. On the other hand, $\phi = \begin{bmatrix} \phi_x\ \phi_y\ \phi_z \end{bmatrix}^T$ are the Euler angles whilst $\omega^b \in \mathbb{R}^3$ is the angular velocity in the body frame.

Rigid Body Dynamics

By regarding the helicopter as a rigid body, equations of motion of a model helicopter can be derived by applying Newton-Euler equation. Thus, the translational and rotational movements of a rigid body are described by the conservation of linear and angular momentum. For translation this yields to

$$m\dot{v} = F \tag{1}$$

where m is the helicopter mass and $F \in \mathbb{R}^3$ is the force applied to the center of mass. All the vectors are given in the inertial frame.

For rotation, it is easier to switch to the body fixed frame. In this case the equation of motion yields to

8   M. Béjar, A. Ollero, and F. Cuesta

$$J\dot{\omega}^b + \omega^b \times J\omega^b = \tau^b \tag{2}$$

where $J \in \mathbb{R}^{3\times3}$ is the inertial matrix and $\tau^b \in \mathbb{R}^3$ the torque applied to the helicopter body.

The transformation from body to inertial frame can be parameterized by Euler angles ϕ and yields to a rotation matrix $R(\phi)$ that makes possible the conversion $F = R(\phi)F^b$. By differentiating $R(\phi)$ with respect to time, state equations of the Euler angles are obtained $\dot{\phi} = \hat{\Psi}(\phi)\omega^b$.

Therefore, the motion equations of a rigid body can be written as:

$$\begin{bmatrix} \dot{s} \\ \dot{v} \\ \dot{\phi} \\ \dot{\omega}^b \end{bmatrix} = \begin{bmatrix} v \\ \frac{1}{m}R(\phi)F^b \\ \Psi(\phi)\omega^b \\ J^{-1}(\tau^b - \omega^b \times J\omega^b) \end{bmatrix} \tag{3}$$

Force and Moment Generation

The force experienced by the helicopter is the sum of the thrust generated by main (F_m) and tail (F_t) rotors and the gravitational force. An additional term F_f for the air resistance of the fuselage is also added:

$$F^b = F_m + F_t + F_f + R^T(\phi)\begin{bmatrix} 0 \\ 0 \\ -mg \end{bmatrix} \tag{4}$$

The torque is composed of the torques generated by the main rotor (τ_m), tail rotor (τ_t) and the torques generated by the forces, since their point of attack is displaced from the center of mass:

$$\tau^b = \tau_m + \tau_t + r_m \times F_m + r_t \times F_t \tag{5}$$

where r_m, r_t are the positions of the main and the tail rotors and g is the acceleration of the gravity.

Aerodynamics

Fuselage Air Resistance

Air resistance of the fuselage (F_f) can be approximated by the 3×1 vector

$$F_f = \begin{bmatrix} K_{f_x} v^b_{a_x} |v^b_{a_x}| \\ K_{f_y} v^b_{a_y} |v^b_{a_y}| \\ K_{f_z} v^b_{a_z} |v^b_{a_z}| \end{bmatrix} \tag{6}$$

where K_{f_i} are aerodynamical parameters, v^b_a is the air velocity in body fixed frame and $|-|$ denotes absolute value.

Dynamics of a General Rotor

In this subsection, equations for a general rotor with collective and cyclic pitch inputs are derived [6] [7]. Subsequently, these equations should be applied to each rotor to obtain the forces (F_m, F_t) and the torques (τ_m, τ_t) indicated in Equations (4) and (5).

Forces and torques generated by a rotor will be calculated as mean values of the forces of a single rotating blade. Due to wind and pitch, these forces will be changing during one cycle. Therefore, integrating these forces along each blade and for one cycle will yield to the mean lift force F_L and the mean drag M_D, both along z^r axe of the rotor disk, and the mean torques M_x, M_y along axes x^r and y^r of the rotor disk, respectively.

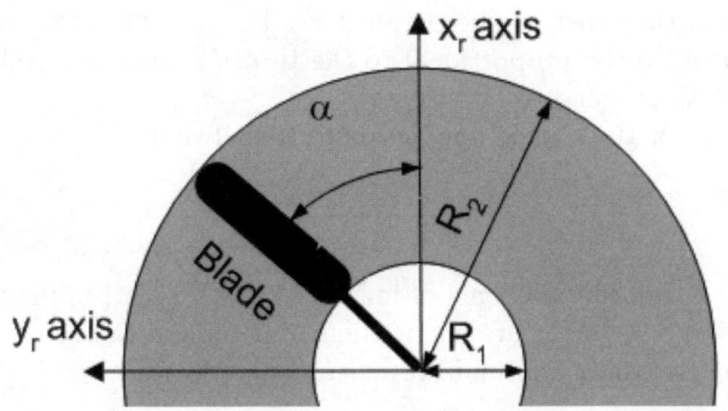

Fig. 6. Rotor disk for integration

Mean lift force (F_L). For the determination of mean lift, all lift components of each rotor element are integrated according to Fig. 6. Blades are assumed to be non-twisted and of constant chord from radius R_1 to R_2.

The mean lift yields to

$$F_L = \int_0^{2\pi} \int_{R_1}^{R_2} F_L' \, dr \, d\alpha \tag{7}$$

where F_L' depends on several values such as the pitch of the blade and the air velocity v_r relative and orthogonal to the blade.

Mean torque (M_x and M_y). Torques are produced by the same forces F_L' as lift. Since these forces are applied asymmetrically, two different torques are produced along axes x^r and y^r. To obtain these toques, forces must be integrated as follows:

$$M_x = \int_0^{2\pi} \int_{R_1}^{R_2} F_L' r^2 sin(\alpha) \, dr \, d\alpha \tag{8}$$

$$M_y = \int_0^{2\pi} \int_{R_1}^{R_2} -F_L' r^2 cos(\alpha) \, dr \, d\alpha \tag{9}$$

Mean drag (M_D). Estimating drag of a rotor is very similar to estimating lift. In this case, the second force exerting on the blade (drag force F'_D) is integrated along each blade and for one cycle. Thus, the mean drag of the rotor can be expressed as

$$M_D = \int_0^{2\pi} \int_{R_1}^{R_2} F'_D r^2 dr d\alpha \tag{10}$$

where the drag force F'_D is assumed to be proportional to the square of the pitch.

Engine Dynamics

For the dynamics of the main engine a first order linear differential equation is considered. As a simplification, it is assumed that the engine (ω_e) and the main rotor (ω_m) have the same angle velocity. This means that the real ratio is identified within the other engine parameters. The torque M_e generated by the engine is assumed to be proportional to the throttle input p_{th} and the friction M_g of the gear is assumed to be viscose.

The equation for the engine now yields to (recall that $\omega_e = \omega_m$):

$$\dot{\omega}_e = \frac{M_e - M_{D_m} - \eta_t M_{D_t} - M_g}{J_m + \eta_t J_t + J_g} \tag{11}$$

where J_m, J_t, J_g are the moments of inertia for main rotor, tail rotor and gear, respectively, M_{D_m} and M_{D_t} are mean drags of main and tail rotor and η_t is the ration between tail rotor and main rotor angular velocities.

Servo Dynamics

For each control servo, the same first order behavior is assumed:

$$\dot{P}i = -\alpha Pi + \alpha PCMDi \quad i = c, x, y, t, th \tag{12}$$

where α is a parameter representing the damping behavior of the servo and $PCMDi$ is the commanded value for the servo i.

3.2   Model Identification

The main difficulties of helicopter model identification are due to the instability and coupling inherent characteristics associated to the system. These inconveniences can be avoided with some modifications of classic identification approaches.

Mathematically, the identification task can be formulated as an optimization problem:

$$\min_{\Theta} F(\Theta, \Gamma), \tag{13}$$

where Θ is the set of parameters to be identified, Γ the captured reference flight (Γ_u inputs, Γ_x states, Γ_y outputs) and F a fitness function which is decreasing for a better approximation of the model.

In order to obtain data for solving this optimization problem, it will be necessary to record the helicopter responses and the corresponding control signals given by a pilot at a reference flight.

Rest of this section is devoted to illustrate an identification tool for autonomous helicopters recently developed by the GRVC.

Fitness Function

To define the fitness function there are different methods whose applicability depends on the characteristics of the model to be identified.

For further discussions we assume a general non-linear model in state-space-form

$$\dot{x} = f(x, u), \tag{14}$$

$$y = h(x), \tag{15}$$

where $x \in \mathbb{R}^n$ is the state vector, $u \in \mathbb{R}^p$ the control vector and $y \in \mathbb{R}^q$ the observation vector.

The parameters Θ are included in the system function $f(x, u)$ and the observation function $h(x)$. To compare a captured reference flight Γ with the identified model response, the model has to be simulated with the same control signal Γ_u and the same initial condition $\Gamma_x(t_0)$.

$$\dot{x} = f(x, \Gamma_u), \quad x(t_0) = \Gamma_x(t_0), \tag{16}$$

$$y = h(x).$$

Output Error Method (OEM)

Fitness is seen as the error between the simulated and the actual trajectories of the outputs measured in the system.

In particular, the parameter fitness can now be written as a weighted mean-square-error:

$$F_{OEM}(\Theta, \Gamma) = \sum_{k=1}^{N} (y(t_k) - \Gamma_y(t_k))^T W (y(t_k) - \Gamma_y(t_k)) \tag{17}$$

where N is the number of data points and W a weighting matrix.

This method is the most widely used, since it has many desirable statistical properties. However it poses difficulties when applied to inherent unstable systems. In this case, numerical integration leads to diverging solutions even if the correct parameters are used, due to the open-loop character of the simulation.

Stabilized Output Error Method (SOEM)

The instability caused by numerical divergence in OEM can be solved by incorporating stabilization using some states that can be measured.

The fitness function to be minimized is the same used in the normal OEM and is given above by equation (17). Only the simulation structure (Fig. 7) is modified by replacing some states with those of the measured trajectory. Thus $\hat{x}^{(l)}$ is the combination of measured states Γ_x and simulated states x.

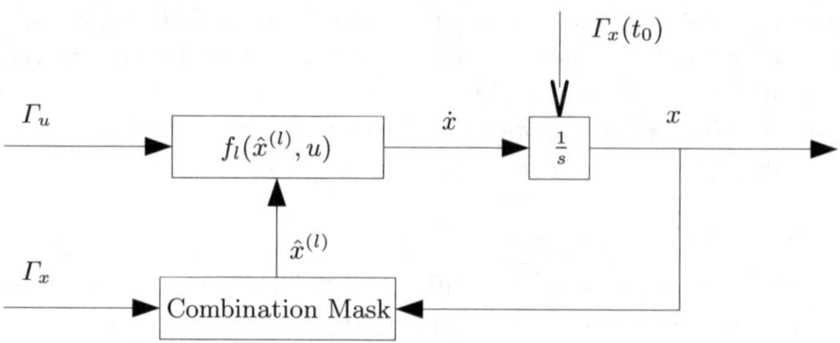

Fig. 7. Structure of a stabilized simulation used with the SOEM

Preprocessing

Parameter Range

Before applying optimization algorithms, the proper choice of the helicopter parameters range is crucial for a successful identification. If the range is chosen too small, the parameters may be outside, and if it is too large the search may take very long or will not converge. This choice of the ranges can be challenging and requires good knowledge of the helicopter model.

Reference Flights

Another important step is the specification of the reference flights. On the one hand, sometimes several parameter combinations could adapt well to a particular reference flight; to overcome this, some flights are captured with a known parameter offset, like mounting a mass with known weight and moments of inertia. On the other hand, trajectories commanded in reference flights should be those that excite as much helicopter dynamics as possible. This last specification can be tested performing a sensitivity analysis for each parameter.

Parameter Sensitivity

Analyzing the sensitivity of the fitness function with respect to each parameter gives an estimate of how every parameter is involved in the behavior of the helicopter and therefore how precise each can be identified. A sensitivity analysis is performed for a specific reference flight, assuming a set of parameters. The data corresponding to the reference flight are used to evaluate the fitness function while varying each parameter around its assumed default values. Due to neglecting the coupling between the parameters, the result is only an estimate, but it helps to get a feeling of how suitable a particular maneuver will be to identify all the parameters.

Fig. 8 shows a sensitivity analysis. The vertical line represents the default value of the parameter and the curve shows the fitness while modifying each parameter.

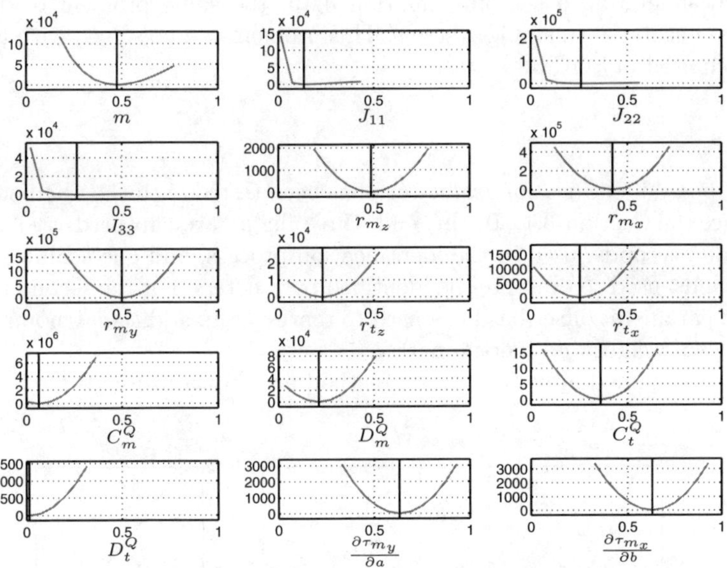

Fig. 8. Sensitivity analysis for a particular manoeuver. Each subgraph shows the variation of the fitness function (gray line) while modifying a parameter around its default value (black vertical line). One possible conclusion is that parameters C_t^Q and D_t^Q have small influence in this manoeuver (and therefore, it would be more complicated to identify them) since their fitness variation is less than in other cases.

Optimization Algorithm

To solve Equation (13), the functions provided by [27] have been implemented. These functions are designed to find the minimum of a constrained nonlinear multivariable function of several variables. In our case, the multivariable function is the fitness described in Equation (17) and the parameters values associated to the minimum are those that make the model response be more similar to the reference flight.

Notice that an observed limitation of these functions is that might only give local solutions. This highlights the importance of the remarks made previously about parameter range choice.

Validating the Model

After having found an optimal set of parameters, it is desirable to see how good this model will predict different behaviors of the real helicopter. This can be done by plotting and comparing the real and the simulated trajectory.

The set of data used in the identification process (reference data) should not be the same that is used in this validation process (evaluation data). This way, parameter adaptation to a particular set of data is avoided.

When simulating a set of evaluation data, the same problem of diverging solutions occurs for unstable systems. This problem can be solved with the same approach used in SOEM.

Results

The proposed identification framework has been tested with the reference model introduced in Section 3.1. To this end, three flights are captured: two reference flights of 4 seconds, used in the identification process, and one evaluation flight of 6 seconds, used to evaluate the identified parameters. For the second reference flight, a parameter offset has been used to represent an additional mounted mass with known weight and inertia matrix.

Results obtained applying SOEM method are shown in Fig. 9.

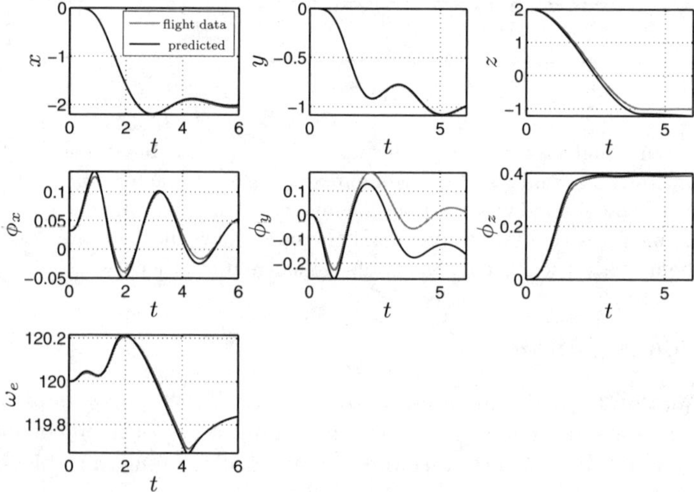

Fig. 9. Comparison between actual evaluation data and simulation data with parameters identified with SOEM

4   Control Techniques

In this section model-based control techniques are considered. Firstly, a general perspective is given. Then, the model described in previous sections is used to design control strategies that illustrate the concepts explained in the mentioned overview.

Although a helicopter is a coupled nonlinear multivariable and underactuated system, simplification of some coupling terms leads to a first simplified scheme of main relations between input-ouput variables of Fig. 5, as shown in Table 1.

Notice that translational variables are expressed in the body coordinate frame defined in Fig. 1. This set of relationships is the base of the typical control

Table 1. Basic input-output relations in a helicopter

Control Input	Translation	Rotation
Pc	z^b	-
Pt	-	ϕ_z
Px	x^b	ϕ_y
Py	y^b	ϕ_x

scheme shown in Fig. 10. This control scheme not only takes into account the main relationships in Table 1 but also considers the most important couplings, such us the lateral and longitudinal movement effect on vertical dynamics.

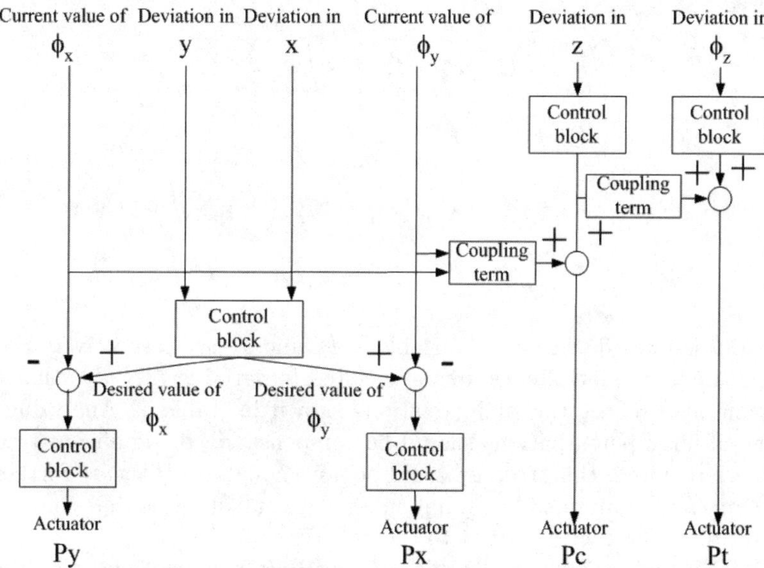

Fig. 10. Basic control scheme

4.1 Model-Based Techniques

Both linear and nonlinear control strategies have been applied to autonomous helicopters. However, even when linear control laws are applied to the inputs defined in Fig. 5, some authors add nonlinear transformations in particular conditions. Thus, in [13] a nonlinear rotation matrix is considered to deal with yaw angles not equal to zero. This rotation matrix converts x and y deviations (global system) into x^b and y^b deviations (local system).

In [36] linear robust multivariable control, fuzzy logic control and nonlinear tracking control are compared in the simulation of two scenarios: vertical climb and simultaneous longitudinal and lateral motion. In order to design the multivariable linear control law, Equations (3) together with the corresponding nonlinear expressions for forces and torques given by (4) and (5), are linearized. Fuzzy control is based on the generation by means of fuzzy logic of the parameters of four separated PID chains, which correspond to each of the inputs of Fig. 5, excluding Pth since engine dynamics are not modeled. The nonlinear approach consists of feedback linearization, assuming the simplification of the coupling effects between forces and torques of Equations (4) and (5). This approach is shown to be more general and to cover wider ranges of flight envelopes. However, it also requires accurate knowledge about the system and is sensitive to model disparities, such as changes in the payload, or to the aero-dynamic thrust-torque model.

In [19] multiloop linear PID techniques obtained good results when applied to the Yamaha R-50. Using variables of Fig. 5, the control laws can be outlined as follows:

$$Pc(t) = -P_{v_z} v_z(t) - P_z e_z(t) - I_z \int e_z(t) dt$$

$$Pt(t) = -P_{\phi_z} \phi_z(t) - I_{\phi_z} \int e_{\phi_z}(t) dt$$

$$Px(t) = -P_{\phi_y} \phi_y(t) - P_{v_x} v_x(t) - P_x e_x(t) - I_x \int e_x(t) dt$$

$$Py(t) = -P_{\phi_x} \phi_x(t) - P_{v_y} v_y(t) - P_y e_y(t) - I_y \int e_y(t) dt \tag{18}$$

where $e_i(t)$ denotes deviation in variable i. P_j and I_j are respectively proportional and integral control constants associated to variable j. Notice that these expressions also reflect the main relations shown in Table 1. Analyzing performance, if large perturbations should be compensated, or significant tracking abilities are required, this strategy could be not enough. In this case further improvements can be obtained by adding nonlinear control terms that compensate significant deviations with respect to the hovering conditions.

At CMU a high-order linear model of the R-50 Yamaha helicopter is used for control in [23] [24]. The controller consists of one multivariable (MIMO) inner loop for stabilization and four separate (SISO) guidance loops for velocity and position control. Several manoeuvre tests have been conducted with the helicopter (square, forward turn, backward turn and nose-out circle). The controller is designed for hovering but its robustness leads the helicopter to perform the manoeuvres efficiently even if the trajectories are not optimal. Videos and further information can be accessed in http://www.roboticflight.org.

The work at [18] aimed at achieving aggressive manoeuvrability at the level of attitude angles, whose dynamics are defined by the fourth element of Equation (3). However the authors do not consider the contribution of tail rotor force in terms of the tail rotor torque, shown in (5). Likewise, engine dynamics block in Fig. 5 are not considered, which implies the assumption of a constant angular velocity ω_m. After these assumptions, the nonlinear model is approximated by a Takagi-Sugeno fuzzy model, which boils down to convex combination of linear submodels. Finally, a gain scheduled output feedback $H\infty$ controller for the approximated model is designed.

Fig. 11 outlines the control scheme adopted in [17] and [32], using the same notation of Fig. 5 for inputs and outputs.

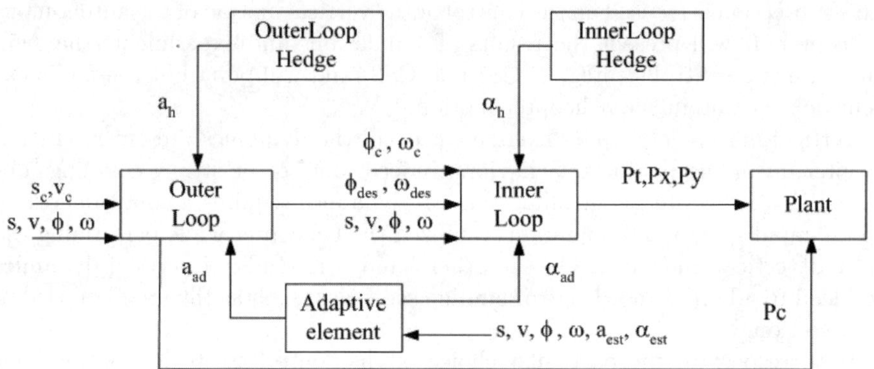

Fig. 11. Adaptive control scheme

The proposed structure is based in feedback linearization and combines the helicopter attitude inner control loop and the outer trajectory control loop. It also applies adaptive techniques (a_{ad}, α_{ad}) to cancel model errors. Furthermore, Pseudo-Control-Hedging (a_h, α_h) is used to prevent unwanted adaptation to actuator limits and dynamics in the inner loop. The commanded references are s_c, v_c, ϕ_c and ω_c whereas ϕ_{des} and ω_{des} are the values imposed to the inner loop by the outer loop.

In general no guarantee of robustness against model uncertainties or disturbances and no adaptive capabilities are provided by many feedback linearization techniques. However, in some cases, nonlinear controller robustness properties are increased using sliding mode and Lyapunov based control [26]. Typically, these techniques trade the controller performance against uncertainty, but require a priori estimates of parameter bounds, which may be difficult to obtain. However, research efforts to design new robust nonlinear control laws are pursued. Then in [15] the vertical motion of a nonlinear model of a helicopter tracks a reference signal, while stabilizing the lateral and longitudinal position.

In [19] the application of nonlinear model predictive control is proposed. At each sample time, the controller computes a finite control sequence which minimizes a quadratic index. This index includes the errors of the outputs (x, y, z and ϕ_z variables in Fig. 5) with respect to desired trajectories, additional state variables, which should be bounded, and the control actions (Pc, Pt, Px and Py in Fig. 5). A gradient descent technique is used to compute the optimal values of the control variable. The method improves the tracking performance at the expenses of heavy computing load.

In [11] the control of underactuated systems including helicopters and Planar VTOL (PVTOL) is studied. Several control techniques are presented including backstepping, energy based controllers and Lyapunov-based controllers.

4.2   Vertical Movement

This subsection is focused on the control of the vertical motion of the autonomous helicopter. It will present the results of simulations and experiments that will illustrate two control strategies (CS1 and CS2) and will point out the nonlinear behavior existing in the helicopter motion.

Vertical movement implies excitement of all the dynamics present in the helicopter, including nonlinear behaviors. Indeed, due to the strong couplings observed in a helicopter, controlling vertical dynamics requires a control effort in all the input variables to maintain stabilized the helicopter while performing the desired vertical maneuver. On the other hand, the study of vertical dynamics will lead to a better knowledge of landing scenarios, such as the one proposed in next section.

With respect to the particular choice of the control strategies, noting that an important point to consider is the applicability of control methods, which in its turn is highly dependent on the capabilities of the on-board hardware. CS1 and CS2 strategies are simpler than other approaches referenced in the previous survey, showing a good performance and robustness with low computational cost. More complex control laws, with an improved performance, have been considered but the main problem is its practical application within the actual platform used in the experiments (fixed point micro-controller).

Control Strategy 1 (CS1)

This strategy is based on a linear control approach described in [6]. The complete control hierarchy consists of several elementary controllers, each one controlling a single scalar state variable. Two of these elementary controllers, those more directly related to vertical dynamics, will be described in the following.

Trajectory and System Model

Let z and ϕ_z in Fig. 5 be the controlled variables and Pc and Pt the associated control variables, as indicated in Fig. 10. For controller design purposes, Equation (3) can be simplified and the following second order model can be assumed for the variables involved in translational vertical dynamics:

$$\ddot{z}(t) = \frac{1}{f_z} Pc(t)$$

$$\ddot{\phi}(t) = \frac{1}{f_{\phi_z}} Pt(t) \tag{19}$$

where f_z and f_{ϕ_z} are system parameters that can be measured.

When fixing a desired trajectory for $z(t)$ and $\phi_z(t)$ to be followed by this system to reach some command values $z|_{ref}(t)$ and $\phi_z|_{ref}(t)$, two free parameters are necessary in each case to adapt to the current state $(z(t_0), \dot{z}(t_0), \phi_z(t_0), \dot{\phi}_z(t_0))$ given as initial condition at the current time t_0. On the other hand, $\ddot{z}(t_0)$ and $\ddot{\phi}_z(t_0)$ must be chosen by the controller according to Equation 19.

In order to allow smooth convergence, the expressions adopted for the desired trajectories of $z(t)$ and $\phi_z(t)$ are:

$$z(t) = z|_{ref}(t) + A_z e^{-K1_z t} + B_z e^{-K2_z t}$$
$$\phi_z(t) = \phi_z|_{ref}(t) + A_{\phi_z} e^{-K1_{\phi_z} t} + B_{\phi_z} e^{-K2_{\phi_z} t} \tag{20}$$

where $K1_i$ and $K2_i$ (for $i = z, \phi_z$) are constants strictly positive that adjust the rate of convergence towards $z|_{ref}(t)$ and $\phi_z|_{ref}(t)$. A_i and B_i (for $i = z, \phi_z$) represent the two free parameters in each case to fix the initial conditions.

Basic control law can be expressed as follows:

$$Pc(t) = f_z \ddot{z}^*(t)$$
$$Pt(t) = f_{\phi_z} \ddot{\phi}_z^*(t) \tag{21}$$

$$\ddot{z}^* = -K1_z K2_z (z - z|_{ref}) - (K1_z + K2_z)(\dot{z} - \dot{z}|_{ref})$$
$$\ddot{\phi}_z^* = -K1_{\phi_z} K2_{\phi_z}(\phi_z - \phi_z|_{ref}) - (K1_{\phi_z} + K2_{\phi_z})(\dot{\phi}_z - \dot{\phi}_z|_{ref}) \tag{22}$$

where Equation (22) is interpreted as the desired accelerations to generate the intended trajectories according to Equation (20). Time dependence has been removed for clarity.

Model Error

It would also be desirable to add some term to eliminate steady state errors. The reason for the necessity of such compensation is that real systems differ from the ideal equations (19). These equations change for real systems with model errors $me_z(t)$ and $me_{\phi_z}(t)$ into:

$$\ddot{z}(t) = \frac{1}{f_z} Pc(t) - me_z(t)$$
$$\ddot{\phi}_z(t) = \frac{1}{f_{\phi_z}} Pt(t) - me_{\phi_z}(t) \tag{23}$$

This lead to the following new expression for the controller output:

$$Pc(t) = f_z[\ddot{z}^*(t) + me_z|_{est}(t)]$$
$$Pt(t) = f_{\phi_z}[\ddot{\phi}_z^*(t) + me_{\phi_z}|_{est}(t)] \tag{24}$$

where $me_z|_{est}(t)$ and $me_{\phi_z}|_{est}(t)$ are estimators of the model error. These estimators are gradually adjusted at selectable rates a_z and a_{ϕ_z} by integrating the deviation between the current desired accelerations, $\ddot{z}^*(t)$ and $\ddot{\phi}_z^{\;*}(t)$, and the current actual observed accelerations $\ddot{z}(t)$ and $\ddot{\phi}_z(t)$:

$$\dot{m}e_z|_{est}(t) = a_z(\ddot{z}^*(t) - \ddot{z}(t))$$
$$\dot{m}e_{\phi_z}|_{est}(t) = a_{\phi_z}(\ddot{\phi}_z^{\;*}(t) - \ddot{\phi}_z(t)) \tag{25}$$

Couplings

Finally, the aforementioned couplings should also be taken into account. To this end, the coupling terms included in Fig. 10 are added to the control laws.

$$Pc(t) = f_z[\ddot{z}^*(t) + me_z|_{est}(t) + c_z(t)]$$
$$Pt(t) = f_{\phi_z}[\ddot{\phi}_z^{\;*}(t) + me_{\phi_z}|_{est}(t) + c_{\phi_z}(t)] \tag{26}$$

where the coupling terms, $c_z(t)$ and $c_{\phi_z}(t)$, are functions of the main coupled variables in each case:

$$c_z(t) = f(\phi_x, \phi_y)$$
$$c_{\phi_z}(t) = f(Pc) \tag{27}$$

Conclusions

Analyzing the structure proposed in this section, it can be stated that it is similar to PID structure. However, there are some differences that make CS1 a better approach than conventional PID implementations. Controllers designed following CS1 are stable by design and oscillation-free, having also step response without overshooting. This result has been proven experimentally in the MAR-VIN helicopter shown in Fig. 3.

Control Strategy 2 (CS2)

The values of the coefficients of the Control Strategy 1 (CS1) previously described ($K1_z$, $K1_{\phi_z}$, $K2_z$, $K2_{\phi_z}$, a_z and a_{ϕ_z}) are fixed. In this second section, a nonlinear improvement of CS1 is shown [2].

The CS1 is possibly one of the simplest approach that shows a good performance and can be used for both hovering and trajectory tracking exhibiting good robustness in most cases. Another advantage is its very low computational cost. Its main drawback is that the control gains are fixed for all the operation range and then low values have to be used due to saturations and large errors, although this could be partially solved by using different controllers for position and velocity control.

Control strategy 2 (CS2) developed by GRVC at the University of Seville proposes a nonlinear strategy, based on CS1 structure, that applies different control laws according to the operation conditions. In fact, CS1 could be considered as a particular case of a more general structure CS2, where gains vary around the fixed values proposed for CS1. This more general structure of CS2 can be defined substituting Equations (22) and (26) by:

$$\frac{Pc}{f_z} = +\varUpsilon_1(z,\phi_z)(z - z|_{ref}) + \varUpsilon_2(z,\phi_z)(\dot{z} - \dot{z}|_{ref})$$

$$+\varUpsilon_3(z,\phi_z)\int(z - z|_{ref})dt + c_z(\phi_x,\phi_y)$$

$$\frac{Pt}{f_{\phi_z}} = +\varUpsilon_4(z,\phi_z)(\phi_z - \phi_z|_{ref}) + \varUpsilon_5(z,\phi_z)(\dot{\phi}_z - \dot{\phi}_z|_{ref})$$

$$+\varUpsilon_6(z,\phi_z)\int(\phi_z - \phi_z|_{ref})dt + c_{\phi_z}(Pc) \tag{28}$$

where $\varUpsilon_i(z,\phi_z)$ are nonlinear feedback functions of z and ϕ_z. Time dependence has been removed for clarity.

Notice that other nonlinear approaches for $\varUpsilon_i$ could be considered in Equation (28), what highlights its generality and flexibility in the study of the vertical dynamics. For instance, $\varUpsilon_i$ could become dependent of other variables of the system, thus reflecting also the couplings mentioned in previous sections.

CS2 can lead to improve overall controller performance, as shown in [2]. Furthermore, this approach can also be used to deal with saturations or even to induce nonlinear behaviors on the helicopter flight for testing issues. Lastly, noting that since CS2 design is based on CS1, it can also be implemented with low computational cost, quite similar to the required by CS1.

Nonlinear Study of the Model

The following will show how CS2 can be used to validate the nonlinear behaviors associated to the model. Thus CS2 is designed to induce those nonlinear behaviors (including multiple equilibriums, limit cycles, etc), that will be validated with experimental flights. To this end, the proposed nonlinear law CS2 is mainly a variation of the CS1 structure in the vertical dynamics. Thus, instead of using fixed gains (Equations 21 and 22) in the computation of both collective input of main rotor and tail rotor control signals $Pc(t)$ and $Pt(t)$, respectively), a nonlinear gain-scheduling approach has been applied taking into account the altitude error $e_z(t)$.

Under CS2, the system exhibit the behavior shown in Fig. 12 corresponding to the "altitude_error vs. vertical_velocity" phase plane. As it can be observed, the system presents five equilibrium points (the origin is the only one stable equilibrium, but it is only locally stable; there are also two unstable equilibrium points and two saddle points between the stable and the unstable equilibria), which are surrounded by a stable limit cycle.

In this way, if the helicopter starts close to the target altitude, i.e., with a small altitude error, it will be able to reach the origin (it achieves null altitude error). However, if the error is a little bit larger (see dashed line in Fig. 12), or a disturbance brings the helicopter far from the target point, it can be out of the attraction basin of the origin and then the helicopter will tend to go away from the origin. Notice that the existence of the stable limit cycle prevents the helicopter from becoming totally unstable. Instead of this, the helicopter will exhibit a permanent stable oscillation. If the error is even larger, the stable limit cycle will attract the helicopter preventing again from instability (see Fig. 12).

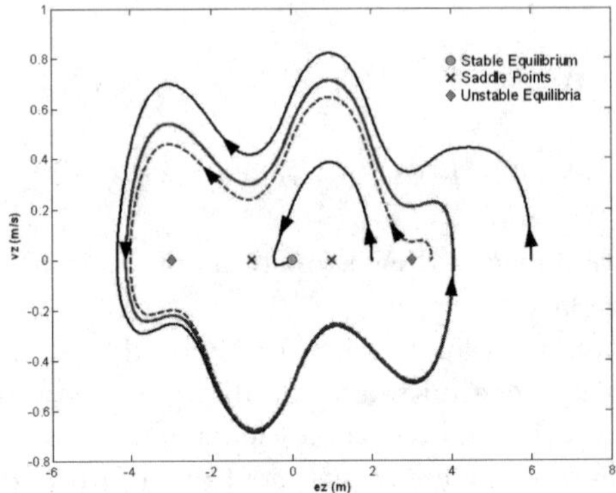

Fig. 12. Altitude_error vs. vertical_velocity phase plane

Clearly, from the point of view of model validation, testing a controller like this one on the real helicopter has several advantages. On the one hand, it makes it possible to compare the response of the real helicopter and the model in the face of non globally stable controllers. On the other hand, it makes it also possible to test the capabilities of the model to reflect the nonlinear behavior of the actual helicopter by comparing the predicted oscillations and the real ones. Moreover, if the model is not good enough, the predicted limit cycle could not exist with the actual helicopter.

Experiments

A series of experiments with CS2 have been performed to validate the aforementioned simulation results. Some of these experiments are depicted below.

In order to analyze the behavior close to the origin an experiment was performed starting at 2.66 meters from the equilibrium point. As can be observed in Fig. 13, black continuous line corresponding to LOCAL STABILITY, when the real helicopter starts at 2.66 m from the origin it tends to the stable equilibrium. However, that stability is only local. A perturbation (gray continuous line corresponding to PERTURBATION) appeared during the real flight bringing the real helicopter out of the attraction basin of the origin, so it evolved to an stable limit cycle of 4 meters error amplitude (black dashed line corresponding to STABLE LIMIT CYCLE). In a similar experiment, with an initial altitude error of 5 meters, i.e. starting out of the limit cycle, it was observed that the helicopter tends to the same stable limit cycle than in the previous experiment.

Conclusions

The final conclusion is that the nonlinear qualitative behavior of the model with CS2 is quite similar to that of the real helicopter with CS2. However, some differences on the quantitative velocity behavior have been found, probably due to non modeled effects of induced velocity. Videos of real experiments can be accessed in http://www.esi.us.es/ fcuesta/videos/helicopter.html.

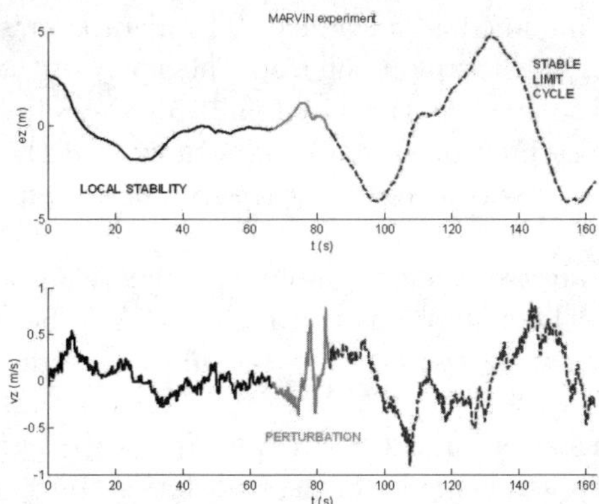

Fig. 13. Real experiment: it starts 2.66m far from the origin and it tends to the equilibrium (black continuous line corresponding to LOCAL STABILITY); a perturbation (gray continuous line corresponding to PERTURBATION) brings the helicopter to the stable limit cycle (black dashed line corresponding to STABLE LIMIT CYCLE).

5 Open Research Lines

There are still many open problems in autonomous helicopter design. To mention some of them, noting the lack of a standard methodology for controller specification and its subsequent automated controller synthesis against multiple and conflicting specifications. Likewise, in navigation and guidance research areas, aspects such as considering vehicle dynamics in path planing or usage of online optimization (according to vehicle constraints like time, fuel, efficiency, etc.), focus the attention of many researchers.

Amongst these open points, it can be highlighted two relevant applications that do imply the design and testing of new techniques and algorithms. These are autonomous landing on oscillating platforms and transporting a single load with several helicopters. In the following the research work in these two problems is reviewed.

Autonomous Landing on Oscillating Platforms

Autonomous landing on the deck of a ship is a matter of importance when sea is rough or other bad weather conditions arise (lack of visibility, etc.). Many accidents have taken place under these circumstances.

This is complex goal, which implies a continuous feedback of ship position and pose. To this end, elaborated position estimation techniques are required. On the other hand, it is also necessary the design of control algorithms that deal with the tracking of the complex trajectories imposed by ship movements and that lead to smooth landing. For this second point, works previously shown on vertical motion can also help in the understanding of the inherent nonlinear problem.

In [35] it is presented a multiple view algorithm that could be used for vision based landing of an unmanned aerial vehicle. The algorithm is based in results of multiple view geometry that exploit the rank deficiency of the so called multiple view matrix. It is also shown how the use of multiple views improves significantly motion and structure estimation. Final results show that the vision-based state estimates are accurate to within 7 cm in each axis of translation and 4° in each axis of rotation.

The work at [31] addresses the design of an autopilot for autonomous landing of a helicopter on a rocking ship, due to rough sea. A tether is used for landing and securing a helicopter to the deck of the ship in rough weather. Two controllers are proposed. In the first (A), the rotation time scale is chosen much shorter than the translation, and the rotation reference signals are created to achieve a desired controlled behavior of the translation. In the second (B), due to coupling of the translation of the helicopter to the rotation through the tether, the translation reference rates are created to achieve a desired controlled behavior of the attitude and altitude. Controller A is proposed for use when the helicopter is far away from the goal, while Controller B is for the case when the helicopter is close to the ship. The proposed control schemes are proved to be robust to the tracking error of its internal loop and results in local exponential stability.

Even though the tether is used to increase the safety of the landing manoeuver, its usage also implies some drawbacks that require further research. As mentioned before, coupling between position and orientation variables, which is normally absent in a helicopter free of a tether, appear in this case. Thus, the variation range of x and y is influenced by attitude control, which could take helicopter out of the landing area of the ship. To control this, the authors point out that cable tension should be as high as possible. However an analysis in more depth, aiming at real implementation, should be considered since excessive tension could provoke the breaking of the tether or damages to the helicopter.

In [16] it is considered the problem of controlling the vertical motion of a nonlinear model of a helicopter, while stabilizing the lateral and horizontal position and maintaining a constant attitude. The vertical reference to be tracked is a sum of a constant and a fixed number of sinusoidal signals:

$$z|_{ref} = z_o + \sum_{i=1}^{n} A_i sin(\Omega_i t + \phi_t) \tag{29}$$

This reference is assumed not to be available to the controller. This represents a possible situation in which the controller is required to synchronize the vehicle motion with that of an oscillating platform, such as the deck of a ship in high seas. The authors design a nonlinear controller which combines recent results on nonlinear adaptive output regulations and robust stabilization of systems in feedforward form by means of saturated controls.

Although [16] outlines a complete set of techniques to carry out the autonomous landing of a helicopter on a ship, some points are still open. On the one hand, the achieved control goal allows vertical position tracking but longitudinal and lateral position, as well as attitude, are only stabilized to a constant configuration. It would be desirable to impose some tracking requirements on attitude. Thus, the effect of strong oscillations in the deck due to rather rough sea could be better avoided. On the other hand, the reduction of the longitudinal and translational stabilization phase would also improve the performance. Notice also that [16] did not consider the modeling of engine dynamics; however the work presented in [3] shows that the technique also works well when they are taken into account. Finally, recall that only simulation results are available until the moment. These simulations show the effectiveness of the method and its ability to cope with uncertainties on the plant and actuator model. However, implementation in a real UAV should be carried out to obtain the definitive validation of the algorithm.

Lastly, noting also that some experimental efforts have been already done. The U.S. Navy and Northrop Grumman Corporation developed some preliminary tests (http://www.northropgrumman.com) with two MQ-8B Fire Scout prototypes. The MQ-8B is the aircraft element of a complete system called the Vertical takeoff and landing Tactical Unmanned Aerial Vehicle (VTUAV) system. After it was launched from the naval air station, the Fire Scout flew to the designated test area, where a ship was waiting for the air vehicle to land and take off under its own control. The flight was monitored from a ship-based control station called a tactical control system, and the air vehicle was guided onto the ship using an unmanned air vehicle common automatic recovery system. These preliminary tests only dealt with moderate conditions of sea. Therefore, further efforts in control algorithms development are needed for the scenario of rough sea mentioned at the beginning of this section.

Lifting and Transporting a Single Load with Several Helicopters

Development of new techniques that lead to helicopters working together in load transport missions would increase the load capacity of single low cost platforms. Civil Security and Disaster Management activities could be reinforced by applying these cooperative approaches.

Main challenges involved in this goal stem from the presence of strong nonlinearities in the complete system, unknown disturbances due to gust of winds and the necessity of high gain controllers, which in its turn implies saturated inputs.

In [34], two H∞ controller designs are presented for a twin lift helicopter system (TLHS). The TLHS configuration consists of two UH-GOA Sikorsky Blackhawk helicopters jointly lifting a heavy payload. The first design presented considers the case in which the tethers connecting each helicopter to the load are equal in length, and the second considers the case in which the tether lengths are unequal. Both designs are based on a seven degree of freedom model linearized about hover. The primary objective of each controller is to minimize the control action and pitching motion required to stabilize the helicopters as they perform elementary maneuvers. A simulation of a typical TLHS command scenario is used to evaluate stability and robustness of the resulting two feedback systems with respect to structured parametric uncertainty.

In [14], a twin lift system is also studied. Because of the special structure of the these systems, controllability, observability, stability and existence of decentralized fixed modes of such systems can be tested on matrices of lower order.

The Technical University of Berlin (TUB) is also conducting a project on the lifting and transporting of a single load by means of several UAVs, involving experimental validation. They plan to apply this system in the self-deployment of the communication network elements involved in AWARE project (http://grvc.us.es/aware). AWARE is a research project whose objective is the design, development and experimentation of a platform for the cooperation among aerial flying objects and a ground sensor-actuator wireless network, including mobile nodes carried by people and vehicles. The project considers the validation in two different applications: Civil Security / Disaster Management and Filming dynamically evolving scenes.

6   Conclusions

This chapter has been devoted to autonomous helicopters. These kind of aerial vehicles have been extensively used for aerial robotic applications such as cinematography, inspection, search and rescue operations and others. Their manoeuvrability and hovering ability explain their use in those contexts.

The chapter surveys the main aspects involved in the design and implementation of an autonomous helicopter: platforms, architectures for control, derivation of physical models and their identification for a particular helicopter and model-based control methods. Along the survey, specific references to works developed by several research groups including the GRVC in the University of Seville were also highlighted. Thus, a model identification tool and different control techniques have been implemented with real helicopters. Furthermore the control strategy CS2 in Section 4 has been used to validate the identified model and the nonlinear behaviors observed in simulation. This model will lead to the implementation of new control laws that will be able to increase performance and robustness.

Some specific points that require further research arise as a consequence of a global analysis of the work reviewed in this chapter. Thus, for example, in the control of vertical motion it could be interesting to design controllers that can compensate the effect of wind gusts on the altitude (due to the Effective Translational Lift). Likewise, the modeling of different induced velocity fields should be carried out to consider the case of a helicopter flying near vertical obstacles, such as the wall of a building. There exists also a lack of research concerning the determination of the wind regimes that can be afforded by a helicopter with a given controller. To this end, nonlinear techniques such as the mentioned CS2 could be used to analyze and improve the performance. Lastly, noting that most of the presented works assume that all the state variables are measurable. Since this assumption is not always true in real implementations, it is necessary the consideration in the close loop of state estimation techniques as well as an analysis of their effect on the performance of the overall control scheme.

As final conclusion, it can be stated that designing and implementing an autonomous helicopter is a complex task that implies efforts in varied work areas. Analyzing the state of the art, it can be concluded that even though there are many works already done in this direction, some aspects are not covered yet.

Acknowledgements

This research has been partially funded by the ongoing projects AEROSENS (Spanish National Research Programme, DPI2005-02293) and AWARE (European Comission, IST-2006-33579).

Likewise, some of the experiments described has been carried out in the framework of the COMETS project (European Commission, IST 2001-34304). The collaborative work and information provided by the TUB partner of the COMETS team is also acknowledged.

Special thanks also to the researchers Marco La Civita (Flying-Cam and formerly CMU) and Kontantin Kondak (TUB) for their valuable contributions to some of the topics discussed in the Chapter.

References

1. Barrientos A, Del Cerro J, Campoy P, García P J (2002) An autonomous helicopter guided by computer vision for inspection of overhead power cables, In: Workshop on Aerial Robotics - IEEE / RSJ International Conference on Intelliget Robots and Systems IROS 2002
2. Bejar M, Cuesta F, Ollero A (2007) On the use of soft computing techniques for helicopter control in environment protection mission scenarios, To appear in Intelligent Automation and Soft Computing
3. Bejar M, Isidori A, Marconi L, Naldi R (2005) Robust vertical/lateral/longitudinal control of an helicopter with constant yaw-attitude In: Proceedings of the IEEE Conference on Decision and Control
4. Buskey G, Wyeth G, Roberts J (2001) Autonomous helicopter hover using an artificial neural network, In: Proceedings of the IEEE International Conference on Robotics and Automation, pages 1635–1640
5. Cavalcante C, Cardoso J, Ramos J G, Nerves O R (1995) Design and tuning of a helicopter fuzzy controller, In: Proceedings of IEEE International Conference on Fuzzy Systems, volume 3, pages 1549–1554
6. Deeg C, Musial M, Hommel G (2004) Control and simulation of an autonomously flying model helicopter, In: IFAC Symposium on Intelligent Autonomous Vehicles

7. Deeg C (2006) Modeling simulation, and implementation of an autonomous flying robot, PhD thesis, Technische Universität Berlin
8. Doherty P, Granlund G, Kuchcinski K, Sandewall E, Nordberg K, Skarman E, Wiklund J (2000) The witas unmanned aerial vehicle project, In: Proceedings of the 14th European Conference on Artificial Intelligence, pages 747–755
9. Eck C, Chapuis J, Geering H P (2001) Software-supported design and evaluation of low-cost navigation units, In: Proceedings of the 8th Saint Peterburg International Conference on Integrated Navigation Systems, pages 163–172
10. Fagg A H, Lewis M A, Montgomery J F, Bekey G A (1993) The usc autonomous flying vehicle: an experiment in real-time behaviour-based control, In: Proceedings of the 1993 IEEE/RSJ International Conference on Intelligent Robots and Systems, pages 1173–1180
11. Fantoni I, Lozano R (2002) Non-linear Control for Underactuated Mechanical Systems, Chapter 13: Helicopter on a platform
12. Gavrilets V, Frazzoli E, Mettler B, Piedmonte M, Feron E (2001) Aggressive maneuvering of small autonomous helicopters: A human-centered approach, The International Journal of Robotics Research, 20(10):795–807
13. González A, Mahtani R, Béjar M, Ollero A (2004) Control and stability analysis of an autonomous helicopter, In: Proceedings of World Automation Congress
14. Huang S, Jing Y, Yang G, Zhang S (1997) The decentralized fixed modes of twin lift systems, In: Proceedings of the American Control Conference, pages 2388–2389
15. Isidori A, Marconi L, Serrani A (2001) Robust nonlinear motion control of a helicopter, In: Proceedings of the 40th IEEE Conference on Decision and Control, pages 4586–4591
16. Isidori A, Marconi L, Serrani A (2003) Robust nonlinear motion control of a helicopter, IEEE Transactions on Automatic Control, 48(3):413–426
17. Johnson E N, Kannan S K (2002) Adaptive flight control for an autonomous unmanned helicopter, AIAA Guidance, Navigation and Control Conference, (AIAA-2002-4439)
18. Kadmiry B, Bergsten P, Driankov D (2001) Autonomous helicopter using fuzzy-gain scheduling, In: Proceedings of the IEEE Conference on Robotic and Automation ICRA, volume 3, pages 2980–2985
19. Kim H J, Shim D H (2003) A flight control system for aerial robots: algorithms and experiments, Control Engineering Practice, 11(12):1351–1515
20. Kim S K, Tilbury D M (2004) Mathematical modelling and experimental identification of a model helicopter, Journal of Robotic Systems, 21(3):95–116
21. Koo T J, Hoffman F, Shim H, Sinopoli B, Sastry S (1998) Hybrid control of model helicopters, In: Proceedings of the IFAC Workshop on Motion Control, pages 285–290
22. Koo T J, Sastry S (1998) Output tracking control design of a helicopter model based on approximate linearization, In: Proceedings of the 37th IEEE Conference on Decision and Control, pages 3635–3640
23. La Civita M, Papageorgiou G, Messner W C, Kanade T (2002) Design and flight testing of a high-bandwidth $\mathcal{H}_\infty$ loop shaping controller for a robotic helicopter, In: Proceedings of the AIAA Guidance, Navigation, and Control Conference, number AIAA-2002-4836, Montery, CA
24. La Civita M, Papageorgiou G, Messner W C, Kanade T (2006) Design and flight testing of a high-bandwidth $\mathcal{H}_\infty$ loop shaping controller for a robotic helicopter, Journal phf Guidance, Control, and Dynamics, 29(2):485–494
25. Gordon Leishman J (2000) Principles of Helicopter Aerodynamics, Cambridge University Press

26. Maharaj D Y (1994)   The application of non-linear control theory to robust behaviour-based control,   PhD thesis, Dept of Aeronautics, Imperial College of Science, Technology and Medicine
27. The MathWorks, Inc. (2004) Matlab Optimization Toolbox User's Guide, revised for version 3.0 (release 14), fifth printing edition
28. Mettler M, Tischler M B, Kanade T (2001) System identification modelling of a small-scale unmanned rotorcraft for flight control design, American Helicopter Society Journal
29. Montgomery J F, Bekey G A (1998) Learning helicopter control through "teaching by showing", In: Proceedings of the 37th IEEE Conference on Decision and Control
30. Montgomery J F, Fagg A H, Bekey G A (1995) The usc afv-i: A behaviour based entry, 1994 Aerial Robotics Competition, IEEE Expert, 10(2):16–22
31. Oh S, Pathak K, Agrawal S K, Pota H R, Garratt M (2006) Approaches for a tether-guided landing of an autonomous helicopter, IEEE Transactions on Robotics, 22(3):536–544
32. Raimundez J C, Camaño J L, Béjar M (2006) Application of adaptive neural-network control to a scale 6dof helicopter In: Proceedings of Artificial Intelligence and Cognitive Science
33. Remuss V, Musial M, Hommel G (2002) Marvin, an autonomous flying robot based on components of the shelf, In: Proceedings of Aerial Robotics Workshop, IROS
34. Reynolds H K, Rodriguez A A (1992) H∞ control of a twin lift helicopter system, In: Proceedings of the 31st IEEE Conference on Decision and Contro, pages 2442–2447
35. Shakernia O, Sharp C S, Vidal R, Shim D H, Ma Y, Sastry S (2002) Multiple view motion estimation and control for landing an unmanned aerial vehicle, In: Proceedings of IEEE International Conference on Robotics and Automation
36. Shim H, Koo T J, Hoffman F, Sastry S (1998) A comprehensive study of control design of an autonomous helicopter, In: Proceedings of the 37th IEEE Conference on Decision and Control, pages 3653–3658

Efficient Quantization in the Average Consensus Problem

Ruggero Carli and Sandro Zampieri

Department of Information Engineering, Università di Padova, Italy
{calirug,zampi}@dei.unipd.it

Summary. In the average consensus a set of linear systems has to be driven to the same final state which corresponds to the average of their initial states. This mathematical problem can be seen as the simplest example of coordination task and in fact it can be used to model both the control of multiple autonomous vehicles which all have to be driven to the centroid of the initial positions, and to model the decentralized estimation of a quantity from multiple measure coming from distributed sensors. In general we can expect that the performance of a consensus strategy will be strongly related to the amount of information the agents exchange each other. This contribution presents a consensus strategy in which the exchanged data are symbols and not real numbers. This is based on a logarithmic quantizer based state estimator. The stability of this technique is then analyzed.

Keywords: Distributed Estimations, Quantization, Distributed Algorithms, Consensus, Multiagent Systems.

1 Introduction

The design of coordination algorithms for multiple autonomous vehicles and of decentralized estimation techniques for handling data coming from distributed sensor networks is attracting large attention in recent years [13, 14, 16, 15, 1, 3, 4]. In fact both in coordinated control and in distributed estimation the agents need to communicate data in order to execute the task. In particular they may need to agree on the value of certain coordination state variables. One expects that, in order to achieve coordination, the variables shared by the agents, converge to a common value, asymptotically. The problem of designing controllers that lead to such asymptotic coordination is called *coordinated consensus*, see for example [2, 3, 7] and references therein. The interest in this type of problems is not limited to the field of mobile vehicles coordination but also in the field of synchronization theory [10, 9].

One of the simplest consensus problems that has been mostly studied in the literature consists in starting from systems described by an integrator and in finding a feedback control driving all the states to the same value [3]. The information exchange is modelled by a directed graph describing in which pair of agents the data transmission is allowed. Many variations of this problems has been considered namely depending on the properties of the data exchange.

C. Bonivento et al. (Eds.): Adv. in Control Theory and Applications, LNCIS 353, pp. 31–49, 2007.
springerlink.com   © Springer-Verlag Berlin Heidelberg 2007

In [13, 14, 7] the problem of designing control strategies for mobile agents leading consensus when the communication graph is time-varying and depending of the agents positions. Robustness to communication link failure [6] and the effects of time delays [3] has also been considered recently. Randomly time-varying networks have also been analyzed in [11].

In [19, 20] the consensus problem is treated in case the agents are allowed to exchange not real numbers by instead only quantized information. This paper continues the analysis proposed in [19]. More precisely we consider the average consensus problem for simple first-order dynamics linear systems which can exchange information according to a fixed strongly connected digital communication network. Hence, besides the decentralized computational aspects induced by the choice of the communication network, we have to face the quantization effects due to the digital links. In order to achieve the consensus, some encoding of the data to be transmitted is necessary. Here we present a encoding/decoding strategy based on the exchange of logarithmically quantized information. Then the stability analysis is provided.

The paper is organized as follows. In Section 2 we provide some basic notions of graph theory and some notational conventions. In Section 3 we formally define the average consensus problem. We then propose a model of the encoder/decoder structure through which the systems exchange information. In Section 4 we introduce the logarithmic quantizer based encoder/decoder. In section 5 we discuss the stability of this technique, first presenting some general theoretical results and then restricting our attention to two particular structures of communication graph allowed us to determine some simple results. Finally we gather our conclusions in Section 6.

2 Preliminaries

Before defining the problem we want to solve, we summarize some notions on graph theory and we provide some notational conventions that will be useful throughout the rest of the paper.

Let $\mathcal{G} = (V, \mathcal{W})$ be a directed graph where $V = (1, \ldots, N)$ is the set of vertices and $\mathcal{W} \subset V \times V$ is the set of arcs. If $(i, j) \in \mathcal{W}$ we say that the arc (i, j) is outgoing from i and incoming in j. In our setup we admit the presence of self-loops. The adjacency matrix A is a $\{0, 1\}$-valued square matrix indexed by the elements in V defined by letting $A_{ij} = 1$ if and only $(j, i) \in \mathcal{W}$. Define the in-degree of a vertex i as $\mathrm{indeg}(i) := \sum_j A_{ij}$ and the out-degree of a vertex j as $\mathrm{outdeg}(j) := \sum_i A_{ij}$. A path in $\mathcal{G}$ consists of a sequence of vertices $i_1 i_2 \ldots \ldots i_r$ such that $(i_\ell, i_{\ell+1}) \in \mathcal{W}$ for every $\ell = 1, \ldots, r - 1$; i_1 (resp. i_r) is said to be the initial (resp. terminal) vertex of the path. A cycle is a path in which the initial and the terminal vertices coincide. A vertex i is said to be connected to a vertex j if there exists a path with initial vertex i and terminal vertex j. A directed graph is said to be connected if, given any pair of vertices i and j, either i is connected to j or j is connected to i. A directed graph is said to be strongly connected if, given any pair of vertices i and j, i is connected to j. A direct graph $\mathcal{G} = (V, \mathcal{W})$ is said to be a *circulant directed graph* if $(i, j) \in \mathcal{W}$ implies that $(i + p, j + p) \in \mathcal{W}$ for any $p \in \mathbb{N}$, where the sum is meant mod N. A graph is said to be undirected if $(i, j) \in \mathcal{W}$ implies that also $(j, i) \in \mathcal{W}$.

Now some notational conventions. Given a matrix $M \in \mathbb{R}^{N \times N}$, $diag\{M\}$ means a diagonal matrix with the same diagonal elements of the matrix M. Given a vector $m \in \mathbb{R}^N$, $diag\{m\}$ means a diagonal matrix having the components of m as diagonal elements. Given a matrix $M \in \mathbb{R}^{N \times N}$, with the symbol $M^\dagger$ we denote the pseudo-inverse of M, with $rk\, M$ and with $\sigma\{M\}$ we indicate respectively the rank and the set of eigenvalues of the matrix.

3   Problem Formulation

Consider $N > 1$ identical systems whose dynamics are described by the following discrete time state equations

$$x_i^+ = x_i + u_i \qquad i = 1, \ldots, N$$

where $x_i \in \mathbb{R}$ is the state of the i-th system, x_i^+ represents the updated state and $u_i \in \mathbb{R}$ is the control input. More compactly we can write

$$x^+ = x + u \tag{1}$$

where $x, u \in \mathbb{R}^N$. The goal is to design an input control u yielding the consensus of the states, namely a control such that all the x_i's become equal asymptotically, i.e.

$$\lim_{t \to \infty} x(t) = \alpha \mathbb{1} \tag{2}$$

where $\mathbb{1} := (1, \ldots, 1)^T$ and α is a scalar depending on $x(0)$. Moreover, we also require that $x(t) = x(0)$ for all $t \in \mathbb{N}$ if $x(0) = \lambda \mathbb{1}$.

An interesting case that has been widely studied in literature (see [3, 12, 18]) corresponds to the case in which x is a static feedback function of u

$$u = Kx, \quad K \in \mathbb{R}^{N \times N} \tag{3}$$

In such case the system (1) is described by the following closed loop system

$$x^+ = (I + K)x. \tag{4}$$

It is easy to see that the consensus problem for system (4) is solved if and only if the following three conditions hold:

(A)   the only eigenvalue of $I + K$ on the unit circle is 1;
(B)   the eigenvalue 1 has algebraic multiplicity one (namely it is a simple root of the characteristic polynomial of $I + K$) and $\mathbb{1}$ is its eigenvector;
(C)   all the other eigenvalues are strictly inside the unit circle.

In the sequel we will restrict to matrices K such that $I + K$ is a nonnegative matrix, namely a matrix with all elements nonnegative. Condition (B) then says that $I + K$ is a stochastic matrix. Conditions (A) and (C) yield the asymptotic behavior

$$(I + K)^t \to \mathbb{1} v^T$$

where $v \in \mathbb{R}^N$ is the unique probability vector such that $v^T(I + K) = v^T$. This implies that

$$x(t) \to v^T x(0) \mathbb{1}.$$

In the special case when $v = N^{-1}\mathbb{1}$ we obtain that the consensus is achieved at the average of the initial conditions. In this case $I + K$ is said to be a doubly stochastic matrix and K a *average consensus controller*.

We observe that the use of control law as in Equation (3) implies the exchange of perfect information through the communication network. More precisely, the fact that the element in position i, j of the matrix K is different from zero, means that the system i needs to know exactly the state of the system j in order to compute its feedback action. This implies that the agent j-th must communicate his state x_j to the system i. A good description of the communication effort required by a specific feedback K is given by the directed graph $\mathcal{G}_K$ with set of vertices $\{1, \ldots, N\}$ in which there is an arc from j to i whenever in the feedback matrix K the element $K_{ij} \neq 0$. The graph $\mathcal{G}_K$ is said to be the *communication graph* associated with K. Conversely, given any directed graph $\mathcal{G}$ with set of vertices $\{1, \ldots, N\}$, a feedback K is said to be *compatible* with $\mathcal{G}$ if $\mathcal{G}_K$ is a subgraph of $\mathcal{G}$ (we will use the notation $\mathcal{G}_K \subseteq \mathcal{G}$). The average consensus problem is said to be solvable on a graph $\mathcal{G}$ if there exists a feedback K compatible with $\mathcal{G}$ solving the average consensus problem. The following result completely characterize those graphs for which the average consensus problem is solvable.

Proposition 1. *Let $\mathcal{G}$ be a directed graph and assume that $\mathcal{G}$ contains all loops (i, i). The following conditions are equivalent:*

(A) The average consensus problem is solvable on $\mathcal{G}$.
(B) $\mathcal{G}$ is strongly connected.

Furthermore, if the above conditions are satisfied, any K such that $I + K$ is doubly stochastic and $\mathcal{G}_{I+K} = \mathcal{G}$, solves the average consensus problem.

Now in our setup we assume that the communication network is constituted only of digital links. This implies that the exchange of perfect information between the systems is not allowed. In fact, through a digital channel, the i-th agent can only send to the j-th agent symbolic data that will be used by the j-th agent to built at most an estimate of the i-th agent's state. Here we consider a control law which has the same form of (3) where, in place of the exact knowledge of the states of the systems, we substitute estimates calculated according to the symbols sent through the communication network.

More precisely, first we assume we have a fixed strongly connected graph $\mathcal{G}$ and a matrix K such that $I + K$ is doubly stochastic and $\mathcal{G}_{I+K} = \mathcal{G}$. The control input u_i has then the following form

$$u_i = K_{ii}x_i + \sum_{\substack{j=1 \\ j \neq i}}^{N} K_{ij}\hat{x}_{ij} \,, \tag{5}$$

where $\hat{x}_{ij}$ is the estimate of the state x_j which has been obtained by the agent i.

Now we proceed to explain how the estimate $\hat{x}_{ij}$ is obtained. Suppose that the j-th agent sends to the i-th agent, through a digital channel, at each time instant t, a symbol $s_{ij}(t)$ belonging to a finite or denumerable alphabet $\mathcal{S}_{ij}$. It is assumed that each symbol transmitted is received without error. In general, see [21], the structure of the coder by which the j-th agent produces the symbol to be sent to the i-th agent can be described by the following equations

$$\begin{cases} \xi_{ij}(t+1) = F_{ij}(\xi_{ij}(t), s_{ij}(t)) \\ \quad s_{ij}(t) = Q_{ij}(\xi_{ij}(t), x_j(t), u_j(t)) \end{cases} \tag{6}$$

where $s_{ij}(t) \in \mathcal{S}_{ij}, \xi_{ij}(t) \in \Xi_{ij}, Q_{ij} : \Xi_{ij} \times \mathbb{R} \times \mathbb{R} \to \mathcal{S}_{ij}$, and $F_{ij} : \Xi \times \mathcal{S}_{ij} \to \Xi_{ij}$ and where also the set Ξ_{ij} is finite or denumerable. The decoder, placed at the system i, coincides with the system

$$\begin{cases} \xi_{ij}(t+1) = F_{ij}(\xi_{ij}(t), s_{ij}(t)) \\ \quad \hat{x}_{ij}(t) = H_{ij}(\xi_{ij}(t), s_{ij}(t)), \end{cases} \tag{7}$$

where $H_{ij} : \Xi_{ij} \times \mathcal{S}_{ij} \to \mathbb{R}$.

In general, we may have different encoders at system j, according to the various systems the system j wants to send its data. For the sake of notational convenience, we assume however, in this paper, that system j uses the same encoder for all data transmissions. Thus, system j will send the same symbol $s_j(t) := s_{ij}(t)$ to all the other systems i which receive information from it. In this case all systems receiving data from j, will obtain the same estimate of x_j, namely we can define a single state estimate $\hat{x}_j := \hat{x}_{ij}$. In this way the previous coder/decoder couple can be represented by the following state estimator with memory

$$\begin{cases} \xi_j(t+1) = F_j(\xi_j(t), s_j(t)) \\ \quad s_j(t) = Q_j(\xi_j(t), x_j(t), u_j(t)) \\ \quad \hat{x}_j(t) = H_j(\xi_j(t), s_j(t)) \end{cases} \tag{8}$$

We point out that all the result presented in this paper can be extended to the more general case.

The main objective of the present paper is to understand whether it is possible to design some smart encoding/decoding strategies such that a control law of the form (5) yields the consensus for the overall system. In the sequel we concentrate our attention on a particular way of exchanging information which fits into the previous scheme: the logarithmic quantized strategy.

4 Logarithmic Quantizers

This strategy is based on the techniques proposed in [17]. In this case we assume the following form for equation (8)

$$\begin{cases} \xi_j(t+1) = \xi_j(t) + s_j(t) \\ \quad\ s_j(t) = q_L\left(x_j(t) - \xi_j(t)\right) \\ \quad\ \hat{x}_j(t) = \xi_j(t) + s_j(t) \end{cases} \qquad (9)$$

where q_L is a logarithmic quantizer depending on a parameter $\delta \in\,]0, 1[$, precisely defined as follows. Suppose that $x \in \mathbb{R}^+$ and that $0 < \delta < 1$ and let $k \in \mathbb{Z}$ be such that $\frac{(1+\delta)^{k-1}}{(1-\delta)^k} \leq x \leq \frac{(1+\delta)^k}{(1-\delta)^{k+1}}$. We then define

$$q_L(x) = \left(\frac{1+\delta}{1-\delta}\right)^k.$$

If $x < 0$, then we define $q_L(x) = -q_L(-x)$. The graph of the logarithmic quantizer is depicted in Figure 1.

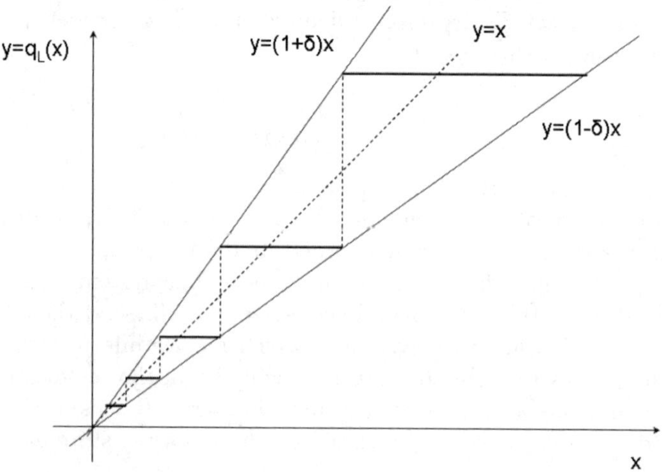

Fig. 1. Logarithmic quantizer

Notice that the logarithmic quantizer is such that

$$|q_L(x) - x| \leq \delta|x|$$

and so the parameter δ can be seen as the precision of the quantizer. Moreover δ determines also the number of quantization intervals we have in any finite subset of $\mathbb{R}$. It is clear that the sets $\mathcal{S}_j$ are denumerable. We impose the initial condition $\xi_j(0) = 0$. One can verify immediately that the estimate $\hat{x}_j(t)$ satisfies the following recursive relation

$$\hat{x}_j(t+1) = \hat{x}_j(t) + q_L\left(x_j(t+1) - \hat{x}_j(t)\right). \qquad (10)$$

Notice that $\xi_j(0) = 0$ implies $\hat{x}_j(0) = q_L(x_j(0))$. Now if we define

$$\varepsilon_j(t) = \frac{q_L(x_j(t+1) - \hat{x}_j(t)) - (x_j(t+1) - \hat{x}_j(t))}{x_j(t+1) - \hat{x}_j(t)}$$

we obtain that

$$\hat{x}_j(t+1) = \hat{x}_j(t) + (1 + \varepsilon_j(t))(x_j(t+1) - \hat{x}_j(t)). \tag{11}$$

where $-\delta \le \varepsilon_j(t) \le +\delta$.

By defining the matrix $E(t) = diag\{\varepsilon_1(t), \ldots, \varepsilon_N(t)\}$ the equations (1), (5) and (11) can be rewritten in the following vector form

$$\begin{bmatrix} x(t+1) \\ \hat{x}(t+1) \end{bmatrix} = \begin{bmatrix} diag\{K\} + I & K - diag\{K\} \\ diag\{K\} + I & K - diag\{K\} \end{bmatrix} \begin{bmatrix} x(t) \\ \hat{x}(t) \end{bmatrix} +$$

$$+ \begin{bmatrix} 0 & 0 \\ E(t)(diag\{K\} + I) & E(t)(K - diag\{K\} - I) \end{bmatrix} \begin{bmatrix} x(t) \\ \hat{x}(t) \end{bmatrix}. \tag{12}$$

In order to analyze the previous system, it is useful to introduce the new variables $y(t) = Kx(t), e(t) = x(t) - \hat{x}(t)$, where $e(t)$ expresses the estimation error. Assumptions (A), (B), and (C) made on K implies that the consensus problem is solved if and only if $y(t) \to 0$. Moreover, observe that $y(t)^T \mathbb{1} = 0$, $\forall t \ge 0$. By straightforward calculations we obtain

$$\begin{bmatrix} y(t+1) \\ e(t+1) \end{bmatrix} = \begin{bmatrix} I + K & K(-K + diag\{K\}) \\ 0 & 0 \end{bmatrix} \begin{bmatrix} y(t) \\ e(t) \end{bmatrix} +$$

$$+ \begin{bmatrix} 0 & 0 \\ -E(t) & -E(t)(I + diag\{K\} - K) \end{bmatrix} \begin{bmatrix} y(t) \\ e(t) \end{bmatrix} \tag{13}$$

From now on, for the sake of the notational convenience we denote $A = I + K$, $B = K(-K + diag\{K\})$, $C = -I$ and $D = -(I + diag\{K\} - K)$. Hence

$$\begin{bmatrix} y(t+1) \\ e(t+1) \end{bmatrix} = \begin{bmatrix} I & 0 \\ 0 & -E(t) \end{bmatrix} \begin{bmatrix} A & B \\ C & D \end{bmatrix} \begin{bmatrix} y(t) \\ e(t) \end{bmatrix}. \tag{14}$$

Finally let

$$\mathcal{A}(t) = \begin{bmatrix} I & 0 \\ 0 & -E(t) \end{bmatrix} \begin{bmatrix} A & B \\ C & D \end{bmatrix} \tag{15}$$

and

$$v(t) = \begin{bmatrix} y(t) \\ e(t) \end{bmatrix} \tag{16}$$

from which

$$v(t+1) = \mathcal{A}(t)v(t). \tag{17}$$

The question we want to address now is if there exist some conditions on the magnitude of δ, which guarantee that the consensus can be reached. This analysis is carried out in the following section where the system (14) is treated as a linear parameter varying (LPV) system [22].

5 Convergence Analysis

We start by rewriting (14) in a more suitable way. Let

$$\mathcal{E} = \left\{ E \in \mathbb{R}^{N \times N} : E = diag\{e_1, \ldots, e_N\}, \ e_i = \pm 1 \ \ 1 \leq i \leq N \right\}$$

Notice that $\mathcal{E}$ contains 2^N elements. Hence we can write $\mathcal{E} = \{E_1, \ldots, E_{2^N}\}$, where we are assuming that some suitable way to enumerate the matrices inside $\mathcal{E}$ is used. We assume that $E_1 = I$. By means of the above definitions we can introduce an another set of matrices

$$\mathcal{R} = \left\{ R_i = \begin{bmatrix} I & 0 \\ 0 & \delta E_i \end{bmatrix} \begin{bmatrix} A & B \\ C & D \end{bmatrix} : E_i \in \mathcal{E} \right\}.$$

The set $\mathcal{R}$ is useful because it is easy to see that the matrix $\mathcal{A}(t)$, defined in (15), belongs to $Co\{\mathcal{R}\}$, $\forall t \geq 0$, where $Co\{\mathcal{R}\}$ denote the convex hull of the set $\mathcal{R}$. In other words there exist $\lambda_1(t), \lambda_2(t), \ldots, \lambda_{2^N}(t)$ nonnegative real numbers λ_i such that $\sum_{i=1}^{2^N} \lambda_i(t) = 1$ and

$$\mathcal{A}(t) = \sum_{i=1}^{2^N} \lambda_i(t) R_i.$$

This problem formulation allows us to analyze (14) by means of Lyapunov approach proposed in [22]. In fact, it is well known in the literature that if we consider the system

$$x(t+1) = F(t)x(t), \ \ x(t) \in \mathbb{R}^n, \ F(t) \in \mathbb{R}^{n \times n},$$

where $F(t) \in Co(\mathcal{F})$ with $\mathcal{F} = \{F_1, \ldots, F_m\}$, a sufficient condition ensuring the stability of the system is the existence of a definite positive matrix $P \in \mathbb{R}^{n \times n}$ such that

$$\frac{1}{2}\left(F_i^T P F_j + F_j^T P F_i\right) - P < 0, \ \ \forall (F_i, F_j) \in \mathcal{F} \times \mathcal{F}, \tag{18}$$

or equivalently

$$\frac{1}{2}\left(F_i P F_j^T + F_j P F_i^T\right) - P < 0, \ \ \forall (F_i, F_j) \in \mathcal{F} \times \mathcal{F}. \tag{19}$$

This last condition is called the dual condition. Our situation is slightly different, since we are addressing the stability of (17) in the hyperplane $\{[y^T \ e^T]^T \in \mathbb{R}^{2N} : y^T \mathbb{1} = 0\}$. However it is possible to provide sufficient conditions similar to (18) and (19) that ensure the stability of (17). They are stated in the following Lemma.

Lemma 1. *Consider the system (17). If there exists a positive semidefinite matrix $P \in \mathbb{R}^{2N \times 2N}$ such that*

$$z^T P z > 0, \tag{20}$$

and

$$z^T \left(\frac{R_i^T P R_j + R_j^T P R_i}{2} - P \right) z < 0, \quad \forall (R_i, R_j) \in \mathcal{R} \times \mathcal{R}, \qquad (21)$$

for each nonzero $z \in \mathbb{R}^{2N}$ such that $[\mathbb{1}^T \, \mathbf{0}^T] z = 0$ ($\mathbf{0}$ denotes the N dimensional column vector with all zeros), then $\lim_{t \to +\infty} v(t) = 0$, $\forall v(0) = [y(0)^T \, e(0)^T]^T$ such that $y(0)^T \mathbb{1} = 0$, and $\forall \{\mathcal{A}(t)\}_{t=0}^{\infty}$.

Equivalently if there exists $P \geq 0 \in \mathbb{R}^{2N \times 2N}$ such that

$$z^T P z > 0, \qquad (22)$$

and

$$z^T \left(\frac{R_i P R_j^T + R_j P R_i^T}{2} - P \right) z < 0, \quad \forall (R_i, R_j) \in \mathcal{R} \times \mathcal{R}, \qquad (23)$$

for each nonzero z such that $z \notin \mathrm{span} \left([\mathbb{1}^T \, \mathbf{0}^T]^T \right)$ then $\lim_{t \to +\infty} v(t) = 0$, $\forall v(0) = [y(0)^T \, e(0)^T]^T$ such that $y(0)^T \mathbb{1} = 0$, and $\forall \{\mathcal{A}(t)\}_{t=0}^{\infty}$.

Proof. We report here the proof only of the first part of the theorem. The dual condition can be proved in an analogous way. Before proceeding, we introduce the following notation that will be useful during the proof. With $\mathbb{1}_{N-1}$ and with 0_{N-1} we denote the $N-1$ dimensional vectors having respectively all the components equal to 1 and all the components equal to 0; with I_{N-1} we indicate the $(N-1) \times (N-1)$ identity matrix. We start by considering the following change of coordinates

$$w(t) = \begin{bmatrix} T^{-1} & 0 \\ 0 & I \end{bmatrix} z(t)$$

where

$$T^{-1} = \begin{bmatrix} 1 & \mathbb{1}_{N-1}^T \\ 0_{N-1} & I_{N-1} \end{bmatrix} \in \mathbb{R}^{N \times N}.$$

Notice that

$$T = \begin{bmatrix} 1 & -\mathbb{1}_{N-1}^T \\ 0_{N-1} & I_{N-1} \end{bmatrix}$$

Let L^{-1} denote the matrix $\begin{bmatrix} T^{-1} & 0 \\ 0 & I \end{bmatrix}$. We have that

$$w(t+1) = \left\{ \sum_{i=1}^{2^N} \lambda_i(t) L^{-1} R_i L \right\} w(t) = \left\{ \sum_{i=1}^{2^N} \lambda_i(t) G_i \right\} w(t)$$

where

$$G_i = L^{-1} R_i L = \begin{bmatrix} I & 0 \\ 0 & \delta E_i \end{bmatrix} \begin{bmatrix} A & B \\ C & D \end{bmatrix}$$

$$= \begin{bmatrix} T^{-1} A T & T^{-1} B \\ E_i T & E_i D \end{bmatrix}$$

$$= \begin{bmatrix} 1 & 0_{N-1}^T \\ * & \tilde{R}_i \end{bmatrix}.$$

Notice that in the last expression $*$ denote a suitable $(N-1)$ dimensional vector whereas $\tilde{R}_i \in \mathbb{R}^{(N-1)\times(N-1)}$ is obtained from R_i by taking off the first row and the first column. Now let $[w_1(t), \ldots, w_{2N}(t)]$ denote the $2N$ components of the vector $w(t)$. By the structure of L it is immediate to see that $w_1(0) = 0$. Moreover, since $w(t) = L^{-1}v(t) = [(T^{-1}y(t))^T \ \ e(t)^T]^T = [(T^{-1}Kx(t))^T \ \ e(t)^T]^T$ and since it can be checked easily that the first row of $T^{-1}K$ has all the components equal to 0, we have also $w_1(t) = 0, \forall t > 0$. Hence, letting $\tilde{w}(t) = [w_2(t), \ldots, w_{2N}(t)]^T \in \mathbb{R}^{2N-1}$ we have that

$$\tilde{w}(t+1) = \left(\sum_{i=1}^{2^N} \lambda_i(t)\tilde{R}_i \right) \tilde{w}(t). \tag{24}$$

Clearly $\lim_{t\to+\infty} v(t) = 0$ if and only if $\lim_{t\to+\infty} \tilde{w}(t) = 0$. Assume now that there exist $\tilde{P} \in \mathbb{R}^{(2N-1)\times(2N-1)}$ such that $\tilde{P} > 0$ and

$$\frac{\tilde{R}_i^T \tilde{P}\tilde{R}_j + \tilde{R}_j^T \tilde{P}\tilde{R}_i}{2} - \tilde{P} < 0.$$

As said before this is a sufficient condition in order to guarantee that (24) is stable. It follows that $\eta^T diag\left\{0, \tilde{P}\right\} \eta > 0$, and

$$\eta^T \left[\frac{G_i^T diag\left\{0, \tilde{P}\right\} G_j + G_j^T diag\left\{0, \tilde{P}\right\} G_i}{2} - diag\left\{0, \tilde{P}\right\} \right] \eta < 0,$$

for any $\eta \in \mathbb{R}^{2N}$ such that $[1, 0, \ldots, 0]\eta = 0$. Hence, if we define

$$P = \left(L^{-1}\right)^T diag\left\{0, \tilde{P}\right\} L^{-1},$$

one can verify, after some algebraic manipulations, that $(L\eta)^T P (L\eta) > 0$ and

$$(L\eta)^T \left(\frac{R_i^T PR_j + R_j^T PR_i}{2} - P \right) L\eta < 0$$

for any nonzero η such that $[1, 0, \ldots, 0]\eta = 0$. In order to conclude the proof it remains to prove that

$$\{L\eta : [1, 0, \ldots, 0]\eta = 0\} = \left\{z \in \mathbb{R}^{2N} : [\mathbb{1}^T \ \mathbb{0}^T] z = 0\right\},$$

but this is quite straightforward.

Now we will show that, under a certain condition on the magnitude of δ, it is possible to exhibit a particular matrix P such that (22) and (23), i.e. the dual conditions in Lemma 1, are satisfied. In order to do so, it is useful to introduce the following set of matrices

$$\mathcal{T} = \left\{T \in \mathbb{R}^{N\times N} : T\mathbb{1} = 0, \ z^T(T - ATA^T)z > 0 \ \ \forall z \in \text{span}(\mathbb{1})^\perp\right\}.$$

The importance of this set is clarified by the following theorem.

Theorem 1. *Consider the system (17) and let T be any matrix in $\mathcal{T}$. Let*

$$\bar{\alpha} = \max\left\{\lambda : \lambda \in \sigma\left((T - ATA^T)^\dagger BB^T\right)\right\}, \tag{25}$$

and for any $\alpha \in R$, let $X_1(\alpha) = \alpha T - \alpha ATA^T - BB^T$, $X_2(\alpha) = \alpha AT + BD^T$, and $X_3(\alpha) = \alpha T + DD^T$. Finally let

$$\bar{\delta}_0 = \max_{\alpha > \bar{\alpha}} \frac{1}{\max\left\{\lambda : \lambda \in \sigma\left(X_3(\alpha) + \frac{1}{4}X_2^T(\alpha)X_1^\dagger(\alpha)X_2(\alpha)\right)\right\}}. \tag{26}$$

Then for all $\delta \leq \bar{\delta}_0$ we have consensus, namely $\lim_{t \to +\infty} e(t) = 0$, and $\lim_{t \to +\infty} x(t) = \gamma \mathbb{1}$, $\forall x(0) \in \mathbb{R}^N$ and $\forall \{\mathcal{A}(t)\}_{t=0}^\infty$.

In order to prove the above results we need the following technical results.

Lemma 2. *Let S and R be semidefinite matrices belonging to $\mathbb{R}^{N \times N}$ such that $S\mathbb{1} = 0$, $R\mathbb{1} = 0$ and $\mathrm{rk}\, S = N - 1$. Let α be a real number. Then the following facts are equivalent*

(i) $z^T(\alpha S - R)z > 0$, $\forall z \notin \mathrm{span}(\mathbb{1})$.
(ii) $\alpha > \max\left\{\lambda : \lambda \in \sigma(S^\dagger R)\right\}$.

Proof. It is easy to decompose S as $S = Q^2$ where also $Q\mathbb{1} = 0$ and $\mathrm{rk}\, Q = N-1$. The following chain of equivalences holds

$$z^T(\alpha Q^2 - R)z > 0 \; \forall z \notin \mathrm{span}(\mathbb{1})$$
$$\Updownarrow$$
$$z^T(\alpha Q^2 - R)z > 0 \; \forall z \in \mathrm{span}(\mathbb{1})^\perp \; \{0\}$$
$$\Updownarrow$$
$$z^T Q(\alpha I - Q^\dagger R Q^\dagger)Qz > 0 \; \forall z \in \mathrm{span}(\mathbb{1})^\perp \setminus \{0\}$$
$$\Updownarrow$$
$$z^T(\alpha I - Q^\dagger R Q^\dagger)z > 0 \; \forall z \in \mathrm{span}(\mathbb{1})^\perp \; \{0\}.$$

The first equivalence is a consequence of the fact that $(\alpha Q - R)\mathbb{1} = 0$ whereas the other ones descend directly from the facts that $Q^\dagger Qz = z$, $\forall z \in \mathrm{span}(\mathbb{1})^\perp$ and $Q\mathrm{span}(\mathbb{1})^\perp = \mathrm{span}(\mathbb{1})^\perp$ and $Q^\dagger \mathrm{span}(\mathbb{1})^\perp = \mathrm{span}(\mathbb{1})^\perp$. Obviously the last condition is satisfied if and only if $\max\left\{\lambda : \lambda \in \sigma(Q^\dagger R Q)\right\} < \alpha$. It is not difficult to prove that $\sigma(Q^\dagger R Q) = \sigma(Q^\dagger Q^\dagger R Q Q) = \sigma(Q^\dagger Q^\dagger R) = \sigma(Q^{\dagger^2} R) = \sigma(S^\dagger R)$. This concludes the proof.

Lemma 3. *Suppose that a symmetric matrix X is partitioned as*

$$X = \begin{bmatrix} X_1 & X_2 \\ X_2^T & X_3 \end{bmatrix}.$$

where X_1 and X_3 are square. Then the following facts are equivalent

1. $z^T X z > 0$, $\forall z \notin \mathrm{span}\left([\mathbb{1}^T\, 0^T]^T\right)$
2. (i) $z_1^T X_1 z_1 > 0$ $\forall z_1 \notin \mathrm{span}(\mathbb{1})$

(ii) $X_3 - X_2^T X_1^\dagger X_2 > 0$

(iii) $Ker\, X_1 \subseteq Ker\, X_2^T$.

Proof. First consider the sufficiency. Since $Ker\, X_1 \subseteq Ker\, X_2^T$ we have that $Im\, X_2 \subseteq Im\, X_1$ and hence $X_2 = X_1 K$ for a suitable matrix K. Now let $Y = -X_1^\dagger X_2$ and calculate

$$\begin{bmatrix} I & 0 \\ Y^T & I \end{bmatrix} \begin{bmatrix} X_1 & X_2 \\ X_2^T & X_3 \end{bmatrix} \begin{bmatrix} I & Y \\ 0 & I \end{bmatrix} = \begin{bmatrix} X_1 & X_1 Y + X_2 \\ Y^T X_1 + X_2^T & Y^T X_1 Y + X_2^T Y + Y^T X_2 + X_3 \end{bmatrix}$$

$$= \begin{bmatrix} X_1 & -X_1 X_1^\dagger X_2 + X_2 \\ -X_2^T X_1^\dagger X_1 + X_2^T & X_2^T X_1^\dagger X_1 X_1^\dagger X_2 - X_2^T X_1^\dagger X_2 - X_2^T X_1^\dagger X_2 + X_3 \end{bmatrix}$$

$$= \begin{bmatrix} X_1 & -X_1 X_1^\dagger X_1 K + X_2 \\ -K^T X_1 X_1^\dagger X_1 + X_2^T & X_2^T X_1^\dagger X_2 - X_2^T X_1^\dagger X_2 - X_2^T X_1^\dagger X_2 + X_3 \end{bmatrix}$$

$$= \begin{bmatrix} X_1 & -X_2 + X_2 \\ -X_2 + X_2^T & X_3 - X_2^T X_1^\dagger X_2 \end{bmatrix}$$

$$= \begin{bmatrix} X_1 & 0 \\ 0 & X_3 - X_2^T X_1^\dagger X_2 \end{bmatrix}$$

Since the right-hand side is semidefinite positive, the positive semidefiniteness of X follows from the exhibited congruence.

Consider now the necessity. By choosing $z = [z_1^T\ \mathbf{0}^T]^T$ with $z_1 \notin \mathrm{span}(\mathbf{1})$ it is immediate to show that $z_1^T X_1 z_1 > 0$, $\forall z_1 \notin \mathrm{span}(\mathbf{1})$.

Suppose now that there exist z_1 such that $v = X_2^T z_1 \neq 0$ and $X_1 z_1 = 0$. Let z_2 be such that $z_2^T v = \gamma \neq 0$. Then

$$[\alpha z_1^T\ z_2^T] \begin{bmatrix} X_1 & X_2 \\ X_2^T & X_3 \end{bmatrix} \begin{bmatrix} \alpha z_1 \\ z_2 \end{bmatrix} = \alpha z_2^T X_2^T z_1 + \alpha z_1^T X_2 z_2 + z_2^T X_3 z_2 = 2\alpha\gamma + z_2^T X_3 z_2.$$

If we choose $\alpha = -\gamma$ whit γ sufficiently large we have that the above quantity is negative contradicting the hypothesis. Hence $Ker X_1 \subseteq Ker X_2^T$. The necessity of $X_3 - X_2^T X_1^\dagger X_2$ follows from the congruence exhibited previously.

Proof (Proof of Theorem 1)

In order to prove the statement of the theorem we show that, for $\delta < \bar\delta_0$, there exists a suitable matrix $P \in \mathbb{R}^{2N \times 2N}$ such that

$$P \begin{bmatrix} \mathbf{1} \\ \mathbf{0} \end{bmatrix} = 0, \tag{27}$$

and

$$z^T P z > 0 \tag{28}$$

and

$$z^T \left(\frac{1}{2} \left(R_i^T P R_j + R_j^T P R_i \right) - P \right) z < 0, \quad \forall\, (R_i, R_j) \in \mathcal{R} \times \mathcal{R}, \tag{29}$$

for each non zero $z \notin \text{span}([\mathbb{1}^T\ \mathbb{0}^T]^T)$. The candidate matrix P has the following form

$$P = \begin{bmatrix} \alpha T & 0 \\ 0 & I \end{bmatrix}$$

where T is any matrix in $\mathcal{T}$ and where α is a suitable positive scalar that we will determine next. It is immediate to see that (27) is satisfied. Moreover one can verify that P has an eigenvalue equal to 0 and all the other eigenvalues positive. Furthermore, the eigenspace associate to the eigenvalue 0 is spanned by the vector $[\mathbb{1}^T\ \mathbb{0}^T]^T$. Hence P satisfies (28).

Now we calculate $\frac{1}{2}\left(R_i P R_j^T + R_j P R_i^T\right) - P$ obtaining the following matrix

$$\begin{bmatrix} \alpha A T A^T + BB^T - \alpha T & \frac{1}{2}\delta(\alpha AT + BD^T)(E_i + E_j) \\ \frac{1}{2}(E_i + E_j)\delta(\alpha T A^T + DB^T) & \frac{1}{2}\delta^2\left\{E_i(\alpha T + DD^T)E_j + E_j(\alpha T + DD^T)E_i\right\} - I \end{bmatrix} \tag{30}$$

By Lemma 3 we have that (30) satisfies (29) if and only if

$$z^T(\alpha T - \alpha A T A^T - BB^T)z > 0, \ \forall z \notin \text{span}(\mathbb{1}), \tag{31}$$

and

$$I - \frac{1}{2}\delta^2\left\{E_i(\alpha T + DD^T)E_j + E_j(\alpha T + DD^T)E_i\right\} -$$
$$- \frac{1}{4}\delta^2(E_i + E_j)(\alpha T A^T + DB^T)(\alpha T - \alpha A T A^T - BB^T)^\dagger \cdot$$
$$(\alpha AT + BD^T)(E_i + E_j) > 0.$$

By Lemma 2, (31) holds if and only if

$$\alpha > \max \sigma\left((T - A T A^T)^\dagger BB^T\right). \tag{32}$$

Now observe that

$$I - \tfrac{1}{2}\delta^2\left\{E_i(\alpha T + DD^T)E_j + E_j(\alpha T + DD^T)E_i\right\} - \tfrac{1}{4}\delta^2(E_i + E_j)\cdot$$
$$\cdot(\alpha T A^T + DB^T)(\alpha T - \alpha A T A^T - BB^T)^\dagger(\alpha AT + BD^T)(E_i + E_j) =$$
$$I - \tfrac{1}{4}\delta^2\left\{4(E_i + E_j)(\alpha T + DD^T)(E_i + E_j) - 2E_j(\alpha T + DD^T)E_j\right.$$
$$-2E_i(\alpha T + DD^T)E_i + (E_i + E_j)(\alpha T A^T + DB^T)\cdot$$
$$\left.\cdot(\alpha T - \alpha A T A^T - BB^T)^\dagger(\alpha AT + BD^T)(E_i + E_j)\right\} >$$
$$I - \tfrac{1}{4}\delta^2\left\{4(E_i + E_j)(\alpha T + DD^T)(E_i + E_j) + (E_i + E_j)(\alpha T A^T + DB^T)\cdot\right.$$
$$\left.\cdot(\alpha T - \alpha A T A^T - BB^T)^\dagger(\alpha AT + BD^T)(E_i + E_j)\right\} >$$
$$I - \tfrac{1}{4}\delta^2\left\{4(\alpha T + DD^T) + (\alpha T A^T + DB^T)(\alpha T - \alpha A T A^T - BB^T)^\dagger\cdot\right.$$
$$\left.\cdot(\alpha AT + BD^T)\right\}.$$

Clearly

$$I - \tfrac{1}{4}\delta^2\left\{4(\alpha T + DD^T) + (\alpha T A^T + DB^T)(\alpha T - \alpha A T A^T - BB^T)^\dagger\cdot\right.$$
$$\left.\cdot(\alpha AT + BD^T)\right\} > 0$$

if and only if

$$\delta < \frac{4}{\max\left\{\lambda : \lambda \in \sigma\left(4(\alpha T + DD^T) + (\alpha T A^T + DB^T)\cdot \\ \cdot(\alpha T - ATA^T - BB^T)^\dagger(\alpha AT + BD^T)\right)\right\}}.$$

This concludes the proof.

In simple words Theorem 1 guarantees that the consensus can be reached by using the same control law that solves the consensus problem when only exchanges of perfect information are assumed, although the systems can share only logarithmically quantized information. This holds provided that a certain condition on the magnitude of δ is satisfied. However for a general matrix K, the expression of $\bar{\delta}_0$ given in (26) is not of immediate interest. In the following, in order to obtain some interesting consequences of this formula, we will restrict our attention on two cases in which K exhibits a particular structure: when K is a generic symmetric matrix and when K is symmetric and circulant.

5.1 K Symmetric

We start this section by recalling the following definition. Let P be any matrix such that $P\mathbb{1} = \mathbb{1}$ and assume that its spectrum $\sigma(P)$ is contained in the closed unit disk centered in 0. Define

$$\rho(P) = \begin{cases} 1 & \text{if } \dim \ker(P - I) > 1 \\ \max\{|\lambda| : \lambda \in \sigma(P) \setminus \{1\}\} & \text{if } \dim \ker(P - I) = 1, \end{cases} \tag{33}$$

which is called the essential spectral radius of P. It is well known in literature [18], that in the case of average consensus controllers, this quantity is responsible of the speed of convergence to the equilibrium point.

Now let K be symmetric. Then the following result holds.

Theorem 2. *Let K be symmetric and let $\bar{\rho} = \underline{\rho}$. Let for any $\epsilon > 0$*

$$\bar{\delta}_1 = \frac{4\epsilon\left(1 - \bar{\rho}\right)^4}{(\epsilon + 4)^2 + 20(\epsilon + 4)(1 - \bar{\rho})^2 + 68(1 - \bar{\rho})^4}. \tag{34}$$

Then for any $\delta \leq \bar{\delta}_1$ we have consensus, namely $\lim_{t \to +\infty} e(t) = 0$, and $\lim_{t \to +\infty} x(t) = \gamma\mathbb{1}, \ \forall x(0) \in \mathbb{R}^N$ and $\forall \{\mathcal{A}(t)\}_{t=0}^{\infty}$.

Proof. Consider the following particular matrix $T \in \mathcal{T}$

$$T = I - \frac{1}{N}\mathbb{1}\mathbb{1}^T.$$

It is easy to see that T commute with any doubly stochastic matrix. We impose that α is such that

$$\alpha(I - A^2)T - BB^T \geq \epsilon T \tag{35}$$

where ϵ is a fixed positive real number. (35) is satisfied if and only if

$$\alpha \geq \max\left\{\sigma\left\{\left[(I - A^2)T\right]^\dagger (\epsilon T + BB^T)\right\}\right\}.$$

Moreover (35) implies $\alpha > \bar{\alpha}$. Notice now that

$$\max\left\{\sigma\left\{\left[(I - A^2)T\right]^\dagger (\epsilon T + BB^T)\right\}\right\} \leq \left\|\left[(I - A^2)T\right]^\dagger\right\| \left\|(\epsilon T + BB^T)\right\|$$

$$\leq \left(\frac{1}{1 - \bar{\rho}}\right)^2 (\epsilon + 4).$$

Hence we can assume that $\alpha = \left(\frac{1}{1-\bar{\rho}}\right)^2 (\epsilon + 4)$. From (26) we obtain that

$$\bar{\delta}_0 \geq \frac{1}{\max\left\{\lambda : \lambda \in \sigma\left(X_3(\alpha) + \frac{1}{4}X_2^T(\alpha)X_1^\dagger(\alpha)X_2(\alpha)\right)\right\}}$$

$$\geq \frac{4}{\max\left\{\lambda : \lambda \in \sigma\left(4(\alpha T + DD^T) + (\alpha TA^T + DB^T)\cdot\right.\right.}$$

$$\overline{\cdot(\alpha T - ATA^T - BB^T)^\dagger(\alpha AT + BD^T))\}}$$

$$\geq \frac{4}{\|4(\alpha T + DD^T)\| + \|\alpha TA + DB^T\|^2 \|(\alpha T - \alpha ATA - BB^T)^\dagger\|}$$

$$\geq \frac{4}{4\alpha\|T\| + \|DD^T\| + (\|\alpha\|\|T\|\|A\| + \|DB^T\|)^2 \|(\alpha T - \alpha ATA - BB^T)^\dagger\|}$$

Notice that condition (35) implies that $\|(\alpha T - \alpha ATA^T - BB^T)^\dagger\| < \frac{1}{\epsilon}$. Moreover we have $\|T\| = 1$, $\|A\| = 1$, $\|DD^T\| \leq 4$ and $\|DB^T\| \leq 8$. Hence

$$\bar{\delta}_0 \geq \frac{4}{4\alpha + 4 + \frac{1}{\epsilon}(\alpha + 8)^2}.$$

By substituting the expression of α in the above expression we obtain (36).

5.2   *K* Symmetric and Circulant

The stability result proposed in the previous section is not the best one can find. Indeed, consider a strongly connected circulant undirected graph $\mathcal{G}(V, \mathcal{W})$, where V and $\mathcal{W}$ are respectively the set of vertices and the set of arcs, and where $|V| = N$. Assume that the in-degree of the graph is $\nu + 1$. We associate to the graph $\mathcal{G}(V, \mathcal{W})$ the matrix K

$$K_{ij} = \begin{cases} \frac{1}{\nu+1} & \text{if } i \neq j \text{ and } i \to j \\ -\frac{\nu}{\nu+1} & \text{if } i = j \\ 0 & \text{otherwise} \end{cases}$$

Since from [18] we have that

$$\rho(I + K) \geq 1 - CN^{-2/\nu},$$

where C is a constant independent of $\mathcal{G}$, then in this case Theorem 5.5 guarantees consensus stability only for all $\delta \leq \bar{\delta}_1$ where however $\bar{\delta}_1$ tends to 0 as N tends to $+\infty$. This seems to suggest that for this class consensus controllers the stability occurs only with logarithmic quantizers having precision which tends to infinity. We show in this section that this is not true as proved by the following theorem.

Theorem 3. *Let K be the matrix defined above. Let for any $\epsilon > 0$*

$$\bar{\delta}_2 = \frac{8}{\nu + 1}. \tag{36}$$

Then for any $\delta \leq \delta_2$ we have consensus, namely $\lim_{t \to +\infty} e(t) = 0$, and $\lim_{t \to +\infty} x(t) = \gamma \mathbb{1}$, $\forall\, x(0) \in \mathbb{R}^N$ and $\forall\, \{\mathcal{A}(t)\}_{t=0}^{\infty}$.

Proof. Now let L be the Laplacian matrix corresponding to $\mathcal{G}\,(V, \mathcal{W})$, i.e.

$$L_{ij} = \begin{cases} -1 & \text{if } i \neq j \;\text{ and } i \to j \\ \nu & \text{if } i = j \\ 0 & \text{otherwise} \end{cases}$$

Finally let $T = L$. It is easy to verify that, letting $A = I - K$ then $B = (A - I)\left(-A + \frac{1}{1+\nu}I\right)$, $D = \frac{2+\nu}{\nu}I - A$ and $T = (\nu + 1)(I - A)$. We recall that any circulant matrix H can be diagonalized by the same matrix F, i.e.

$$F^{-1}HF = diag\left\{\lambda_0(H), \ldots, \lambda_{N-1}(H)\right\},$$

where $\{\lambda_0(H), \ldots, \lambda_{N-1}(H)\}$ is the set of eigenvalues of the matrix R. Then the condition (32) can be rewritten as

$$\bar{\alpha} = \max\left\{0, \max_{1 \leq i \leq N-1}\left\{\frac{\lambda_i^2\,(I - A)\,\lambda_i^2\,(1/(\nu + 1)I - A)}{(\nu + 1)\lambda_i^2(I - A)\lambda_i(I + A)}\right\}\right\}$$

$$= \max_{1 \leq i \leq N-1}\left\{\frac{\lambda_i^2\left(\frac{1}{\nu+1}I - A\right)}{(\nu + 1)\lambda_i(I + A)}\right\}$$

By noticing that

$$\sigma\{A\} \subseteq \left[-1 + \frac{2}{\nu + 1}, 1\right],$$

and

$$\sigma\left\{\left(\frac{1}{\nu + 1}I - A\right)\right\} \subseteq \left[-\frac{\nu}{\nu + 1}, \frac{\nu}{\nu + 1}\right].$$

it is immediate to verify

$$\bar{\alpha} \leq \frac{\nu^2}{2(1 + \nu)^2}.$$

Consider now the expression of $\bar{\delta}$. By reasoning as previously we obtain

$$\bar{\delta}_0 = \max_{\alpha > \bar{\alpha}} \ \max_{1 \leq i \leq N-1} \frac{4\alpha(\nu + 1)\lambda_i(I + A) - 4\left(\lambda_i\left(-\frac{1}{1+\nu}I + A\right)\right)}{\left[\alpha(\nu + 1)\lambda_i(A) + \lambda_i\left(\frac{\nu+2}{\nu+1}I - A\right)\lambda_i\left(-\frac{1}{\nu+1}I + A\right)\right]^2}$$

Let $\alpha = \frac{\nu^2}{2(1+\nu)^2} + \epsilon$ where ϵ is a fixed positive real number. Then

$$\bar{\delta}_0 \geq \frac{\min_{1 \leq i \leq N-1}\left\{4\alpha(\nu + 1)\lambda_i(I + A) - 4\left(\lambda_i\left(-\frac{1}{1+\nu}I + A\right)\right)\right\}}{\max_{1 \leq i \leq N-1}\left\{\left[\alpha(\nu + 1)\lambda_i(A) + \lambda_i\left(\frac{\nu+2}{\nu+1}I - A\right)\lambda_i\left(-\frac{1}{\nu+1}I + A\right)\right]^2\right\}}$$

Observe that

$$\max_{1 \leq i \leq N-1}\left\{\left[\alpha(\nu + 1)\lambda_i(A) + \lambda_i\left(\frac{\nu + 2}{\nu + 1}I - A\right)\lambda_i\left(-\frac{1}{\nu + 1}I + A\right)\right]^2\right\}$$
$$< \frac{\alpha(\nu + 1)^3 + \nu^2 + 3\nu + 1}{(\nu + 1)^2}.$$

Moreover

$$\min_{1 \leq i \leq N-1}\left\{4\alpha(\nu + 1)\lambda_i(I + A) - 4\left(\lambda_i\left(-\frac{1}{1 + \nu}I + A\right)\right)\right\} \geq$$
$$4\left(\frac{\nu^2}{2(\nu + 1)^2} + \epsilon\right)\frac{2(\nu + 1)}{\nu + 1} - \left(1 - \frac{1}{\nu + 1}\right)^2 = 8\epsilon.$$

Hence

$$\bar{\delta}_0 > \frac{8\epsilon}{\epsilon(\nu + 1) + \frac{\nu^3 + 3\nu^2 + 6\nu + 2}{2(\nu + 1)^2}},$$

for all $\epsilon > 0$. By taking $\epsilon \to \infty$ we can argue that $\bar{\delta}_2 = \frac{8}{\nu + 1}$.

6   Conclusions

In this paper we presented a new approach to the consensus problem, where we considered only quantized exchanges of information. In particular we considered a encoding-decoding strategy based on logarithmic quantizers. We restricted our attention on the average consensus controllers and we proved that the consensus problem is solvable even if the systems can share only logarithmically quantized information. Obviously the use of logarithmic quantizers introduces an error starting which prevents in general to obtain an consensus at the average of the initials conditions. The distance $\bar{\delta}$ from the average the systems will reach the consensus will be the object of our future investigations. An another field of future research will be to find encoding and decoding methods which work also for digital noisy channels.

References

1. Mazo M., Speranzon A, Johansson K H, Hu X (2004) Multi-robot tracking of a moving object using directional sensors, Proceedings of 2004 International Conference on Robotics and Automation (ICRA)
2. Fax J A, Murray R M (2004) Information flow and cooperative control of vehicle formations, IEEE Transactions on Automatic Control, 49: 1465–1476
3. Olfati-Saber R, Murray R M (2004) Consensus problems in networks of agents with switching topology and time-delays, IEEE Transactions on Automatic Control, 49: 1520–1533
4. Olfati-Saber R (2005) Distributed Kalman filter with embedded consensus filters, Proceedings of 44th IEEE Conference on Decision and Control-European Control Conference, 8179–8184
5. Carli R, Fagnani F, Speranzon A, Zampieri S (2006) Communication constraints in coordinated consensus problem, Proceedings of 2006 American Control Conference
6. Cortes J, Martinez S, Bullo F (2006) Robust rendezvous for mobile autonomous agents via proximity graphs in arbitrary dimensions, IEEE Transactions on Automatic Control, 51: 1289:1298
7. Jadbabaie A, Lin J, Morse A S (2003) Coordination of groups of mobile autonomous agents using nearest neighbor rules, IEEE Transactions on Automatic Control, 48:988–1001
8. D'Andrea R, Dullerud G E, Distributed control design for spatially interconnected systems, IEEE Transactions on Automatic Control, 48:1478–1495
9. Marodi M, D'Ovidio F, Vicksek T (2002) Synchronization of oscillators with long range interaction: Phase transition and anomalous finite size effects, Phisical Review E, 66
10. Strogatz S H (2000) From Kuramoto to Crawford: exploring the onset of synchronization of in populations of coupled oscillators, Phisical D: Nonlinear Phenomena, 143:1–20
11. Hatano Y, Meshabi M (2004) Agreement of random networks, Proceedings of 43rd IEEE Conference on Decision and Control
12. Carli R, Fagnani F, Speranzon A, Zampieri S (2006) Communication constraints in coordinated consensus problem, Proocedings of the 2006 American Control Conference
13. Tanner H G, Jadbabaie A, Pappas G J (2003) Stable flocking of mobile agents, part i: fixed topology, Proocedings of the 42th IEEE Conference on Decision and Control
14. Tanner H G, Jadbabaie A, Pappas G J (2003) Stable flocking of mobile agents, part ii: dynamic topology. Proocedings of the 42th IEEE Conference on Decision and Control
15. Beard R W, Lawton J, Hadaegh F Y (2001) A coordination architecture for spacecraft formation control, IEEE Transactions on Control Systems Technology, 9:777–790
16. Bhatta P., Leonard N E (2002) Stabilization and coordination of underwater gliders, Proceedings of 41th IEEE Conference on Decision and Control
17. Elia M, Mitter S J (2001) Stabilization of linear systems with limited information, IEEE Transactions on Automatic and Control, 46:1384–1400
18. Carli R, Fagnani F, Speranzon A, Zampieri S (2005) Communication constraints in the state agreement problem, In Technical Report N.32, Politecnico di Torino, Submitted to Automatica

19. Carli R, Fagnani F, Zampieri S (2006) On the state agreement with quantized information, Proceedings of 17th International Symposium on Mathematical Theory of Networks and Systems
20. Kashyap A, Basar T, Srikant R (2006), Consensus with quantized information updates, Proceedings of 45th IEEE Conference on Decision and Control
21. Nair G, Fagnani F, Zampieri S, Evans R (2006) Feedback control under data rate constraints: an overview, Proceedings of 45th IEEE Conference on Decision and Control
22. Blanchini F, Miani S (2003) Stabilization of LPV systems: state feedback, state estimation and duality, SIAM Journal on Control and Optimization, 32:76-97

Human-Robot Interaction Control
Using Force and Vision

Agostino De Santis, Vincenzo Lippiello, Bruno Siciliano,
and Luigi Villani

Dipartimento di Informatica e Sistemistica, Università di Napoli Federico II
{agodesa,vincenzo.lippiello,siciliano,lvillani}@unina.it

Summary. The extension of application domains of robotics from factories to human environments leads to implementing proper strategies for close interaction between people and robots. In order to avoid dangerous collision, force and vision based control can be used, while tracking human motion during such interaction.

Keywords: Physical Human-Robot Interaction, Interaction Control, Impedance Control, Visual Control, Extended Kalman Filter.

1 Introduction

Physical human-robot interaction (pHRI) is an interesting topic for small-scale industrial robotics, where a user may need to share the workspace with a robot, as well as for service robotics. In the elderly-dominated scenario of most industrialized countries, service robotics is a solution for automatizing common daily tasks, also due to the lack or high cost of human expertise.

The size of an industrial robot, or the necessary autonomous behaviour of a service robot, can result in dangerous situations for humans co-existing in the robot operational domain. Therefore, physical issues must be carefully considered, since "natural" or unexpected behaviours of people during interaction with robots can result in injuries, which may be severe, when considering the current mechanical structure of robots available on the market. In this special perspective, an improved analysis of the problems related to the physical interaction with robots leads to rediscuss most of the topics of mechanical design, planning, and control of robots [1].

The physical viewpoint is mainly focused on the risks of collisions or excessive force exchange occurring between the robot and its user: a too high energy-to-power ratio may be transferred by the robot, resulting in serious human damages. Severity indices of injuries may be used to evaluate the safety of robots in pHRI. These should take into account the possible damages occurring when a manipulator collides with a human head, neck, chest or arm. Several standard indices of injury severity exist in other, non-robotic, domains. The automotive industry developed empirical/experimental formulas that correlate human body's acceleration to injury severity, while the suitability of such formulas is still an open issue in robotics.

C. Bonivento et al. (Eds.): Adv. in Control Theory and Applications, LNCIS 353, pp. 51–70, 2007.
springerlink.com

One possible issue to consider, in order to increase robot safety, is the proper use of the two main "senses": vision and touch. Vision and force based control for physical interaction may include collision avoidance, control of close interaction, fusion with other sensory modes, which all may lead to improving available robots' performance, without necessarily considering a novel mechanical design.

Possibly, the need for safety suggests complementing the proposed control system with the adoption of compliant components in the structure. Compliance can be introduced at the contact point by a soft covering of the whole arm with visco-elastic materials or by adopting compliant transmissions at the robot joints. Increasing in this way the robot mechanical compliance while reducing its overall apparent inertia has been realized through different elastic actuation/transmission arrangements which include: relocation of actuators close to the robot base and transmission of motion through steel cables and pulleys, combination of harmonic drives and lightweight link design, and use of parallel and distributed macro-mini [2] or variable-impedance [3] actuation. Other improvements for anticipating and reacting to collisions can be achieved through the use of combinations of external/internal robot sensing, electronic hardware and software safety procedures, which intelligently monitor, supervise, and control manipulator operation.

Modern actuation strategies, as well as force/impedance control schemes, seem to be anyway crucial in human-robot interaction. On the other hand, a more complete set of external sensory devices can be used to monitor task execution and reduce the risks of unexpected impacts. However, even the most robust architecture is endangered by system faults and human unpredictable behaviours. This suggests improving both passive and active safety for robots in anthropic domains.

This work focuses on techniques for augmenting safety by means of control systems. Human-like capabilities in close interaction can be considered as mimicking sensing and actuation of humans. This leads to consider fully integrated vision and force based control. Thanks to the visual perception, the robotic system may achieve global information on the surrounding environment that can be used for task planning and obstacle avoidance. On the other hand, the perception of the force applied to the robot allows adjusting the motion so that the local constraints imposed by the environment during the interaction are satisfied. The safety and dependability of a robotic system are strictly connected to the availability of sensing information on the external environment. Moreover, vision system may substitute the complex infrastructure needed for "intelligent environments" [4] to detect and track people in the operational domain.

Because of such complementary nature, it should be natural to believe that vision and force could be used in an integrated and synergic way to design suitable planning and control strategies for robotic systems. In the last years, several papers on this subject have been presented. Some of them combine force of vision in the same feedback control loop, such as hybrid visual/force control [5], shared and traded control [6, 7] or visual impedance control [8, 9].

2 Safety by Means of Control

The first important criterion to limit injuries due to collisions is to reduce the
weight of the moving parts of the robot. A prototypical example along this
direction is the design of the DLR-III Lightweight Robot [10], which is capable
of operating a payload equal to its own weight (13,5 kg). Advanced light but stiff
materials were used for the moving links, while motor transmission/reduction
is based on harmonic drives, which display high reduction ratio and efficient
power transmission capability. In addition, there is the possibility of relocating
all the relevant weights (mostly, the motors), at the robot base, like it was done
for the Barrett Whole Arm Manipulator (WAM) [11]. This is a very interesting
cable-actuated robot, which is also backdrivable, i.e., by pushing on the links, it
is possible to force motion of all mechanical transmission components, including
the motors' rotors. In the case of a collision, the lighter links display lower inertia
and thus lower energy is transferred during the impact.

On the other hand, compliant transmissions tend to decouple mechanically
the larger inertias of the motors from those of the links. The presence of compliant elements may thus be useful as a protection against unexpected contacts
during pHRI. More in general, a lightweight design and/or the use of compliant
transmissions introduce link [12] and, respectively, joint [13] elasticity. In order to
preserve performance while exploiting the potential offered by lightweight robot
arms, one must consider the effects of structural link flexibility. Distributed link
deformation in robot manipulators arises in the presence of very long and slender
arm design (without special care on materials); notice that "link rigidity" is always an ideal assumption and may fail when increasing payload-to-weight ratio.

In the presence of compliant transmissions, deformation can be assumed to
be instead concentrated at the joints of the manipulator. Neglected joint elasticity or link flexibility limits static (steady-state error) or dynamic (vibrations,
poor tracking) task performance. Problems related to motion speed and control bandwidth must be also considered. Flexible modes of compliant systems
prevent obtaining control bandwidths greater than a limit; in addition, attenua-
tion/suppression of vibrations excited by disturbances can be difficult to achieve.
Intuitively, compliant transmissions tend to respond slowly to torque inputs on
the actuator and to oscillate around the goal position, so that it can be expected that the promptness of an elastically actuated arm is severely reduced if
compliance is high enough to be effective on safety.

From the control point of view, there is a basic difference between link and
joint elasticity. In the first case, we have non-colocation between input commands
and typical outputs to be controlled; for flexible joint robots, the co-location of
input commands and structural flexibility suggests to treat this case separately.

In order to introduce safety tactics for available robots, we can mainly act
on control; interaction control strategies can be grouped in two categories: those
performing indirect force control and those performing direct force control. The
main difference between the two categories is that the former achieve force control indirectly via a motion control loop, while the latter offers the possibility of

controlling the contact force to a desired value, thanks to the closure of a force feedback loop.

Force/impedance control [14] is important in pHRI because a compliant behaviour of a manipulator leads to a more natural physical interaction and reduces the risks of damages in case of unwanted collisions. Similarly, the capability of sensing and controlling exchanged forces is relevant for cooperating tasks between humans and robots. To the category of indirect force control belongs impedance control, where the position error is related to the contact force through a mechanical impedance of adjustable parameters. A robot manipulator under impedance control is described by an equivalent mass-spring-damper system, with the contact force as input (impedance may vary in the various task space directions, typically in a nonlinear and coupled way).

The interaction between the robot and a human results then in a dynamic balance between these two "systems". This balance is influenced by the mutual weight of the human and the robot compliant features. In principle, it is possible to decrease the robot compliance so that it dominates in the pHRI and vice versa. Cognitive information could be used for dynamically setting the parameters of robot impedance, considering task-dependent safety issues. Certain interaction tasks, however, do require the fulfilment of a precise value of the contact force. This would be possible, in theory, by tuning the active compliance control action and by selecting a proper reference location for the robot.

If force measurements are available (typically through a robot wrist sensor), a direct force control loop could be also designed. Note that, a possible way to measure contact forces occurring in any part of a serial robot manipulator is to provide the robot with joint torque sensors. The integration of joint torque control with high performance actuation and lightweight composite structure can help merging the competing requirements of safety and performance. In all cases, the control design should prevent to introduce in the robot system more energy than strictly needed to complete the task. This rough requirement is related to the intuitive consideration that robots with large kinetic and potential energy are eventually more dangerous for a human in case of collision. An elegant mathematical concept satisfying this requirement is passivity.

As already mentioned, compliant transmissions can negatively affect performance during normal robot operation in free space, in terms of increased oscillations and settling times. However, more advanced motion control laws can be designed which take joint elasticity of the robot into account. Moreover, in robots with variable impedance actuation, the simultaneous and decoupled control of both the link motion and the joint stiffness is also possible in principle, reaching a trade-off between performance and safety requirements.

3 Modelling

For pHRI it is necessary to model or track human motion, to get a model of robot motion and of the objects to interact with. Consider a robot in contact with an object, a wrist force sensor and a camera mounted on the end-effector (eye-in-hand configuration) or fixed in the workspace (eye-to-hand configuration). In the following, some modelling assumption concerning the environemnt, the robot and the camera are illustrated.

3.1 Robot

The case of a n-joints robot manipulator is considered, with $n \geq 3$. The tip position $\boldsymbol{p}_q$ can be computed via the direct kinematics equation:

$$\boldsymbol{p}_q = \boldsymbol{k}(\boldsymbol{q}), \tag{1}$$

where $\boldsymbol{q}$ is the $(n \times 1)$ vector of the joint variables. Also, the velocity of the robot's tip $\boldsymbol{v}_{P_q}$ can be expressed as

$$\boldsymbol{v}_{P_q} = \boldsymbol{J}(\boldsymbol{q})\dot{\boldsymbol{q}}$$

where $\boldsymbol{J} = \partial\boldsymbol{k}(\boldsymbol{q})/\partial\boldsymbol{q}$ is the robot Jacobian matrix. The vector $\boldsymbol{v}_{P_q}$ can be decomposed as

$$^o\boldsymbol{v}_{P_q} = {}^o\dot{\boldsymbol{p}}_q + \boldsymbol{\Lambda}(^o\boldsymbol{p}_q)^o\boldsymbol{\nu}_o, \tag{2}$$

with $\boldsymbol{\Lambda}(\cdot) = [\boldsymbol{I}_3 \; -\boldsymbol{S}(\cdot)]$, where $\boldsymbol{I}_3$ is the (3×3) identity matrix and $\boldsymbol{S}(\cdot)$ denotes the (3×3) skew-symmetric matrix operator. In Eq. (2), $^o\dot{\boldsymbol{p}}_q$ is the relative velocity of the tip point P_q with respect to the object frame while $^o\boldsymbol{\nu}_o = [^o\boldsymbol{v}_{O_o}^T \; {}^o\boldsymbol{\omega}_o^T]^T$ is the velocity screw characterizing the motion of the object frame with respect to the base frame in terms of the translational velocity of the origin $\boldsymbol{v}_{O_o}$ and of the angular velocity $\boldsymbol{\omega}_o$; all the quantities are expressed in the object frame.

When the robot is in contact with the object, the normal component of the relative velocity $^o\dot{\boldsymbol{p}}_q$ is null, i.e.:

$$^o\boldsymbol{n}^T(^o\boldsymbol{p}_q)^o\dot{\boldsymbol{p}}_q = 0. \tag{3}$$

3.2 Human User

Positioning of critical parts of a human body may be addressed, like for robots, considering the kinematics of the structure. However, joint measures are not available on the human body; therefore, exteroceptive sensing by means of cameras is used, obtaining the position in the space of some relevant features (hands, head etc.). This leads to finding a simplified kinematic model, to be updated in real time, with the novel "skeleton algorithm" [15]. This algorithm considers a skeleton, composed of segments, as a simplified model of a human (or a robot or even an object), exploiting the simple geometric structures in order to evaluate analytically the distances between the segments, which can be used for collision avoidance, considering all the points of the articulated structure of humans and robots which may collide. For every link of the skeleton of a human figure, the closest point to the robot or the object to be avoided is computed. The distance information between the two closest points of human and obstacle can be used to avoid a collision, using "spheres" located in the selected closest points as protective hulls: these spheres can have a finite or infinite radius and can be the source of repelling forces shaped as effects of virtual springs or potential fields. Summarizing, the steps of the algorithm are:

- Create a skeleton of the human body, by using vision, and of the robot, by using direct kinematics in order to find the extremal point of the segments.
- Compute analytically the distances between the different segments, finding also the two closest points for each pair of links.
- Define intensity and shape of repelling forces between these two points and use them as reference values in the position/force control system.

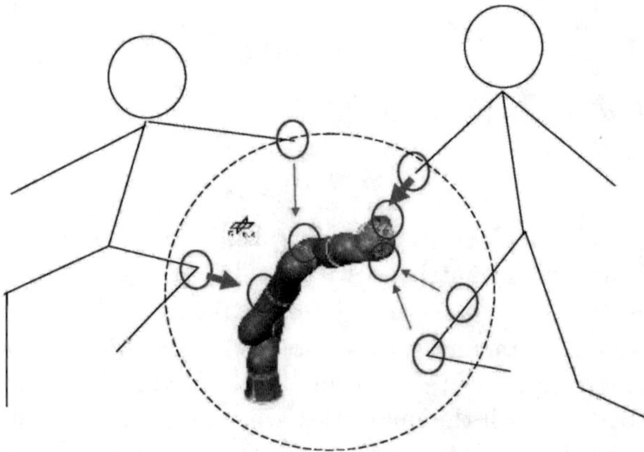

Fig. 1. Exemplification of the skeleton algorithm for the DLR lightweight arm

Almost all structures can be encapsulated by a finite skeleton with spheres, as sketched in Fig. 1 for the DLR arm. The position of the closest point on each link (continuous curves) varies continuously, preserving continuity of reference values for any kind of control scheme. The key point of the proposed approach is that only the two closest points (on each link) of the structure are considered each time, leading to a simple generation of the Cartesian desired velocity (or force) for only one of these points, which eventually is transformed in the corresponding joint trajectory via proper inverse kinematics (or kinetics). Any point on the structure can be considered as a control point. To simplify the problem, there is also the possibility to choose only a subset of control points, e.g., the articulation of the robot [16]. Moreover, it is possible to use an inverse kinematics, an impedance control or whatever is desired, since the algorithm just adds with continuity repelling forces or velocity, preserving stability of the control loops used for the system.

3.3   Camera

A frame $O_c\text{--}x_c y_c z_c$ attached to the camera (either in eye-in-hand or in eye-to-hand configuration) is considered. By using the classical pin-hole model, a point P of the object with coordinates $^c\boldsymbol{p} = \begin{bmatrix} x & y & z \end{bmatrix}^{\mathrm{T}}$ with respect to the camera frame is projected onto the point of the image plane with coordinates

$$\begin{bmatrix} X \\ Y \end{bmatrix} = \frac{\lambda_c}{z} \begin{bmatrix} x \\ y \end{bmatrix} \tag{4}$$

where λ_c is the focal length of the lens of the camera.

Let $\boldsymbol{H}_c$ denote the homogeneous transformation matrix representing the pose of the camera frame referred to the base frame. For eye-to-hand cameras, the matrix $\boldsymbol{H}_c$ is constant, and can be computed through a suitable calibration procedure, while for eye-in-hand cameras this matrix depends on the camera current pose $\boldsymbol{x}_c$ and can be computed as:

$$\boldsymbol{H}_c(\boldsymbol{x}_c) = \boldsymbol{H}_e(\boldsymbol{x}_e)^e\boldsymbol{H}_c$$

where $\boldsymbol{H}_e$ is the homogeneous transformation matrix of the end effector frame e with respect to the base frame, and $^e\boldsymbol{H}_c$ is the homogeneous transformation matrix of camera frame with respect to end effector frame. Notice that $^e\boldsymbol{H}_c$ is constant and can be estimated through suitable calibration procedures, while $\boldsymbol{H}_e$ depends on the current end-effector pose $\boldsymbol{x}_e$ and may be computed using the robot kinematic model. The relevant frames and the transformation matrices are illustrated in Fig. 2, where the more general case of multiple mobile and fixed cameras is depicted.

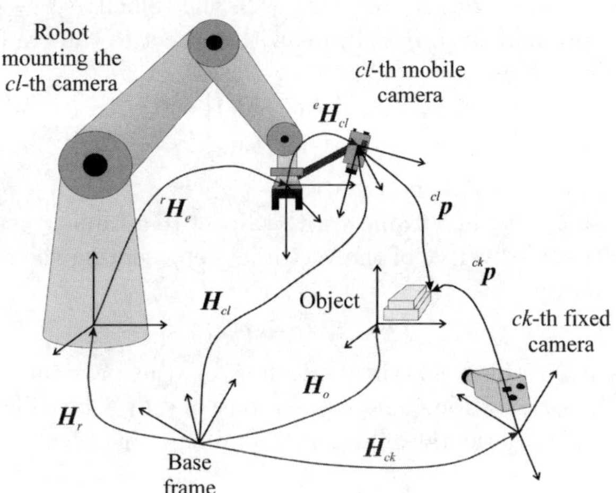

Fig. 2. Relevant camera and object frames

Therefore, the homogeneous coordinate vector of P with respect to the camera frame can be expressed as

$$^c\tilde{p} = {}^c\boldsymbol{H}_o(\boldsymbol{x}_o, \boldsymbol{x}_c){}^o\tilde{p} \tag{5}$$

where $^c\boldsymbol{H}_o(\boldsymbol{x}_o, \boldsymbol{x}_c) = {}^c\boldsymbol{H}^{-1}(\boldsymbol{x}_c)\boldsymbol{H}_o(\boldsymbol{x}_o)$. Notice that $\boldsymbol{x}_c$ is constant for eye-to-hand cameras; moreover, the matrix $^c\boldsymbol{H}_o$ does not depend on $\boldsymbol{x}_c$ and $\boldsymbol{x}_o$ separately but only on the relative pose of the object frame with respect to the camera frame.

The velocity of the camera frame with respect to the base frame can be characterized in terms of the translational velocity of the origin $\boldsymbol{v}_{O_c}$ and of angular velocity $\boldsymbol{\omega}_c$. These vectors, expressed in camera frame, define the velocity screw $^c\boldsymbol{\nu}_c = [{}^c\boldsymbol{v}_{O_c}^T \ {}^c\boldsymbol{\omega}_c^T]^T$. Analogously to (2), the absolute velocity of the origin O_o of the object frame can be computed as

$$^c\boldsymbol{v}_{O_o} = {}^c\dot{\boldsymbol{o}}_o + \boldsymbol{\Lambda}({}^c\boldsymbol{o}_o){}^c\boldsymbol{\nu}_c, \tag{6}$$

where $^c\boldsymbol{o}_o$ is the vector of the coordinates of O_o with respect to camera frame and $^c\dot{\boldsymbol{o}}_o$ is the relative velocity of O_o with respect to camera frame; all the quantities are expressed in camera frame. On the other hand, the absolute angular velocity $^c\boldsymbol{\omega}_o$ of the object frame expressed in camera frame can be computed as

$$^c\boldsymbol{\omega}_o = {}^c\boldsymbol{\omega}_{o,c} + {}^c\boldsymbol{\omega}_c \tag{7}$$

where $^c\boldsymbol{\omega}_{o,c}$ represents the relative angular velocity of the object frame with respect to the camera frame. The two equations (6) and (7) can be rewritten in the compact form

$$^c\boldsymbol{\nu}_o = {}^c\boldsymbol{\nu}_{o,c} + \boldsymbol{\Gamma}({}^c\boldsymbol{o}_o){}^c\boldsymbol{\nu}_c \tag{8}$$

where $^c\boldsymbol{\nu}_o = [{}^c\boldsymbol{v}_{O_o}^T \ {}^c\boldsymbol{\omega}_o^T]^T$ is the velocity screw corresponding to the absolute motion of the object frame, $^c\boldsymbol{\nu}_{o,c} = [{}^c\dot{\boldsymbol{o}}_o^T \ {}^c\boldsymbol{\omega}_{o,c}^T]^T$ is the velocity screw corresponding to the relative motion of the object frame with respect to camera frame, and the matrix $\boldsymbol{\Gamma}(\cdot)$ is defined as

$$\boldsymbol{\Gamma}(\cdot) = \begin{bmatrix} \boldsymbol{I}_3 & -\boldsymbol{S}(\cdot) \\ \boldsymbol{O}_3 & \boldsymbol{I}_3 \end{bmatrix},$$

where $\boldsymbol{O}_3$ denotes the (3×3) null matrix.

The velocity screw $^r\boldsymbol{\nu}_s$ of a frame s with respect to a frame r can be expressed in terms of the time derivative of the vector $\boldsymbol{x}_s$ representing the pose of frame s through the equation

$$^r\boldsymbol{\nu}_s = {}^r\boldsymbol{L}(\boldsymbol{x}_s)\dot{\boldsymbol{x}}_s \tag{9}$$

where $^r\boldsymbol{L}(\cdot)$ is a Jacobian matrix depending on the particular choice of coordinates for the orientation. The expressions of $^r\boldsymbol{L}(\cdot)$ for different kinds of parametrization of the orientation can be found, e.g., in [17].

3.4   Object

The position and orientation of a frame attached to a rigid object O_o–$x_o y_o z_o$ with respect to a base coordinate frame O–xyz can be expressed in terms of the coordinate vector of the origin $\boldsymbol{o}_o = \begin{bmatrix} x_o & y_o & z_o \end{bmatrix}^T$ and of the rotation matrix $\boldsymbol{R}_o(\boldsymbol{\varphi}_o)$, where $\boldsymbol{\varphi}_o$ is a $(p \times 1)$ vector corresponding to a suitable parametrization of the orientation. In the case that a minimal representation of the orientation is adopted, e.g., Euler angles, it is $p = 3$, while it is $p = 4$ if unit quaternions are used. Hence, the $(m \times 1)$ vector $\boldsymbol{x}_o = \begin{bmatrix} \boldsymbol{o}_o^T & \boldsymbol{\varphi}_o^T \end{bmatrix}^T$ defines a representation of the object pose with respect to the base frame in terms of $m = 3 + p$ parameters.

The homogeneous coordinate vector $\tilde{\boldsymbol{p}} = \begin{bmatrix} \boldsymbol{p}^T & 1 \end{bmatrix}^T$ of a point P of the object with respect to the base frame can be computed as $\tilde{\boldsymbol{p}} = \boldsymbol{H}_o(\boldsymbol{x}_o){}^o\tilde{\boldsymbol{p}}$, where ${}^o\tilde{\boldsymbol{p}}$ is the homogeneous coordinate vector of P with respect to the object frame and $\boldsymbol{H}_o$ is the homogeneous transformation matrix representing the pose of the object frame referred to the base frame:

$$\boldsymbol{H}_o(\boldsymbol{x}_o) = \begin{bmatrix} \boldsymbol{R}_o(\boldsymbol{\varphi}_o) & \boldsymbol{o}_o \\ \boldsymbol{0}_3^T & 1 \end{bmatrix},$$

where $\boldsymbol{0}_3$ is the (3×1) null vector.

It is assumed that the geometry of the object is known and that the interaction involves a portion of the external surface which satisfies the continuously differentiable scalar equation $\varphi({}^o\boldsymbol{p}) = 0$.

The unit vector normal to the surface at the point ${}^o\boldsymbol{p}$ and pointing outwards can be computed as:

$${}^o\boldsymbol{n}({}^o\boldsymbol{p}) = \frac{(\partial\varphi({}^o\boldsymbol{p})/\partial\,{}^o\boldsymbol{p})^T}{\|(\partial\varphi({}^o\boldsymbol{p})/\partial\,{}^o\boldsymbol{p}\|}, \tag{10}$$

where ${}^o\boldsymbol{n}$ is expressed in the object frame.

Notice that the object pose $\boldsymbol{x}_o$ is assumed to be unknown and may change during the task execution. As an example, a compliant contact can be modelled assuming that $\boldsymbol{x}_o$ changes during the interaction according to an elastic law.

A further assumption is that the contact between the robot and the object is of point type and frictionless. Therefore, when in contact, the tip point P_q of the robot instantaneously coincides with a point P of the object, so that the tip position ${}^o\boldsymbol{p}_q$ satisfies the constraint equation:

$$\varphi({}^o\boldsymbol{p}_q) = 0. \tag{11}$$

Moreover, the (3×1) contact force ${}^o\boldsymbol{h}$ is aligned to the normal unit vector ${}^o\boldsymbol{n}$.

4 Use of Vision, Force and Joint Measurements

When the robot moves in free space, the unknown object pose and the position of the head of a human user can be estimated online by using the data provided by the camera; when the robot is in contact to the object, also the force measurements and the joint position measurements are used. Joint values are used for evaluating the position of the links for collision avoidance. In the following, the equations mapping the measurements to the unknown position and orientation of the object are derived.

4.1   Vision

Vision is used to measure the image features, i.e., any structural feature that can be extracted from an image, corresponding to the projection of a physical feature of the object onto the camera image plane. An image feature can be characterized by a set of scalar parameters f_j that can be grouped in a vector $\boldsymbol{f} = [f_1 \cdots f_k]^T$, where k is the dimension of the image feature parameter space. The mapping from the position and orientation of the object to the corresponding image feature vector can be computed using the projective geometry of the camera and can be written in the form

$$\boldsymbol{f} = \boldsymbol{g}_f({}^c\boldsymbol{H}_o(\boldsymbol{x}_o, \boldsymbol{x}_c)), \tag{12}$$

where only the dependence from the relative pose of the object frame with respect to camera frame has been explicitly evidenced.

For the estimation of the object pose, it is required the computation of the Jacobian matrix

$$\boldsymbol{J}_f = \frac{\partial \boldsymbol{g}_f}{\partial \boldsymbol{x}_o}.$$

To this purpose, the time derivative of (12) can be computed in the form

$$\dot{\boldsymbol{f}} = \frac{\partial \boldsymbol{g}_f}{\partial \boldsymbol{x}_o}\dot{\boldsymbol{x}}_o + \frac{\partial \boldsymbol{g}_f}{\partial \boldsymbol{x}_c}\dot{\boldsymbol{x}}_c, \tag{13}$$

where the second term on the right-hand side is null for eye-to-hand cameras. On the other hand, the time derivative of (12) can be expressed also in the form

$$\dot{\boldsymbol{f}} = \boldsymbol{J}_{o,c}{}^c\boldsymbol{\nu}_{o,c} \tag{14}$$

where the matrix $\boldsymbol{J}_{o,c}$ is the Jacobian mapping the relative velocity screw of the object frame with respect to the camera frame into the variation of the image feature parameters. The expression of $\boldsymbol{J}_{o,c}$ depends on the choice of the image features; examples of computation can be found in [17].

By taking into account the velocity composition (8), Eq. (14) can be rewritten in the form

$$\dot{\boldsymbol{f}} = \boldsymbol{J}_{o,c}{}^c\boldsymbol{\nu}_o - \boldsymbol{J}_c{}^c\boldsymbol{\nu}_c \tag{15}$$

where $\boldsymbol{J}_c = \boldsymbol{J}_{o,c}\boldsymbol{\Gamma}({}^c\boldsymbol{o}_o)$ is the Jacobian corresponding to the contribution of the absolute velocity screw of the camera frame, known in the literature as interaction matrix [18]. In view of (9), the comparison of (15) with (13) yields

$$\boldsymbol{J}_f = \boldsymbol{J}_{o,c}{}^c\boldsymbol{L}(\boldsymbol{x}_o). \tag{16}$$

4.2   Force and Joint Measurements

In the case of frictionless point contact, the measure of the force $\boldsymbol{h}$ at the robot tip during the interaction can be used to compute the unit vector normal to the object surface at the contact point ${}^o\boldsymbol{p}_q$, i.e.,

$$n_h = \frac{h}{\|h\|}. \tag{17}$$

On the other hand, the vector n_h can be expressed as a function of the object pose x_o and of the robot position p_q in the form

$$n_h = R_o\,{}^o n({}^o p_q) = g_h(x_o, p_q), \tag{18}$$

being ${}^o p_q = R_o^T(p_q - o_o)$.

For the estimation of the object pose, it is required the computation of the Jacobian matrix

$$J_h = \frac{\partial g_h}{\partial x_o}.$$

To this purpose, the time derivative of (18) can be expressed as

$$\dot{n}_h = \frac{\partial g_h}{\partial x_o}\dot{x}_o + \frac{\partial g_h}{\partial p_q}\dot{p}_q. \tag{19}$$

On the other hand, the time derivative of (18) can be computed also in the form

$$\dot{n}_h = \dot{R}_o\,{}^o n({}^o p_q) + R_o\,{}^o N({}^o p_q)\,{}^o\dot{p}_q, \tag{20}$$

where ${}^o N({}^o p_q) = \partial{}^o n/\partial{}^o p_q$ depends on the surface curvature and ${}^o\dot{p}_q$ can be computed from (2). Hence, by comparing (19) with (20) and taking into account (9) and the equality $\dot{R}_o\,{}^o n({}^o p_q) = -S(n_h)\omega_o$, the following expression can be found:

$$J_h = -[N\;S(n_h) - NS(p_q - o_o)]L(x_o), \tag{21}$$

where $N = R_o\,{}^o N({}^o p_q)R_o^T$.

The joint positions q are used not only to evaluate the configuration of the robot, which can possibly collide with a user, but also to evaluate the position of the point P of the object when in contact to the robot's tip point P_q, using the direct kinematics equation (1). In particular, it is significant computing the scalar

$$\delta_{hq} = n_h^T p_q = g_{hq}(x_o, p_q), \tag{22}$$

using also the force measurements via (17).

For the estimation of the object pose it is required the computation of the Jacobian matrix

$$J_{hq} = \frac{\partial g_{hq}}{\partial x_o}.$$

As in the previous subsection, the time derivative of δ_{hq} can be expressed as

$$\dot{\delta}_{hq} = \frac{\partial g_{hq}}{\partial x_o}\dot{x}_o + \frac{\partial g_{hq}}{\partial p_q}\dot{p}_q. \tag{23}$$

On the other hand, the time derivative of δ_{hq} can be computed also as

$$\dot{\delta}_{hq} = \dot{n}_h^T p_q + n_h^T R_o({}^o\dot{p}_q + \Lambda({}^o p_q){}^o\nu_o)$$

62   A. De Santis et al.

where the expression of the absolute velocity of the point P_q in (2) has been used. Using identity (3), the above equation can be rewritten as

$$\dot{\delta}_{hq} = \boldsymbol{p}_q^T \dot{\boldsymbol{n}}_h + \boldsymbol{n}_h^T \boldsymbol{\Lambda}(\boldsymbol{p}_q - \boldsymbol{o}_o)\boldsymbol{\nu}_o. \tag{24}$$

Hence, by comparing (23) with (24) and taking into account (20), (21) and (9), the following expression can be found

$$\boldsymbol{J}_{hq} = \boldsymbol{p}_q^T \boldsymbol{J}_h + \boldsymbol{n}_h^T \boldsymbol{\Lambda}(\boldsymbol{p}_q - \boldsymbol{o}_o)\boldsymbol{L}(\boldsymbol{x}_o). \tag{25}$$

5 Vision-Based Pose Estimation

5.1 Human Operator's Pose Estimation

In order to use the skeleton algorithm [15] for collision avoidance, simple fixed cameras are employed to detect the positions of face and hands of an operator in the operational space of the robot. In assembly tasks in cooperation with the robot, the operator does not move fast, simplifying the tracking by means of cameras. In preliminary experiments, markers are used to help the detection and tracking. The detected positions of the human operator are to be tracked in order to keep a safety volume around him/her, repelling the robot when it approaches too much. Cameras mounted on the robot can be used as well. Potential fields or optimization techniques are then to be designed, in order to create modifications to the robot's trajectory aimed at avoiding dangerous approaches. Simple virtual springs or more complex modifications to trajectories, using null-space motion if possible, can be adopted also while using an interaction control with an object, which is considered in the following. The shape of the computed repelling force or velocity must preserve continuity of reference values for the robot controllers.

5.2 Object Pose Estimation

In this section, the problem of the estimation of the pose vector $\boldsymbol{x}_o$ of the object with respect to the base frame using visual, force and joint position measurements. The proposed solution is based on the EKF [19].

To this purpose, a discrete-time state space dynamic model has to be considered, describing the object motion. The state vector of the dynamic model is chosen as $\boldsymbol{w} = \begin{bmatrix} \boldsymbol{x}_o^T & \dot{\boldsymbol{x}}_o^T \end{bmatrix}^T$. For simplicity, the object velocity is assumed to be constant over one sample period T_s. This approximation is reasonable in the hypothesis that T_s is sufficiently small. The corresponding dynamic modeling error can be considered as an input disturbance $\boldsymbol{\gamma}$ described by zero mean Gaussian noise with covariance $\boldsymbol{Q}$. The discrete-time dynamic model can be written as

$$\boldsymbol{w}_k = \boldsymbol{A}\boldsymbol{w}_{k-1} + \boldsymbol{\gamma}_k, \tag{26}$$

where $\boldsymbol{A}$ is the $(2m \times 2m)$ block matrix

$$\boldsymbol{A} = \begin{bmatrix} \boldsymbol{I}_m & T_s\boldsymbol{I}_m \\ \boldsymbol{O}_m & \boldsymbol{I}_m \end{bmatrix}.$$

The output of the Kalman filter, in the case that all the available data can be used, is the vector of the measurements at time kT_s

$$\boldsymbol{\zeta}_k = \left[\boldsymbol{\zeta}_{f,k}^T \ \boldsymbol{\zeta}_{h,k}^T \ \zeta_{hq,k}^T\right]^T,$$

where $\boldsymbol{\zeta}_{f,k} = \boldsymbol{f}_k + \boldsymbol{\mu}_{f,k}$, $\boldsymbol{\zeta}_{h,k} = \boldsymbol{h}_k + \boldsymbol{\mu}_{h,k}$, and $\zeta_{hq,k} = \delta_k + \mu_{hq,k}$, being $\boldsymbol{\mu}$ the measurement noise. The measurement noise is assumed to be zero mean Gaussian noise with covariance $\boldsymbol{\Pi}$.

In view of (12), (18), and (22), the output model of the Kalman filter can be written in the form:

$$\boldsymbol{\zeta}_k = \boldsymbol{g}(\boldsymbol{w}_k) + \boldsymbol{\mu}_k,$$

where $[\boldsymbol{\mu}_{f,k}^T \ \boldsymbol{\mu}_{h,k}^T \ \mu_{hq,k}^T]^T$ and

$$\boldsymbol{g}(\boldsymbol{w}_k) = \left[\boldsymbol{g}_f^T(\boldsymbol{w}_k) \ \boldsymbol{g}_h^T(\boldsymbol{w}_k) \ \boldsymbol{g}_{hq}^T(\boldsymbol{w}_k)\right]^T \tag{27}$$

where only the explicit dependence on the state vector $\boldsymbol{w}_k$ has been evidenced.

Since the output model is nonlinear in the system state, the EKF must be adopted, which requires the computation of the Jacobian matrix of the output equation

$$\boldsymbol{C}_k = \left.\frac{\partial \boldsymbol{g}(\boldsymbol{w})}{\partial \boldsymbol{w}}\right|_{\boldsymbol{w}=\hat{\boldsymbol{w}}_{k,k-1}} = \left[\frac{\partial \boldsymbol{g}(\boldsymbol{w})}{\partial \boldsymbol{x}_o} \ \boldsymbol{O}\right]_{\boldsymbol{w}=\hat{\boldsymbol{w}}_{k,k-1}},$$

where $\boldsymbol{O}$ is a null matrix of proper dimensions corresponding to the partial derivative of $\boldsymbol{g}$ with respect to the velocity variables, which is null because the function $\boldsymbol{g}$ does not depend on the velocity.

The Jacobian matrix $\partial \boldsymbol{g}(\boldsymbol{w})/\partial \boldsymbol{x}_o$, in view of (16), (21) and (25), has the expression

$$\frac{\partial \boldsymbol{g}(\boldsymbol{w})}{\partial \boldsymbol{x}_o} = \left[\boldsymbol{J}_f^T \ \boldsymbol{J}_h^T \ \boldsymbol{J}_{hq}^T\right]^T.$$

The equations of the recursive form of the EKF are standard and are omitted here for brevity.

6 Interaction Control

The proposed algorithm can be used to estimate on-line the pose of an object in the workspace; hence it allows the computation of the constraint (11) with respect to the base frame in the form

$$\varphi(\boldsymbol{R}_o^T(\boldsymbol{p}_q - \boldsymbol{o}_o)) = 0.$$

This information can be suitably exploited to implement any kind of interaction control strategy. In this work, an impedance control is adopted, according to a position-based control scheme [14].

In detail, a position and orientation control is adopted for the robot end-effector, and a pose trajectory for a desired frame d is specified in terms of $\boldsymbol{p}_d$ and $\boldsymbol{R}_d$. To manage the interaction with the environment, a compliant frame r

is introduced, specified in terms of $\boldsymbol{p}_r$ and $\boldsymbol{R}_r$. Then, a mechanical impedance between the desired and the compliant frame is considered, so as to keep limited the values of the interaction force $\boldsymbol{h}$ and moment $\boldsymbol{m}$. In other words, the desired position and orientation, together with the measured contact force and moment, are input to the impedance equation which, via a suitable integration, generates the position and orientation of the compliant frame to be used as a reference for the pose control of the robot end effector.

As far as the compliant frame is concerned, the position $\boldsymbol{p}_r$ can be computed via the translational impedance equation

$$M_p\Delta\ddot{\boldsymbol{p}}_{dr} + D_p\Delta\dot{\boldsymbol{p}}_{dr} + K_p\Delta\boldsymbol{p}_{dr} = \boldsymbol{h}, \tag{28}$$

where $\Delta\boldsymbol{p}_{dr} = \boldsymbol{p}_d - \boldsymbol{p}_r$, and $\boldsymbol{M}_p$, $\boldsymbol{D}_p$ and $\boldsymbol{K}_p$ are positive definite matrices representing the mass, damping, and stiffness characterizing the impedance.

The orientation of the reference frame $\boldsymbol{R}_r$ is computed via a geometrically consistent impedance equation similar to (28), in terms of an orientation error based on the (3×1) vector $^r\boldsymbol{\epsilon}_{dr}$, defined as the vector part of the unit quaternion that can be extracted from $^r\boldsymbol{R}_d = \boldsymbol{R}_r^T\boldsymbol{R}_d$. The corresponding mass, damping and inertia matrices are $\boldsymbol{M}_o$, $\boldsymbol{D}_o$ and $\boldsymbol{K}_o$ respectively. More details about the geometrically consistent impedance based on the unit quaternion can be found in [14].

Notice that, when the robot moves in free space, the proposed scheme is equivalent to a position-based visual servoing [20]. Hence, it can be classified as a position-based visual impedance control.

7 Case Studies

7.1 Interaction with an Object

A planar object surface is considered, described by the equation

$$^o\boldsymbol{n}^T{}^o\boldsymbol{p} = 0,$$

assuming that the origin O_o of the object frame is a point of the plane and the axis z_o is aligned to the normal $^o\boldsymbol{n}$. During the interaction with the robot, the normal vector $\boldsymbol{n}$ remains constant in the base frame while the plane is elastically compliant along $\boldsymbol{n}$ according to a simple elastic law. The contact force of the object on the robot's tip at $\boldsymbol{p}_q$ is given by

$$\boldsymbol{h} = \begin{cases} k\boldsymbol{n}\boldsymbol{n}^T(\boldsymbol{p}_o - \boldsymbol{p}_q) & \text{if} \quad \boldsymbol{n}^T(\boldsymbol{p}_o - \boldsymbol{p}) \geq 0 \\ \boldsymbol{0}_3 & \text{if} \quad \boldsymbol{n}^T(\boldsymbol{p}_o - \boldsymbol{p}) < 0 \end{cases}$$

where $\boldsymbol{p}_q$ is on the plane when $\boldsymbol{h} \neq \boldsymbol{0}_3$ while $\boldsymbol{p}_o$ is a constant vector representing the position of a point of the plane when $\boldsymbol{h} = \boldsymbol{0}_3$. The scalar k, representing the stiffness of the surface, has been set to 10000 N/m.

An industrial robot Comau SMART-3 S is considered. The robot has a six-degree-of-freedom anthropomorphic geometry (see [14] for the kinematic and dynamic model).

The end-effector tool is a rigid stick of 25 cm length ending with a circular disk of 5 cm radius. The end-effector frame has its origin at the center of the disk and its approach axis normal to the disk surface and pointing outwards. During the interaction, when the disk surface and the plane are not parallel, the robot's tip point P_Q is assumed to be the instantaneous contact point of the external contour of the disk with the plane. In the case that the disk and the plane are parallel and in contact, the instantaneous contact point is chosen as the center of the disk.

The robot has a force/position sensor mounted at the wrist. Neglecting the weight and inertia of the tool, the force at the robot's tip point P_Q and that at the origin of the end-effector frame are the same, while a moment is present at the origin of the end-effector frame due to the contact with the external contour of the disk. Notice that both the impedance equation and the pose control law are formulated for the end-effector frame.

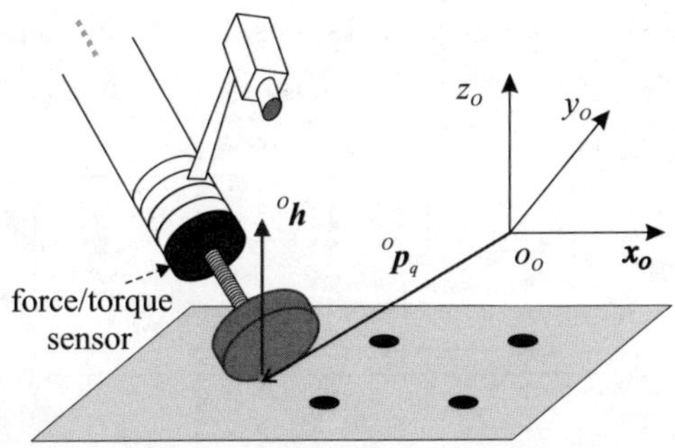

Fig. 3. Sketch of the end-effector in contact with the plane

A camera is mounted on the robot end effector. It is assumed that the intrinsic parameters of the camera are affected by a 2% error, while the extrinsic parameters are known. The object features are 4 landmark points lying on the plane at the corners of a square of 10 cm side.

A sketch of the end-effector in contact with the plane is reported in Fig. 3.

The impedance parameters are chosen as: $M_p = 9I_3$, $D_p = 5I_3$ and $K_p = 700I_3$, $M_o = 0.4I_3$, $D_o = 5I_3$ and $K_o = 2I_3$; a 2 ms sampling time has been selected for the impedance and the pose controller.

The desired task is planned in the object frame and consists in a straight-line motion of the end-effector along the z_o-axis keeping a fixed orientation with the disk surface parallel to the x_oy_o-plane. The final position is:

$$^op_f = {}^op_i - {}^on({}^on^T{}^op_i - \delta),$$

66   A. De Santis et al.

where $^o\boldsymbol{p}_i = [0.5\ 0\ 0]^T$ m is the initial position of the end effector and $\delta = 0.033$ m is chosen to have a normal force of about 22 N at the equilibrium, with the available estimate of the environment stiffness. A trapezoidal velocity profile time-law is adopted, with a cruise velocity of 0.023 m/s. The absolute trajectory is computed from the desired relative trajectory using the current object pose estimation.

In the EKF, the non-null elements of the matrix $\boldsymbol{\Pi}$ have been set equal to $625 \cdot 10^{-12}$ for $\boldsymbol{f}$, 10^{-7} for $\boldsymbol{n}_h$ and $6.5 \cdot 10^{-5}$ for δ_{hq}. The state noise covariance matrix has been selected so as to give a rough measure of the errors due to the simplification introduced on the model (constant velocity), by considering only velocity disturbance, i.e.

$$\boldsymbol{Q} = \mathrm{diag}\{0, 0, 0, 0, 0, 0, 0, 5, 5, 0.5, 10^2, 10^3, 10^3, 10^3\} \cdot 10^{-12}.$$

Notice that the unit quaternion has been used for the orientation in the EKF, to avoid any occurrence of representation singularities. Moreover a 20 ms sampling time has been set for the estimation algorithm, corresponding to the typical camera frame rate of 50 Hz.

Two different case studies are presented, to show the effectiveness of the use of force and joint position measurements, besides visual measurements.

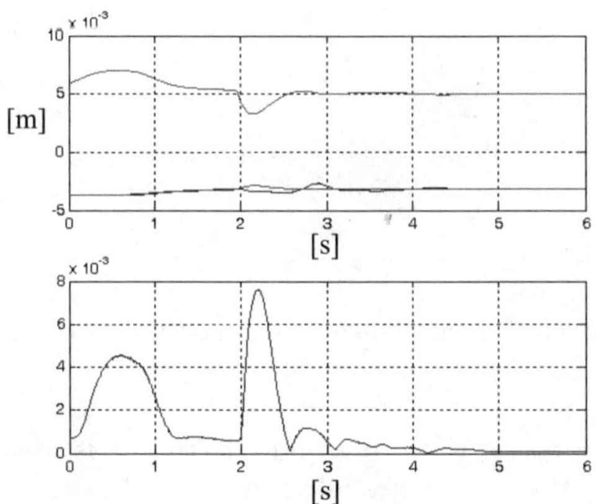

Fig. 4. Pose estimation error in the first case study. Top: position error; bottom: orientation error.

In the first case study only the visual measurements are used. The object pose estimation errors are reported in Fig. 4. The position error is computed as the difference between the real position of the origin of the objet frame and the estimated position; the orientation error is defined as the norm of the vector part of the quaternion that can be extracted from the rotation matrix representing the mutual orientation of the real object frame with respect to the estimated frame. The task starts at time $t_o = 0$ s, when an estimate of the object pose is available from visual measurements; notice that the initial value of the pose estimation error in non null, due to the camera calibration error. From t_o to $t_1 \simeq 2\,\mathrm{s}$ the error varies slowly due to the robot motion. At time t_1 the disk comes into contact with the plane; the abrupt change of robot velocity causes an increment of the estimation error that, after a transient, becomes constant and approximatively equal to the initial value.

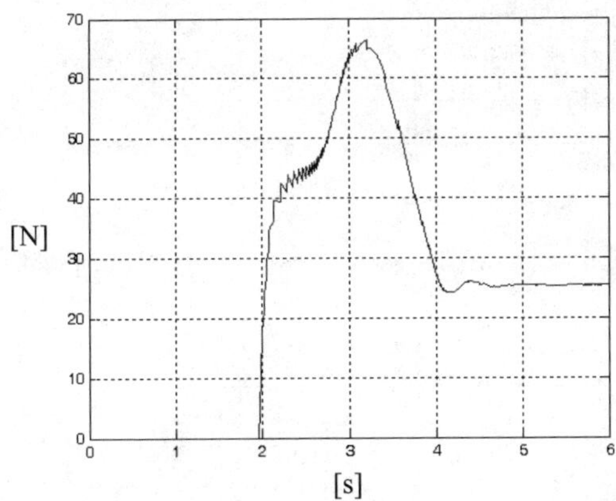

Fig. 5. Contact force in the first case study

The time history of the contact force in the object frame is reported in Fig. 5. Notice that the contact force is null during the motion in free space and becomes different from zero after the contact at time t_1. The impedance control keeps the force limited during the transient while, at steady state, the force reaches a value of about 26 N, which is different from the desired value due to the presence of the estimation error along to the z_o-axis.

The same task is repeated using also the contact force and the joint position measurements for object pose estimation; the results are reported in Fig. 6 and Fig. 7. Before the contact (i.e. before time t_1), the results are the same as in the previous case study. After the contact, the benefit of using additional measurements in the EKF produces a significant reduction of the pose estimation error, especially for the z_o component and for the orientation. Moreover, the peak of the contact force is lower than before and the force value at steady state is near to the expect value of 22 N.

7.2 Vision-Based Head Avoidance

During a task involving interaction with an object, there is the possibility that a human operator is present in the workspace. In such a case, the robot has to reconfigure in order to avoid the body of the operator, tracked by a camera. In a simple case, it is possible to consider the head and the arms of a person present in

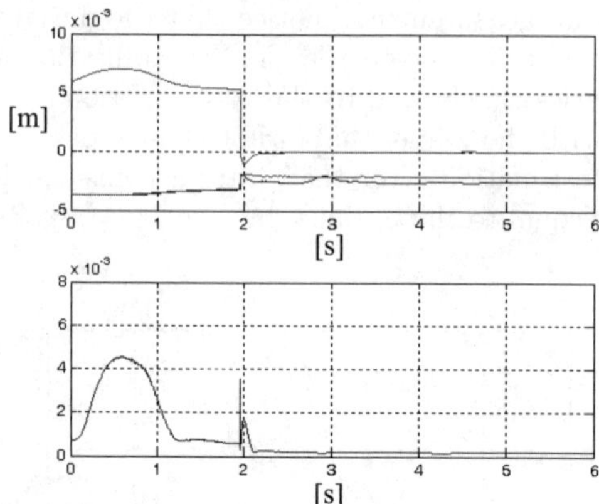

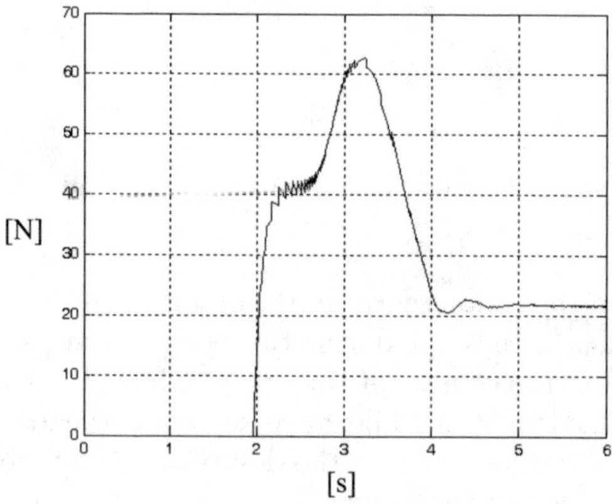

Fig. 6. Pose estimation error in the second case study. Top: position error; bottom: orientation error.

Fig. 7. Contact force in the second case study

the workspace as a source of a repelling elastic force. A volume is created around the head and the arms: the robot is pushed with continuous reference values given to force or velocity for a point on each link which is the closest to the considered "safety volume". Results of an experiment with the Comau SMART 3S industrial robot are reported in Fig. 8. The planned trajectory (dotted line) is abandoned for the presence of the arm (segment parallel to the axis x, with $y = 1$ and $z = 0.5$). The bold trajectory is the path followed with an elastic constant $K = 0.5$ for planning the desired velocity v of the closest points with the formula $v = K(d_{ref} - d)$ for $d > d_{ref}$, where d_{ref} is the radius of the protective sphere and d is the distance between the robot links and the center of such a sphere. The thin path in Fig. 8 is tracked for $K = 0.5$. This simple case study shows the robustness of the skeleton algorithm, which gives continuous references to different inverse kinematics schemes (one for each link of a robot) in order to push a robot in a certain direction during any kind of operation and with any kind of motion/force controller.

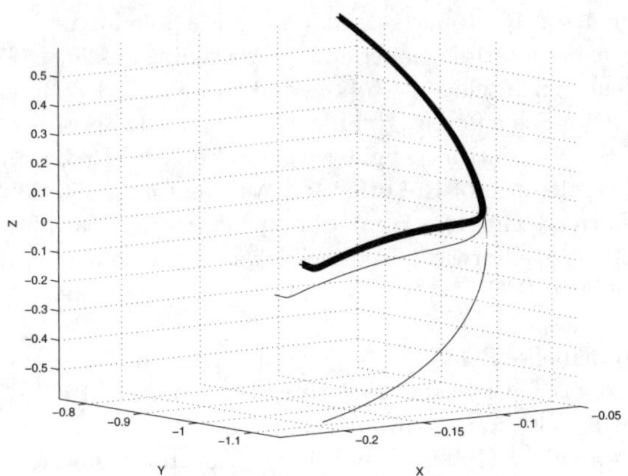

Fig. 8. Trajectory modifications for collision avoidance

8   Conclusion

The integration of force and visual control to achieve safe human-robot interaction has been discussed. A position-based visual impedance control scheme has been presented, employing a pose estimation algorithm on the basis of visual, force and joint position data. The addition of collision avoidance facilities with the so-called skeleton algorithm gives the opportunity of sharing the workspace with a human operator.

References

1. De Santis A, Siciliano B, Villani L (2006) The Atlas of Physical Human-Robot Interaction, Final Report of the EURON Perspective Research Project PHRIDOM
2. Zinn M, Khatib O, Roth B, Salisbury J K (2004) Playing it safe [human-friendly robot], IEEE Robotics and Automation Magazine, 11(2):12–21
3. Bicchi A, Tonietti G (2004) Fast and "soft-arm" tactics, IEEE Robotics and Automation Magazine 11(2):22–33
4. Hashimoto H (2005) Intelligent interactive spaces - integration of IT and robotics, In: Proceedings of 2005 IEEE Workshop on Advanced Robotics and its Social Impacts, 85–90
5. Hosoda K, Igarashi K, Asada M (1998) Adaptive hybrid control for visual and force servoing in an unknownenvironment, IEEE Robotics and Automation Magazine 5(4):39–43
6. Nelson BJ, Morrow JD, Khosla PK (1995) Improved force control through visual servoing, In: Proceedings of 1995 American Control Conference, 380–386

7. Baeten J, De Schutter J (2004) Integrated Visual Servoing and Force Control. The Task Frame Approach, Springer, Berlin Heidelberg New York
8. Morel G, Malis E, Boudet S (1998) Impedance based combination of visual and force control, In: Proceedings of 1998 IEEE International Conference on Robotics and Automation, 1743–1748
9. Olsson T, Johansson R, Robertsson A (2004) Flexible force-vision control for surface following using multiple cameras, In: Proceedings of 2004 IEEE/RSJ International Conference on Intelligent Robots and System, 798–803
10. Hirzinger G, Albu-Schaeffer A, Hahnle M, Schaefer I, Sporer N (2001) On a new generation of torque controlled light-weight robots, In: Proceedings of 2001 IEEE International Conference of Robotics and Automation, 3356–3363
11. http://www.barretttechnology.com/robot/products/arm/armfram.htm
12. De Luca A (2000) Feedforward/feedback laws for the control of flexible robots, In: Proceedings of 2000 IEEE International Conference on Robotics and Automation, 233–240
13. De Luca A, Lucibello P (1998) A general algorithm for dynamic feedback linearization of robots with elastic joint, In: Proceedings of 1998 IEEE International Conference on Robotics and Automation, 504–510
14. Siciliano B, Villani L (1999) Robot Force Control, Kluwer, Dordrecht Boston London
15. De Santis A, Albu-Schaeffer A, Ott C, Siciliano B, Hirzinger G (2007), The skeleton algorithm for real-time collision avoidance of a humanoid manipulator interacting with humans, Submitted to IEEE Transactions on Robotics
16. De Santis A, Pierro P, Siciliano B (2006) The virtual end-effectors approach for human-robot interaction, In: Lenarčič J, Roth B (eds) Advances in Robot Kinematics, Springer, Berlin Heidelberg New York, 133–144
17. Lippiello V, Siciliano B, Villani L (2006) 3D pose estimation for robotic applications based on a multi-camera hybrid visual system, In: Proceedings of 2006 IEEE International Conference on Robotics and Automation, 2732–2737
18. Espiau B, Chaumette F, Rives P (1992) A new approach to visual servoing in robotics, IEEE Transactions on Robotics and Automation, 8:313–326
19. Lippiello V, Villani L (2003) Managing redundant visual measurements for accurate pose tracking, Robotica, 21:511–519
20. Wilson W J, Hulls C C W, Bell G S (1996) Relative end-effector control using Cartesian position based visual servoing, IEEE Transactions on Robotics and Automation, 12:684–696

A Dissipation Inequality for the Minimum Phase Property of Nonlinear Control Systems

Christian Ebenbauer[1] and Frank Allgöwer[2]

[1] Laboratory for Information and Decision Systems, Massachusetts Institute of Technology, USA
`ebenbauer@mit.edu`
[2] Institute for Systems Theory and Automatic Control, University of Stuttgart, Germany
`allgower@ist.uni-stuttgart.de`

Summary. The minimum phase property is an important notion in systems and control theory. In this paper, a characterization of the minimum phase property of nonlinear control systems in terms of a dissipation inequality is derived. It is shown that this dissipation inequality is equivalent to the classical definition of the minimum phase property in the sense of Byrnes and Isidori, if the control system is affine in the input and the so-called input-output normal form exists.

Keywords: Minimum Phase Property, Dissipation Inequalities, Nonlinear Control Systems.

1 Introduction

Bode introduced the notion of minimum phase property in his seminal paper [4] more than 60 years ago. Today, the minimum phase property plays an important role in systems analysis and control design [12, 14, 13, 27]. For example, the notion of the minimum phase property can be used to describe fundamental performance limitations in feedback design (e.g., [4, 20, 15, 9, 8, 18, 23, 24, 2, 26, 1]) and thus allows, roughly speaking, to distinguish between easy and difficult control problems. For linear time-invariant single-input-single-output systems, the minimum phase property is characterized for example by all zeros of the transfer function being in the open left half plane. The notion of zeros was generalized by Byrnes and Isidori (cf. e.g [12]) to nonlinear control systems. For nonlinear control systems, loosely speaking, a system is said to be minimum phase if it has asymptotically stable zero output constrained dynamics (zero dynamics), which is obtained when the output of the system is kept identically equal to zero. For the special class of nonlinear control systems that are affine in the input and that posses a well-defined input-output normal form in the sense of [12], a rigorous definition of the minimum phase property can be given. In the following, this situation is referred to as the minimum phase property in the sense of Byrnes-Isidori. The minimum phase property is then equivalent to the situation that an equilibrium point, let's say $x = 0$, is asymptotically stable under the constraint that the output $y(t) = 0$, $t \geq 0$. In general, however, a

C. Bonivento et al. (Eds.): Adv. in Control Theory and Applications, LNCIS 353, pp. 71–83, 2007.
springerlink.com © Springer-Verlag Berlin Heidelberg 2007

precise definition of the minimum phase property for general nonlinear control systems is not an easy task. The reason for this is that the zero dynamics may not be well-defined, and even if this were the case, it makes no sense to speak about stability without saying something about equilibrium points (or sets). Beside this, it may be difficult to check if a control system is minimum phase or not. In the literature (cf. e.g. [12]), there exist at least two strategies for a minimum phase analysis: The first one makes use of a transformation of the control system into the input-output normal form, if the normal form exists. The second one is based on simply setting $y(t), \dot{y}(t), \ldots$ to zero, i.e, by setting the output and its higher order Lie-derivatives to zero and by calculating the remaining dynamics, which is equivalent to the zero dynamics. The second strategy is more general, since it also works when a transformation into the input-output normal form does not exist.

In this paper, a new third possibility is given to characterize the minimum phase property, namely in terms of a dissipation inequality. It is shown that the definition of the minimum phase property in the sense of Byrnes-Isidori for affine control systems with a well-defined input-output normal form is equivalent to the fact that a certain dissipation inequality is satisfied. Hence the minimum phase property, which has its origin in the frequency domain world and in geometric control, is expressed in terms of a Lyapunov-based language in this new approach. Moreover, the dissipation inequality can be easily applied to general nonlinear control systems that are not necessarily affine in the input. In addition to the preliminary work [5], an additional result on the smoothness of a certain function involved in the derived dissipation inequality is established. This result plays an important role when using the dissipation inequality in a constructive or in a computational way.

The only known results where the minimum phase property is expressed in terms of a Lyapunov-based language, i.e., in terms of a dissipation inequality are [16] and [5]. In [16] another alternative (stronger) notion of minimum phase property is given, based on output-input-stability which is in the spirit of Sontag's "input-to-state stability" philosophy. In particular a dissipation inequality is used in [16], which is a sufficient condition for the minimum phase property. The dissipation inequality there is, however, not a necessary condition, since the notion used there is motivated by introducing additional robustness in the minimum phase property. Therefore, the dissipation inequality derived there does not fully coincide with the well-established notion of minimum phase property in the sense of Byrnes-Isidori. In the preliminary work [5], a dissipation inequality is derived which is necessary and sufficient for the minimum phase property and which is slightly different from the dissipation inequality derived below. However, in contrast to [5], the results established in Section 3, in particular Theorem 2 allow an additional smoothness statement of a function that appears in the dissipativity characterization of the minimum phase property.

The structure of the paper is as follows: In Section 2, results from the literature are revisited and the class of control systems to be considered, the input-output normal form, and the definition of the minimum phase property in the sense of Byrnes-Isidori is given. In Section 3, the dissipation inequality which characterizes the minimum phase property is derived. Some numerical examples demonstrate the results of this paper in Section 4. Finally, Section 5 concludes with a discussion and summary.

2 Preliminaries

The class of control systems studied in this paper is of the form

$$\dot{x} = f(x) + G(x)u$$
$$y = h(x), \tag{1}$$

where $f : \mathbb{R}^n \rightarrow \mathbb{R}^n$, $G : \mathbb{R}^n \rightarrow \mathbb{R}^{n \times p}$, $h : \mathbb{R}^n \rightarrow \mathbb{R}^p$ and $x \in \mathbb{R}^n$ is the state, $u \in \mathbb{R}^p$ is the input, and $y \in \mathbb{R}^p$ is the output.

The main assumption on the control system (1) is that an input-output normal form exists, i.e., the following assumption is made:

Assumption 1. *The functions f, G, h in (1) are assumed to be sufficiently smooth with $f(0) = 0$, $h(0) = 0$ and furthermore, it is assumed that there exists a local change of coordinates $[\xi, \eta]^T = \Phi(x)$ with $\Phi(0) = 0$, Φ sufficiently smooth, such that the control system (1) with the same number of inputs and outputs can be represented in input-output normal form ([12], p.224):*

$$\dot{\xi}_1^i = \xi_2^i$$
$$\vdots$$
$$\dot{\xi}_{r_i-1}^i = \xi_{r_i}^i$$
$$\dot{\xi}_{r_i}^i = b_i(\xi, \eta) + \sum_{j=1}^{p} a_{ij}(\xi, \eta)u_j \tag{2}$$
$$\dot{\eta} = q(\xi, \eta) + P(\xi, \eta)u$$
$$y_i = \xi_1^i,$$

where $\xi = [\xi_1^1 \ldots \xi_{r_1}^1, \xi_1^2 \ldots]^T$, $i = 1 \ldots p$. Moreover, it is assumed that $q(0, \eta) - P(0, \eta)A(0, \eta)^{-1}b(0, \eta)$ is sufficiently smooth, with the square invertible (decoupling) matrix $A(\xi, \eta) = (a_{ij}(\xi, \eta))$, $i, j = 1 \ldots p$ and a vectorial relative degree $r = [r_1, \ldots, r_p]$. Note that the output zeroing feedback $u = k_z(\xi, \eta)$ is unique [12] and is given by

$$u = k_z(\xi, \eta) = -A(\xi, \eta)^{-1}b(\xi, \eta) \tag{3}$$

with $b(\xi, \eta) = [b_1(\xi, \eta) \ldots b_p(\xi, \eta)]^T$.

For example, if the control system (1) is a single-input-single-output system with f, G, h sufficiently smooth and if the relative degree is well-defined, then a local change of coordinates exists that transforms the control system into the given form. The multi-input-multi-output case is more involved [12]. However, control systems that are minimum phase in the sense of Byrnes-Isidori exhibit stable behavior under the constraint that the output is identically zero. More precisely:

Definition 1. *The control system (1) under the Assumption 1 is said to possess the minimum phase property with respect to the equilibrium point $x = 0$, if $x = 0$ is asymptotically stable under the constraint $y(t) = 0$, $t \geq 0$. In other words, the zero dynamics*

$$\dot{\eta} = q(0, \eta) - P(0, \eta)A(0, \eta)^{-1}b(0, \eta) \tag{4}$$

of the control system (1), respectively (2), is asymptotically stable at $\eta = 0$.

Further definitions and notations. A function $V : \mathbb{R}^n \to \mathbb{R}$ is called positive definite, if $V(0) = 0$, $V(x) > 0$ for all nonzero x. V is called radially unbounded, if $V(x) \to \infty$ whenever $\|x\| \to \infty$. A continuously differentiable, positive definite, radially unbounded function V is called a Lyapunov function candidate. For a function $V : \mathbb{R}^n \to \mathbb{R}$, the row vector $\frac{\partial V}{\partial x}(x) = \nabla V(x) = [V_{x_1}(x) \dots V_{x_n}(x)]$ denotes the derivative of V with respect to x.

3   A Dissipation Inequality for the Minimum Phase Property

In the following, the main results of the paper are derived. In particular, a characterization of the minimum phase property for the control system (1) under the Assumption 1 is given in terms of a dissipation inequality, Theorem 1, and an additional smoothness result for the dissipation inequality, Theorem 2, is derived. To define the dissipation inequality for the minimum phase property, the following so-called derivative array (cf. e.g. [10]) is used:

Definition 2. *The derivative array $H_r : \mathbb{R}^n \times \mathbb{R}^q \to \mathbb{R}^{r_1 + \dots + r_p + p}$ of the output function $y = h(x) = [h_1(x) \dots h_p(x)]^T$ in (1) is defined by the first r Lie-derivatives of the output, i.e.,*

$$H_r(x, u) = \begin{bmatrix} h_1(x) \\ \dot{h}_1(x) \\ \vdots \\ h_1^{(r_1)}(x) \\ h_2(x) \\ \vdots \\ h_p^{(r_p)}(x) \end{bmatrix} \tag{5}$$

with $\dot{h}_i(x) = \frac{\partial h_i}{\partial x}(x)(f(x) + G(x)u) = L_f h_i(x) + L_G h_i(x)u$ etc., i.e., the Lie-derivatives of h_i with respect to (1) up to degree r_i. Notice that H_r is a function of x and u, since $h_i^{(r_i)}$ depends on u.

Using the derivative array, the first result in this section is a characterization of the minimum phase property in terms a dissipation inequality.

Theorem 1. *The control system* (1) *under the Assumption 1 has the minimum phase property according to Definition 1 if and only if there exists a Lyapunov function* $V : \mathbb{R}^n \to \mathbb{R}$ *and a function* $\rho : \mathbb{R}^n \times \mathbb{R}^p \to \mathbb{R}^{r_1 + \cdots + r_p + p}$ *such that the dissipation inequality*

$$\nabla V(x)(f(x) + G(x)u) < H_r(x, u)^T \rho(x, u) \tag{6}$$

is satisfied for all u *and all nonzero* x *in a neighborhood of* $x = 0$.

Proof. The first part of the proof of Theorem 1 shows an explicit construction of the functions V, ρ, in case the control system (1) is minimum phase. The second part shows that if the dissipation inequality (6) is satisfied, then the minimum phase property follows.

Part 1 ((1) is minimum phase $\overset{(2)}{\Rightarrow}$ (6) is satisfied): In the following, it is assumed that the control system (1) is represented in the input-output normal form (2). Since (1) is minimum phase, the zero dynamics of (1) is asymptotically stable and is given by

$$\dot{\eta} = q(0, \eta) - P(0, \eta)A(0, \eta)^{-1}b(0, \eta) = z(0, \eta), \tag{7}$$

which follows by substituting the output zeroing feedback

$$k_z(\xi, \eta) = -A(\xi, \eta)^{-1}b(\xi, \eta) \tag{8}$$

into (2) and by setting $\xi = 0$. Let W be a continuously differentiable Lyapunov function of (7). The existence of such an Lyapunov function is guaranteed due to Massera's converse Lyapunov theorem [17, 25, 13]. Massera's theorem assumes a locally Lipschitz right-hand side of the differential equation for the existence of a smooth differential Lyapunov function. Since this is assumed in Assumption 1, W exists. Define now a Lyapunov function candidate

$$V(\xi, \eta) = U(\xi) + W(\eta), \tag{9}$$

where U is an arbitrary Lyapunov function candidate, i.e., a positive definite, radially unbounded, continuously differentiable scalar-valued function. The derivative of V along the trajectories of (2) is given by:

$$\dot{V}(\xi, \eta) = \nabla U(\xi)\dot{\xi} + \nabla W(\eta)\dot{\eta}. \tag{10}$$

Next, two cases are distinguished: *Case 1:* H_r is zero in (6), i.e., $\xi_1^i = \ldots = \xi_{r_i}^i = \dot{\xi}_{r_i}^i = 0$, $(u = k_z(\xi, \eta))$, $i = 1 \ldots p$. In this case define the value of ρ to be zero, i.e.,

$$\rho(\xi, \eta, u) = 0. \tag{11}$$

What remains to show that (6) is satisfied is that $\nabla W(\eta)\dot{\eta} < 0$ holds for some neighborhood around $\eta = 0$. But this is the case, since asymptotic stability of the zero dynamics is assumed. *Case 2:* H_r is not zero in (6), i.e., there exists $\xi^i_j \neq 0$ or $\xi^i_{r_i} \neq 0$ $(u \neq k_z(\xi, \eta))$. In this case define the value of ρ such that

$$\rho(\xi, \eta, u) = H_r(\xi, \eta, u) \cdot \tilde{\rho}(\xi, \eta, u),$$

$$\tilde{\rho}(\xi, \eta, u) > \frac{\nabla U(\xi)\dot{\xi} + \nabla W(\eta)\dot{\eta}}{\|H_r(\xi, \eta, u)\|^2}. \tag{12}$$

The value of ρ is finite since $H_r(\xi, \eta, u) \neq 0$. With the definition of ρ by (11), (12) and V according to (9), the dissipation inequality (6) is satisfied (in (ξ, η)-coordinates). Note that the dissipation inequality in the original coordinates can be obtained by the inverse transformation $x = \Phi^{-1}(\xi, \eta)$.

Part 2 ((6) is satisfied $\overset{(2)}{\Rightarrow}$ (1) is minimum phase): To show this, consider the zero dynamics, i.e., consider the dynamics which is defined by initial conditions (ξ_0, η_0) with $\xi_0 = 0$ and by the output zeroing feedback $u = k_z(\xi, \eta)$. Under these initial conditions and under the output zeroing feedback $u = k_z(\xi, \eta)$, $y(t) = 0$ for all $t \geq 0$. Hence $H_r(0, \eta(t), u(t)) = 0$, $t \geq 0$ because of $\xi(t) = 0$, $t \geq 0$. Thus the dissipation inequality (6) turns into $\dot{V}(0, \eta(t)) < 0$ and therefore V is a Lyapunov function and the equilibrium point $\eta = 0$ of the zero dynamics is asymptotically stable.

Theorem 1 establishes a symmetric statement between the minimum phase property and a dissipation inequality (6). To understand the dissipation inequality (6) is not difficult. However, a few points have to be explained. Firstly, the role of the derivative array H_r: The zero dynamics is the dynamics such that the output is identically zero. This dynamics evolves on the zero dynamics manifold, which is implicitly defined by $\|H_r(x, u)\| = 0$, since H_r is identically zero, if $y(t) = 0$, $t \geq 0$. Remember that ρ in the proof of Theorem 1 has the form $\rho(\xi, \eta, u) = H_r(\xi, \eta, u)\tilde{\rho}(\xi, \eta, u)$, which turns the inequality (6) into

$$\nabla V(x)(f(x) + G(x)u) < \|H_r(x, u)\|^2 \tilde{\rho}(x, u) \tag{13}$$

with $x = [\xi\ \eta]^T$. Thus, stability on the manifold $\|H_r(x, u)\| = 0$ has to be studied, i.e., a Lyapunov function is needed subject to the constraint $\|H_r(x, u)\| = 0$. $\|H_r(x, u)\| > 0$ is not of interest. This situation is compactly expressed in inequality (6) ((13)), where ρ plays the role of a penalization function. Geometrically speaking, inequality (6) guarantees negative definiteness of the derivative of V only on a subset, namely on the set where $\|H_r(x, u)\| = 0$. For $\|H_r(x, u)\| > 0$, one can find a function ρ such that the left side is dominated by the right side of the dissipation inequality (6). Algebraically speaking, the right side of the inequality (6) is the ideal generated by $\|H_r(x, u)\|$, i.e, the left side is negative definite modulo $\|H_r(x, u)\| > 0$.

Summarizing, the main ingredients to arrive at the dissipation inequality (6) are the so-called derivative array, which defines the hidden constraints and which finally defines the zero dynamics manifold, as well as a penalization argument, a well-known argument from optimization theory. Furthermore, in contrast to the ISS-like minimum phase characterization introduced in [16], Theorem 1 is necessary and sufficient to express the minimum phase property as defined by Byrnes and Isidori. Hence, Theorem 1 represents a complete characterization of the minimum phase property. It is also worthwile to remark that from the dissipation inequality (6) it can be very clearly seen that the notion of the minimum phase property is feedback invariant, since it must hold for all u. In Theorem 1, no statement is made about the degree of smoothness of ρ. Even no statement on the existence of a continuous ρ is made. However, for computational purposes for example, a guarantee of the existence of a smooth ρ would be desirable. The next theorem shows that under Assumption 1, there exists indeed a smooth ρ. In particular the following proof of Theorem 2 is constructive and an explicit function ρ is constructed. Due to simplicity of exposition, the construction is carried out for the case $p = 1$, i.e., Theorem 2 is stated for the single-input-single-output case. The construction for the multi-input-multi-output is more tedious but goes along the same lines.

Theorem 2. *If the control system* (1) *under the Assumption 1 with $p = 1$ has the minimum phase property according to Definition 1, then there exists a smooth Lyapunov function $V : \mathbb{R}^n \to \mathbb{R}$ and a smooth function $\rho : \mathbb{R}^n \times \mathbb{R} \to \mathbb{R}^{r+1}$ such that the dissipation inequality* (6) *is satisfied for all u and all nonzero x in a neighborhood of $x = 0$.*

Proof. As in the proof of Theorem 1, it is assumed that the control system (1) is represented in the input-output normal form (2). Smoothness of the Lyapunov function W for the zero dynamics (7) follows from Massera's converse Lyapunov theorem [17, 25]. In particular, it is assumed in Assumption 1 that the zero dynamics is sufficiently smooth, hence a sufficiently smooth W exists. To show that also ρ is sufficiently smooth, a smooth Lyapunov function candidate of the form

$$V(\xi, \eta) = \frac{1}{2}\xi^T \xi + W(\eta) \tag{14}$$

is chosen, i.e., $U(\xi) = \frac{1}{2}\xi^T\xi$ in (9). Hence, the dissipation inequality (6) for the control system (2), with $p = 1$, is given by:

$$\begin{aligned}
\xi_1\xi_2 + \ldots + \xi_{r-1}\xi_r + \xi_r(b(\xi, \eta) + a(\xi, \eta)u) + \nabla W(\eta)(q(\xi, \eta) + p(\xi, \eta)u) \\
< \xi_1\rho_1(\xi, \eta, u) + \ldots + \xi_r\rho_r(\xi, \eta, u) + (b(\xi, \eta) + a(\xi, \eta)u)\rho_{r+1}(\xi, \eta, u)
\end{aligned} \tag{15}$$

with $\xi = [\xi_1 \ldots \xi_r]^T$. In a first step, ρ is chosen as

$$\begin{aligned}
\rho_i(\xi, \eta, u) &= \xi_{i+1} + \tilde{\rho}_i(\xi, \eta, u), \\
\rho_r(\xi, \eta, u) &= (b(\xi, \eta) + a(\xi, \eta)u) + \tilde{\rho}_r(\xi, \eta, u)
\end{aligned} \tag{16}$$

$i = 1 \ldots r - 1$, with $\tilde{\rho}_i$ as new auxiliary functions. Hence (15) turns into

$$\begin{aligned}
\nabla W(\eta)(q(\xi, \eta) + p(\xi, \eta)u) < \xi_1\tilde{\rho}_1(\xi, \eta, u) + \ldots + \xi_r\tilde{\rho}_r(\xi, \eta, u) \\
+ (b(\xi, \eta) + a(\xi, \eta)u)\rho_{r+1}(\xi, \eta, u).
\end{aligned} \tag{17}$$

In a second step, u is replaced by

$$u = -\frac{1}{a(\xi,\eta)}(b(\xi,\eta) + v) \tag{18}$$

with v as a new input. Therefore, one obtains from (17)

$$\nabla W(\eta)\left(q(\xi,\eta) - \frac{p(\xi,\eta)}{a(\xi,\eta)}(b(\xi,\eta) + v)\right)$$
$$< \xi_1\tilde{\rho}_1(\xi,\eta,v) + \ldots + \xi_r\tilde{\rho}_r(\xi,\eta,v) + v\rho_{r+1}(\xi,\eta,v). \tag{19}$$

Due to the substitution (18), inequality (19) can be satisfied if and only if (17) can be satisfied. In a next step, ρ is chosen as

$$\rho_{r+1}(\xi,\eta,v) = -\nabla W(\eta)\frac{p(\xi,\eta)}{a(\xi,\eta)} \tag{20}$$

and after rewriting (19), one arrives at

$$\nabla W(\eta)z(0,\eta) + \nabla W(\eta)\left(z(\xi,\eta) - z(0,\eta)\right)$$
$$< \xi_1\tilde{\rho}_1(\xi,\eta,v) + \ldots + \xi_r\tilde{\rho}_r(\xi,\eta,v), \tag{21}$$

where the expression that corresponds to the zero dynamics is given by

$$z(\xi,\eta) = q(\xi,\eta) - p(\xi,\eta)\frac{b(\xi,\eta)}{a(\xi,\eta)}. \tag{22}$$

Since the control system is assumed to be minimum phase, the inequality

$$\nabla W(\eta)z(0,\eta) < 0 \tag{23}$$

holds locally. Thus, it is sufficient to show that

$$\nabla W(\eta)\left(z(\xi,\eta) - z(0,\eta)\right) + \xi^T\xi \le \xi_1\tilde{\rho}_1(\xi,\eta,v) + \ldots + \xi_r\tilde{\rho}_r(\xi,\eta,v) \tag{24}$$

can be satisfied. Assumption 1 implies that the function z is sufficiently smooth and therefore continuously differentiable. By applying a mean-value theorem for vector-valued functions, the so-called Hadamard lemma [22, 3], the difference $z(\xi,\eta) - z(0,\eta)$ can be written as

$$z(\xi,\eta) - z(0,\eta) = Z(\xi,\eta)\xi \tag{25}$$

with a continuous (smooth) matrix-valued function Z defined by

$$Z(\xi,\eta) = \int_0^1 \frac{\partial z}{\partial x}((1-\theta)\xi,\eta)d\theta. \tag{26}$$

Hence, the inequality (24) can be written as

$$\nabla W(\eta)Z(\xi,\eta)\xi + \xi^T\xi \le \xi_1\tilde{\rho}_1(\xi,\eta,v) + \ldots + \xi_r\tilde{\rho}_r(\xi,\eta,v), \tag{27}$$

from which the smooth functions $\tilde{\rho}_i$ easily follow such that (27) holds. For example, choose the $\tilde{\rho}_i$'s such that

$$\nabla W(\eta)Z(\xi,\eta) + \xi^T = [\tilde{\rho}_1(\xi,\eta,v)\ldots\tilde{\rho}_r(\xi,\eta,v)]. \tag{28}$$

Therefore, inequality (24) is satisfied and thus also the desired dissipation inequality (15). Finally, note again that the dissipation inequality in the original coordinates can be obtained by the inverse transformation $x = \Phi^{-1}(\xi,\eta)$ of the input-output normal form transformation.

Notice that the converse statement of Theorem 2 follows immediately from the proof of Theorem 1. Furthermore, Definition 1 and all established results are of local nature with respect to the equilibrium point $x = 0$. From the proofs of Theorem 1 and 2, however, global results can be easily established. Moreover, the results in this paper can be easily extended to the more general input-output normal form in [12] (p.310). In particular, this is one advantage of the established dissipation inequality since it is in principle also applicable to general control systems, as summarized next.

Remark 1. The affine structure of the control system (1) can be easily replaced in (6) by a general, nonaffine control system, i.e.,

$$\nabla V(x)f(x,u) < H_r(x,u)^T\rho(x,u), \tag{29}$$

which leads to a possible extension of the minimum phase property to nonaffine control systems. In this case, however, the zero dynamics might not be well-defined and the output zeroing feedback is not unique anymore. For generalized notions for the minimum phase property to control systems that are not affine in the control input, one may also consult [19, 16]. Since the minimum phase property is basically a matter of stability on manifolds, one may consult [6] which provides a general Lyapunov-based approach for such questions.

Remark 2. It is well-known that the minimum phase property is an important notion for describing fundamental performance limitations in feedback design. In particular, well-known is the Bode integral of the inverse sensitivity function (Bode T-integral), which relates the minimum phase property with limitations in tracking problems [20, 15, 9, 8, 18, 23, 24, 2, 26, 1]. One may ask the question, in how far the derived dissipation inequality in Section 3 reflects this fact. One possible answer is given in [7]. The idea persuited there is to search for a new output such that a given nonminimum phase control system becomes minimum phase and such that the new minimum phase output is the closest one to the true nonminimum phase output in the L_2-sense. In particular, for linear time-invariant control systems it can be shown using duality theory from convex optimization and by utilizing the established dissipation inequality for the minimum phase property, that this leads to an alternative deriviation of the Bode integral for the inverse sensitivity. More details on that as well as its relation to cheap control [24] can be found in [7].

Summarizing, in this section a new characterization of the minimum phase property for control systems which posses a well-defined input-output normal form is derived. Moreover, the established characterization is suitable for computational purposes (cf. Example 2, Section 4) and also applicable to control systems where an input-output normal form does not exist or where the relative degree is not well-defined (cf. Example 1, Section 4).

4   Examples

In the following, two examples are given which illustrate the results in the previous sections. In particular, attention is paid to the following two aspects: computability and generalizability.

Example 1

This example illustrates that the dissipation inequality (6) can also be applied in case the control system does not have a well-defined relative degree and is not affine in the input. Consider the nonaffine control system

$$
\begin{aligned}
\dot{x}_1 &= -x_1 + x_3 e^u \\
\dot{x}_2 &= x_3 \\
\dot{x}_3 &= x_2 u \\
y &= x_2,
\end{aligned}
\tag{30}
$$

which has relative degree two except for $x_2 = 0$. Applying (6) with $V = \frac{1}{2}(x_1^2 + x_2^2 + x_3^2)$ yields

$$
\begin{aligned}
- x_1^2 &+ x_1 x_3 e^u + x_2 x_3 + x_2 x_3 u \\
&< x_2 \rho_1(x, u) + x_3 \rho_2(x, u) + x_2 u \rho_3(x, u).
\end{aligned}
\tag{31}
$$

For example, by choosing $\rho_1(x, u) = x_3 + x_2, \rho_2(x, u) = x_1 e^u + x_2 u + x_3, \rho_3(x, u) = 0$, one obtains $-x_1^2 - x_2^2 - x_3^2 < 0$. Thus global asymptotic stability of the zero dynamics is established, i.e., the control system (30) is (globally) minimum phase.

Example 2

This example illustrates that the dissipation inequality (6) for the minimum phase property is particularly suited for a minimum phase test for control systems with polynomial nonlinearities. In general, it is very difficult to search for a Lyapunov function V and a function ρ such that (6) holds However, recently established methods from computational real algebraic geometry based on semidefinite programming and the sum of squares decomposition allow to verify the dissipation inequality (6) very efficiently in case all the functions involved are of polynomial type (consult for example [21, 11] and references therein).

In the special case of linear time-invariant control systems, (6) can be written as an linear matrix inequality. The following example demonstrates this fact, without going into the computational details. Consider the control system

$$
\begin{aligned}
\dot{x}_1 &= -x_1 + x_1 x_2 + x_1 x_2^2 u \\
\dot{x}_2 &= -x_2 + x_4 - x_1 x_4 \\
\dot{x}_3 &= x_1^2 + x_2 + x_3 + u \\
\dot{x}_4 &= -x_4 + x_1 x_2 + x_1 x_2 x_3 \\
y &= x_3,
\end{aligned}
\tag{32}
$$

which has relative degree one. By using semidefinite programming and sum of squares techniques, the following quadratic Lyapunov function

$$
\begin{aligned}
V = {}& 5.11x_1^2 + 3.82x_2^2 - 0.31x_2 x_3 + 2.35x_3^2 \\
& - 0.07x_1 x_3 + 0.07x_2 x_4 - 1.26x_3 x_4 + 4.94x_4^2
\end{aligned}
\tag{33}
$$

was found. Furthermore, a function ρ was found with monomials of degree one to four. Therefore, it was possible to prove in a computationally efficient way that the control system (32) is globally minimum phase.

5 Conclusions

The paper has two contributions. The first contribution is a characterization of the minimum phase property of nonlinear control systems in terms of a dissipation inequality. This allows to describe the notion of the minimum phase property, which was originally developed in nonlinear geometric control [12], with the help of a Lyapunov-based argument, in case the control systems possesses an input-output normal form. The main idea is the use of a so-called derivative array and a penalizing function, which allows to characterize the stability of the zero dynamics in terms of a dissipation inequality without an explicit knowledge of the equations that define the zero dynamics.

The second contribution of this paper shows that if the control system is sufficiently smooth, then the functions that appear in the derived dissipation inequality (6) can also be chosen smooth. Moreover, demonstrated on an example, it has been shown that the derived dissipation inequality that characterizes the minimum phase property is in particular suitable for a minimum phase analysis using efficient numerical algorithms. It has also been shown by an example that the dissipation inequality can be very easily applied to control systems that are not affine in the input and thus allow a way to generalize the notion of the minimum phase property very easily. Another advantage of the proposed dissipation inequality is the conceptual simplicity.

There are several interesting points for future reseach. Since the penalizing function ρ is motivated from optimization theory, one can also consider ρ as a Lagrange multiplier or as a dual variable. This may be of particular interest in connection with performance limitations and further investigation in this direction is needed. Finally, the dissipation inequality (6) is similar to a generalized phase or passivity condition [12, 14, 13], due to the appearence of the inner product in the dissipation inequality. This similarity may be useful to extend passivity-based results to minimum phase control systems with a higher relative degree.

References

1. Aguiar A P, Hespanha J P, Kokotović P V (2005) Path-following for non-minimum phase systems removes performance limitations, IEEE Transactions on Automatic Control, 50:234–239
2. Åström K J (2000) Limitations on control system performance, European Journal of Control, 6:2–20
3. Aulbach B (2004) Gewöhnliche Differenzialgleichungen, Elsevier Spektrum Akademischer Verlag, 2nd edition
4. Bode H W (1940) Relations between attenuation and phase in feedback amplifier design, Bell System Technical Journal, 19:421–454
5. Ebenbauer C, Allgöwer F (2004) Minimum-phase property of nonlinear systems in terms of a dissipation inequality, In: Proceedings of the American Control Conference (ACC), Boston, USA, pages 1737–1742
6. Ebenbauer C, Allgöwer F (2005) Stability analysis of constrained control systems: an alternative approach, Systems and Control Letters, Accepted for publication
7. Ebenbauer C, Allgöwer F (2006) A dissipation inequality for the minimum phase property of nonlinear control systems and performance limitations, , Submitted for publication
8. Engel S (1988) Optimale lineare Regelung: Grenzen der erreichbaren Regelgüte in linearen zeitinvarianten Regelkreisen, Band 18, Fachbereiche Messen-Steuern-Regeln
9. Freudenberg J S, Looze D P (1985) Right half plane poles and zeros and design tradeoffs in feedback systems, IEEE Transactions on Automatic Control, 30:555–565
10. Griepentrog E (1992) Index reduction methods for differential-algebraic equations, In: Proceedings of the Berliner Seminar on Differential-Algebraic Equations Seminar Notes, Mathematik-92-1, ISSN: 0863-0976
11. Henrion D, Garulli A (2004) Positive Polynomials in Control, Springer Lecture Notes in Control and Information Sciences, Springer Verlag
12. Isidori A (1994) Nonlinear Control Systems, Springer Verlag, 3rd edition
13. Khalil H K (2002) Nonlinear Systems, Prentice Hall, 3rd edition
14. Kokotović P V, Arcak M (2001) Constructive nonlinear control: A historical perspective, Automatica, 37:637–662
15. Kwakernaak H, Sivan R (1972) Linear Optimal Control Systems, John Wiley and Sons
16. Liberzon D, Morse S, Sontag E D (2002) Output-input stability and minimum-phase nonlinear systems, IEEE Transactions on Automatic Control, 34:422–436
17. Massera J L (1956) Contributions to stability theory, Annals of Mathematics, 64:182–206
18. Middleton R H (1991) Trade-offs in linear control system design, Automatica, 27:281–292

19. Nešić D, Skafidas E, Mareels I M Y, Evans R J (1999) Minimum phase properties for input non-affine nonlinear systems, IEEE Transactions on Automatic Control, 44:868–872
20. Newton G C, Gould L A, Kaiser J F (1957) Analytical Design of Linear Feedback Controls, John Wiley and Sons
21. Parrilo P A, Lall S (2003) Semidefinite programming relaxations and algebraic optimization in control, European Journal of Control, 9(2-3)
22. Petrovski I G (1966) Ordinary Differential Equations, Prentice Hall
23. Qiu L, Davidson E J (1993) Performance limitations of nonminimum phase systems in the servomechanism problem, Automatica, 29:337–349
24. Seron M M, Braslavsky J H, Kokotović P V, Mayne D Q (1999) Feedback limitations in nonlinear systems: from Bode integrals to cheap control, IEEE Transactions on Automatic Control, 44:829–833
25. Sontag E D (1998) Mathematical Control Theory, Springer Verlag
26. Su W, Qui L, Chen J (2003) Fundamental performance limitations in tracking sinusoidal signals, IEEE Transactions on Automatic Control, 48:1371–1380
27. Svaricek F (2006) Nulldynamik linearer und nichtlinearer Systeme: Definitionen, Eigenschaften und Anwendungen, Automatisierungstechnik, 54(7):310–322

Input Disturbance Suppression for Port-Hamiltonian Systems: An Internal Model Approach

Luca Gentili, Andrea Paoli, and Claudio Bonivento

Center for Research on Complex Automated Systems (CASY)
"Giuseppe Evangelisti" - DEIS - Department of Electronic Computer Science
and Systems, University of Bologna, Italy
{lgentili,apaoli,cbonivento}@deis.unibo.it

Summary. In this paper an internal model based approach to periodic input disturbance suppression for port-Hamiltonian systems is presented; more specifically, an adaptive solution able to deal with unknown periodic signal belonging to a given class is introduced.

After an introductive section, the adaptive internal model design procedure is presented in order to solve the input disturbance problem. This theoretical machinery is specialized for the energy-based port-Hamiltonian framework in order to prove the global asymptotical stability of the solution.

Finally, in order to clearly point out the effectiveness of the presented design procedure a tracking problem is solved for a robotic manipulator affected by torque ripples.

Keywords: Port-Hamiltonian systems, Internal Model Control, Adaptive Control, Input Disturbance Suppression, Robot Manipulator.

1 Introduction

Input disturbance suppression is a very important topic in control theory as it represents the case in which malfunctions on the systems can be modeled as signals superimposed to the input channels; in real case it is possible to assume that the malfunction effect belongs to a known class of signals while their parameters (amplitude and even phase and frequencies in case of periodic signals) are unknown. For example malfunctioning on rotating systems driven by a power electronic part (e.g. electrical drives, magnetic levitation systems etc.) leads to asymmetries reflecting in spurious harmonics in the electrical variables (see [21], [3], [20], [16], [15]).

In this paper a comprehensive port-Hamiltonian systems (pHs) framework to deal with input disturbance suppression problems is considered. The main idea is to cast the problem into a regulation one and to solve it with an adaptive internal model based regulator. The design procedure turns out to be able to obtain a fault tolerant behavior: the asymptotic regulation is assured even in presence of a fault, and hence in presence of the resulting disturbances. The theoretical machinery exploited in order to prove the global asymptotical stability of

C. Bonivento et al. (Eds.): Adv. in Control Theory and Applications, LNCIS 353, pp. 85–98, 2007.
springerlink.com

the solution is the nonlinear regulation theory, specialized for the energy-based port-Hamiltonian formalism in order to take advantage of its peculiar properties. In [10] pHs were introduced as a generalization of Hamiltonian systems, described by Hamilton's canonical equations, which may represent general physical systems, i.e. mechanical, electric and electro-mechanical systems, nonholomic systems and their combinations (see [14] for further references).

In Section 2, a general exogenous input disturbance problem is considered for a generic pHs. The regulation problem is stated and an adaptive internal model based design procedure able to globally asymptotically solve this problem is introduced under proper assumption regarding the system into account.

It is worth to remark that the design procedure presented in this section is able to deal with disturbances that can be modeled as functions of time within a finitely-parametrized family: i.e. exogenous constant and sinusoidal disturbances characterized by unknown amplitude, phase and frequency. The hypothesis of not perfect knowledge of the characteristic frequencies introduces a complex issue to deal with: in the last years this problem has been pointed out and addressed using different design techniques (see [13], [9], [17], [11], [18] and references therein). In this work a solution to this issue, relying on simple Lyapunov based consideration, is presented.

In order to enlighten the practical effectiveness of the solution presented, in Section 3, a tracking control problem is solved for a robotic manipulator affected by torque ripples. The same example studied in [2] is here suitably modified in the solution according to the general framework presented.

2 Adaptive Input Disturbance Suppression Control for Port-Hamiltonian Systems

In this section we present a design approach to solve a disturbance suppression problem for pHs. The class of disturbance considered consists in addictive actuators disturbances modeled as exogenous input signals belonging to the class of constant and sinusoidal disturbances characterized by unknown amplitude, phase and frequency.

Consider a pHs with an exogenous disturbance $\delta(t)$ acting through the input channel:

$$\dot{x} = (J(x) - R(x))\frac{\partial H}{\partial x} + gu - g\delta \tag{1}$$

where $x \in \mathbb{R}^n$, $u \in \mathbb{R}^m$, $H : \mathbb{R}^n \to \mathbb{R}$ is the energy function (Hamiltonian function), $J(x)$ is a skew symmetric matrix ($J(x) = -J^{\mathrm{T}}(x)$), $R(x)$ is a symmetric semi-positive definite matrix ($R(x) = R^{\mathrm{T}}(x)$) and $g \in \mathbb{R}^{n \times m}$.

The disturbance $\delta(t)$ is generated by a neutrally stable exosystem defined by

$$\begin{cases} \dot{w} = Sw \\ \delta = \Gamma w \end{cases} \tag{2}$$

with $s = 2k + 1$, $w \in \mathbb{R}^s$; $\Gamma \in \mathbb{R}^{s \times m}$ is a known matrix and S is defined by

$$S = \text{diag}\{S_0, S_1, \ldots, S_k\} \tag{3}$$

where $S_0 = 0$,

$$S_i = \begin{bmatrix} 0 & \omega_i \\ -\omega_i & 0 \end{bmatrix} \qquad \omega_i > 0 \qquad i = 1, \ldots, k \, . \tag{4}$$

The initial condition of the exosystem is $w(0) \in \mathcal{W}$, with $\mathcal{W} \subseteq \mathbb{R}^s$ bounded compact set.

In this discussion the dimension s of matrix S is known but all characteristic frequencies ω_i are unknown but ranging within known compact sets, i.e. $\omega_i^{\min} \leq \omega_i \leq \omega_i^{\max}$. In this set up the lack of knowledge of the exogenous disturbance reflects into the lack of knowledge of the initial state $w(0)$ of the exosystem and of the characteristic frequencies. Any disturbance obtained by linear combination of a constant term and sinusoidal signals with unknown frequencies, amplitudes and phases are therefore considered.

The problem to address is to regulate the system to the origin in despite of the presence of exogenous disturbances; it will be remarked later that the solution presented is even able to supply estimates of the disturbances acting on the system.

All above assumptions allow to cast the problem of disturbance suppression as a nonlinear regulation problem (see [8], [4], [7], [1]) complicated by the lack of knowledge of the matrix S and suggests to look for a controller which embeds an *internal model* of the exogenous disturbances augmented by an adaptive part in order to estimate the characteristic frequencies of the disturbances.

The hypothesis of not perfect knowledge of the characteristic frequencies introduces a key issue to deal with (see [13], [9], [17], [11], [18] and references therein); here it is shown how, under some hypotheses regarding the pHs (1), this problem can be overcome introducing an adaptation law, designed exploiting the properties of the pHs structure, able to globally asymptotically stabilize the feedback system.

As discussed in the introduction, the regulator to be designed will embed the internal model of the exogenous disturbance: this internal model unit is designed according to the procedure proposed in [11] (*canonical internal model*). Given a symmetric, negative definite Hurwitz matrix F and any matrix G such that the couple (F, G) is controllable, denote by Y the unique nonsingular matrix solution of the Sylvester equation[1]

$$YS - FY = G\Gamma \tag{5}$$

and define $\Psi := \Gamma Y^{-1}$.

---

[1] Existence and uniqueness of the matrix Y follow from the fact that S and F have disjoint spectrum. The fact that Y is nonsingular can be easily proved using observability of the pairs (S, Γ) and controllability of the pair (F, G).

Let us introduce the adaptive internal model unit as

$$\begin{cases} \dot{\xi} = (F + G\hat{\Psi})\xi + N(x) \\ \dot{\hat{\Psi}}_{ij} = \varphi_{ij}(\xi, x), \quad \begin{array}{l} i = (1, \cdots, m) \\ j = (1, \cdots, s) \end{array} \end{cases} \tag{6}$$

where $\hat{\Psi}_{ij}$ represents the ij-th element of matrix $\hat{\Psi}$, and set the control law as

$$u = \hat{\Psi}\xi + u_{\mathrm{st}} \tag{7}$$

where $N(x)$ and u_{st} are additional terms that will be designed later. The adaptation laws $\varphi_{ij}(\xi, x)$ will be designed in order to assure that, asymptotically, the internal model unit will provide a control able to accommodate all disturbances.

Defining the changes of coordinate

$$\chi = \xi - Yw - Ax$$

$$\tilde{\Psi}_{ij} = \hat{\Psi}_{ij} - \Psi_{ij}, \quad \begin{array}{l} i = (1, \cdots, m) \\ j = (1, \cdots, s) \end{array} \tag{8}$$

where matrix A is chosen according to $Ag = G$, system (1) with controller (6) becomes

$$\begin{cases} \dot{x} = (J(x) - R(x))\dfrac{\partial H}{\partial x} + g\tilde{\Psi}\xi + g\Psi\xi + gu_{\mathrm{st}} - g\Gamma w \\ \dot{\chi} = (F + G\hat{\Psi})\xi + N(x) - YSw - A\dot{x} \\ \dot{\hat{\Psi}}_{ij} = \varphi_{ij}(\xi, x), \quad \begin{array}{l} i = (1, \cdots, m) \\ j = (1, \cdots, s) \, . \end{array} \end{cases} \tag{9}$$

Note that

$$\dot{x} = (J(x) - R(x))\frac{\partial H}{\partial x} + g\tilde{\Psi}\xi + g\Psi(\xi - Yw - Ax) + g\Psi Ax + gu_{\mathrm{st}}$$

$$= (J(x) - R(x))\frac{\partial H}{\partial x} + g\tilde{\Psi}\xi + g\Psi\chi + g\hat{\Psi}Ax - g\tilde{\Psi}Ax + gu_{\mathrm{st}} \, ,$$

hence, choosing $u_{\mathrm{st}} = -\hat{\Psi}Ax$, it is possible to write

$$\dot{x} = (J(x) - R(x))\frac{\partial H}{\partial x} + g\tilde{\Psi}(\xi - Ax) + g\Psi\chi \, .$$

Considering now the following two vectors containing every element of matrix Ψ and $\hat{\Psi}$

$$\begin{aligned} \Phi &= \begin{pmatrix} \Psi_{11} & \cdots & \Psi_{1s} & \cdots & \Psi_{m1} & \cdots & \Psi_{ms} \end{pmatrix}^{\mathrm{T}} \\ \hat{\Phi} &= \begin{pmatrix} \hat{\Psi}_{11} & \cdots & \hat{\Psi}_{1s} & \cdots & \hat{\Psi}_{m1} & \cdots & \hat{\Psi}_{ms} \end{pmatrix}^{\mathrm{T}} \end{aligned} \tag{10}$$

and defining $\tilde{\Phi} = \hat{\Phi} - \Phi$, it is possible to design a matrix $\Pi(x, \xi)$ such that

$$\Pi(x,\xi)\tilde{\Phi} = g\tilde{\Psi}(\xi - Ax) = g\tilde{\Psi}(\chi + Yw)$$

and write

$$\dot{x} = (J(x) - R(x))\frac{\partial H}{\partial x} + \Pi(x,\xi)\tilde{\Phi} + g\Psi\chi \ . \tag{11}$$

Let us concentrate now on the χ-dynamic in order to suitably design the update term $N(x)$:

$$\dot{\chi} = (F + G\hat{\Psi})\xi + N(x) - YMz - G\Gamma w - A\left[(J(x) - R(x))\frac{\partial H}{\partial x} + g\hat{\Psi}\xi\right.$$
$$\left. -g\Gamma w - g\hat{\Psi}Ax\right] =$$
$$= F\chi + FAx + N(x) - A(J(x) - R(x))\frac{\partial H}{\partial x} + Ag\hat{\Psi}Ax \ .$$

Choosing

$$N(x) = -FAx + A(J(x) - R(x))\frac{\partial H}{\partial x} - Ag\hat{\Psi}Ax \tag{12}$$

the obtained χ-dynamic is

$$\dot{\chi} = F\chi \ . \tag{13}$$

As all dynamics of (9) have been investigated, it is now possible to design an adaptation law for $\hat{\Psi}^{\mathrm{T}}$. In order to obtain a system fitting the pHs framework, this adaptation law must be chosen to satisfy the skew-symmetric property:

$$\dot{\hat{\Phi}} = -\Pi(x,\xi)^{\mathrm{T}}\frac{\partial H}{\partial x} \ .$$

With this in mind it is immediate to write the $\tilde{\Phi}$-dynamic as

$$\dot{\tilde{\Phi}} = \dot{\hat{\Phi}} - \dot{\Phi} = -\Pi(x,\xi)^{\mathrm{T}}\frac{\partial H}{\partial x} \ . \tag{14}$$

Consider now equations (2), (11), (13) and (14). The overall new system identifies an interconnection described by:

$$\dot{\bar{x}} = [\bar{J}(\bar{x}) - \bar{R}(\bar{x})]\frac{\partial H_x(\bar{x})}{\partial \bar{x}} + \Lambda \tag{15}$$

with

$$\bar{x} = \begin{pmatrix} x & \chi & \tilde{\Phi} & w \end{pmatrix}^{\mathrm{T}} \ ,$$

and characterized by the Hamiltonian function $H_x(x)$ defined by

$$H_x(\bar{x}) = H(x) + \frac{1}{2}\chi^{\mathrm{T}}\chi + \frac{1}{2}\tilde{\Phi}^{\mathrm{T}}\tilde{\Phi} + \frac{1}{2}w^{\mathrm{T}}w \ ,$$

the skew-symmetric interconnection matrix $\bar{J}(\bar{x})$ defined by

$$\bar{J}(\bar{x}) = \begin{pmatrix} J & 0 & \Pi & 0 \\ 0 & 0 & 0 & 0 \\ -\Pi^{\mathrm{T}} & 0 & 0 & 0 \\ 0 & 0 & 0 & S \end{pmatrix} \ ,$$

the damping matrix $\bar{R}$ defined as

$$\bar{R} = \begin{pmatrix} R(x) & 0 & 0 & 0 \\ 0 & -F & 0 & 0 \\ 0 & 0 & 0 & 0 \\ 0 & 0 & 0 & 0 \end{pmatrix},$$

and finally Λ as

$$\Lambda = \begin{pmatrix} g\Psi\chi & 0 & 0 & 0 \end{pmatrix}^{\mathrm{T}}.$$

At this point the main result of the paper can be stated.

Theorem 1. *Consider the pHs (1), affected by exogenous signals generated by the autonomous system (2), (3), (4).*

If the following hypotheses H1 and H2 hold:

H1: *there exists two numbers $\eta_x \in \mathbb{R}^-$ and $\eta_\Psi \in \mathbb{R}$ and a matrix $Q \in \mathbb{R}^{n \times n}$ such that for all $\chi \in \mathbb{R}^s$ the following holds*

$$-\frac{\partial^{\mathrm{T}} H}{\partial x} R(x) \frac{\partial H}{\partial x} + \frac{\partial^{\mathrm{T}} H}{\partial x} g\Psi\chi \leq \eta_x \|Qx\|^2 + \eta_\Psi \|Qx\| \|\chi\|; \qquad (16)$$

H2: *the origin of (1) is the largest invariant set of the auxiliary system*

$$\dot{x} = (J(x) - R(x)) \frac{\partial H}{\partial x} + g\sigma$$

$$\dot{\sigma} = -g^{\mathrm{T}} \frac{\partial H}{\partial x}$$

characterized by

$$\frac{\partial^{\mathrm{T}} H}{\partial x} R(x) \frac{\partial H}{\partial x} = 0.$$

Define the controller (adaptive internal model unit)

$$\begin{cases} \dot{\xi} = (F + G\hat{\Psi})\xi - FAx + A(J(x) - R(x))\dfrac{\partial H}{\partial x} - Ag\hat{\Psi}Ax \\[2mm] \dot{\hat{\Phi}} = -\Pi(x,\xi)^{\mathrm{T}} \dfrac{\partial H}{\partial x} \\[2mm] u = \hat{\Psi}\xi - \hat{\Psi}Ax, \end{cases} \qquad (17)$$

where A is chosen according to $Ag = G$, $\Pi(x,\xi)$ is a suitably defined updating term designed such that

$$\Pi(x,\xi)\tilde{\Phi} = g\tilde{\Psi}(\xi - Ax),$$

F is a suitably defined symmetric Hurwitz matrix designed according the constructive proof presented in the following and G is a suitably defined matrix such that the couple (F,G) is controllable.

__Then__ controller (17) is able to asymptotically stabilize the origin of system (1), zeroing the effect of the exogenous disturbances.

Proof. Consider system (15) (obtained connecting (1) with (17)) and the following Lyapunov function:

$$V = H_x(x) \, .$$

Simple computations show that the time derivative of this Lyapunov function is defined by

$$\dot{V} = -\frac{\partial^{\mathrm{T}} H}{\partial x} R \frac{\partial H}{\partial x} + \frac{\partial^{\mathrm{T}} H}{\partial x} g \Psi \chi + \chi^{\mathrm{T}} F \chi \, .$$

As (16) holds, there exist real numbers $\eta_x \in \mathbb{R}^-$ and $\eta_\Psi \in \mathbb{R}$, a matrix $Q \in \mathbb{R}^{n \times n}$ and $\eta_F \in \mathbb{R}^-$, such that

$$\dot{V} \leq \eta_x \|Qx\|^2 + \eta_F \|\chi\|^2 + \eta_\Psi \|Qx\| \|\chi\| \, .$$

Using a Young's inequality argumentation it is possible to state that

$$\dot{V} \leq \eta_x \|Qx\|^2 + \eta_F \|\chi\|^2 + \frac{\eta_\Psi}{2} \varepsilon \|Qx\|^2 + \frac{\eta_\Psi}{2\varepsilon} \|\chi\|^2 \, ,$$

for a certain value of ε. Choosing $\varepsilon = -\eta_{k_p}/\eta_\Psi$, it comes out that

$$\dot{V} \leq \eta_x \|Qx\|^2 + \left(\eta_F - \frac{\eta_\Psi^2}{2\eta_x} \right) \|\chi\|^2 \, , \tag{18}$$

hence, choosing matrix F such that

$$\eta_F < \frac{\eta_\Psi^2}{2\eta_x} \, ,$$

it turns out that $\dot{V} \leq 0$ and, for LaSalle invariance principle, system's trajectory are asymptotically captured by the largest invariant set characterized by $\dot{V} = 0$. Considering this, by (18) and hypothesis H2, the system (1) asymptotically converge to the origin proving the statement.

Remark 1. It is worth to remark that, though the main hypotheses H1 and H2 could appear rather conservative, they refer to a system of the form (1) that could be not the original plant but the port-Hamiltonian formulation of the original system already controlled to attain specific tasks; this a priori control action could be suitably designed such that the resulting system satisfies conditions H1 and H2. In particular, it could be easily shown that hypothesis H1 is always verified if the system into account is characterized by a quadratic Hamiltonian function. To enlighten the effectiveness of the property remarked here, in section 3 a n-degree of freedom robotic manipulator is taken into account: the original system does not satisfy both conditions but, with a suitably defined control action able to perform even a tracking objective and a proper change of coordinates, the resulting system turns out in the form (1) satisfying H1 and H2; hence the control algorithm introduced can be used to solve the input disturbance suppression problem.

A further control procedure able to impose particular shape and properties to the controlled system is the well known IDA-PBC (see [19], [12] and reference therein for a survey about this control strategy): this control strategy is able to design a suitable port-Hamiltonian controller such that the interconnected system (original plant and IDA-PBC controller) assumes the form of a desired reference pHs. It is easy to realize that one of the characteristic step of the IDA-PBC control strategy is just the definition of a target system, usually described by the classical port-Hamiltonian structure, characterizing the resulting dynamic after that the controller is designed and connected: this target system could be assumed of the form (1), imposing moreover that all the assumptions in Theorem 1 are satisfied.

An interesting research topic, that is still under investigation, regards the conditions to impose to the original system and to the original problem, such that, for example, an IDA-PBC control strategy makes it possible to cast the problem in the presented framework.

Remark 2. Following the discussion of remark 1, it is important to stress the fact that, in some case, system (1) can be time varying. A typical situation is when (1) represents the error system in a tracking problem; in this case the time depending trajectories could appear in the skew-symmetric matrix $J(x,t)$. Clearly LaSalle invariance principle cannot be used to show convergence; nevertheless the approach can be used again and the asymptotic convergence proved by means of Barbalat Lemma when its hypotheses are satisfied (see [22]). This is the case, for example, of having $Q = I$ in Theorem 1.

Remark 3. In some cases, the arise of periodical disturbances superimposed to the control variable represents the effect of a fault occurred in the system: for example electrical motors as well as magnetic levitation systems can be subject to some asymmetries (e.g. due to some electrical or mechanical faults) that cause the arise of spurious harmonics in the electrical variables (see [3], [20], [16], [15]). In these cases, the design procedure introduced is able to obtain a fault tolerant behavior: the asymptotic regulation is assured even in presence of a fault, and hence in presence of the resulting disturbances superimposed to the control inputs.

In this viewpoint the design procedure can be cast into the so-called *implicit Fault Tolerant Control* framework introduced in [3]. According to this approach the control reconfiguration does not relay on an explicit Fault Detection and Isolation design but is achieved by a proper design of a dynamic controller which is *implicitly* fault tolerant to all the possible faults whose model is embedded in the regulator by means of an *internal model.*

It is interesting to see that, thanks to Theorem 1, even the Fault Detection and Isolation phase can be carried out by testing the state of the internal model unit which automatically activates to offset the presence of disturbances representing the effect of a fault. Let us remember the definition of error variable χ: its asymptotic convergence to the origin implies that the internal model state ξ tends to the disturbance Yw. Hence this phase, which is usually the starting point for the design of the FTC system is now a consequence of the reconfiguration phase.

In the next section an interesting example is presented to point out the main properties of the input disturbance suppression design procedure presented in this paper. More precisely, a n degree of freedom manipulator is controlled in presence of torque disturbances assuring in the meantime a tracking property.

3 An Example: Application to a Robot Manipulator

Consider an n degree of freedom fully-actuated robot manipulator with generalized coordinates $q = (q_1, \cdots, q_n)^{\mathrm{T}}$. If $p = M(q)\dot{q} = (p_1, \cdots, p_n)^{\mathrm{T}}$ are the generalized momenta, with $M(q)$ the inertia matrix, symmetric and positive definite for all q, an explicit port-Hamiltonian representation of this system can be obtained defining the whole state $(q, p)^{\mathrm{T}}$, the Hamiltonian function as the total energy of the system (sum of kinetic energy and potential energy)

$$H(q,p) := \frac{1}{2} p^{\mathrm{T}} M^{-1}(q) p + P(q)$$

and, finally, the matrices

$$J = \begin{pmatrix} 0 & I_n \\ -I_n & 0 \end{pmatrix}, \quad R(q) = \begin{pmatrix} 0 & 0 \\ 0 & D(q) \end{pmatrix}, \quad \gamma = \begin{pmatrix} 0 \\ g \end{pmatrix} = \begin{pmatrix} 0 \\ I_n \end{pmatrix}$$

with $D(q) = D^T(q) \geq 0$ taking into account the dissipation effects. Let us call ν the input effort representing the actuation torques. These positions lead to the following port-Hamiltonian model

$$\begin{bmatrix} \dot{q} \\ \dot{p} \end{bmatrix} = [J - R(q)] \begin{bmatrix} \dfrac{\partial H}{\partial q} \\ \dfrac{\partial H}{\partial p} \end{bmatrix} + \gamma(\nu + \delta(t)).$$

This system is affected by an external torque ripple $\delta(t)$ acting through the control input channel (i.e. the torque applied to the system will be the sum of the control torque and the external disturbance $\nu + \delta(t)$); the main objective to be pursued by this system is to track a known trajectory while compensating this disturbance, detecting and isolating in the meanwhile its entity.

The tracking control is developed following the main idea introduced in [6], but the characteristic change of error-coordinates is suitably modified in order to obtain an error system still described as a pHs and satisfying the conditions imposed by Theorem 1. Note that in our previous paper [2] those conditions were not satisfied and the design procedure was not able to overcome constant torque disturbances.

3.1 Tracking Control

In this subsection a state feedback tracking control algorithm is presented to make the robot manipulator, in absence of fault disturbances, tracks a known target trajectory (defined in generalized coordinates by $(q^\star(t), p^\star(t))$).

94   L. Gentili, A. Paoli, and C. Bonivento

To define new error variables, consider the following change of coordinates

$$\bar{q} = q - q^{\star}(t)$$
$$\bar{p} = p - M(q)\dot{q}^{\star}(t) + M(q)k_q\bar{q}$$

(19)

where k_q is a positive definite symmetric design gain matrix.

Computing time derivatives of new error coordinates we obtain:

$$\dot{\bar{q}} = M^{-1}(q)\bar{p} - k_q\bar{q}$$

$$\dot{\bar{p}} = -\frac{1}{2}p^{\mathrm{T}}\frac{\partial M^{-1}(q)}{\partial q}p - D(q)M^{-1}(q)p - \frac{\partial P(q)}{\partial q}$$

$$+\nu + v(t) - \Pi(q, p, q^{\star}(t), \dot{q}^{\star}(t), \ddot{q}^{\star}(t))$$

(20)

where

$$\Pi(q, p, q^{\star}(t), \dot{q}^{\star}(t), \ddot{q}^{\star}(t)) = \frac{\partial^{\mathrm{T}} M(q)}{\partial q}M^{-1}(q)p\dot{q}^{\star} + M(q)\ddot{q}^{\star}$$

$$-\frac{\partial^{\mathrm{T}} M(q)}{\partial q}M^{-1}(q)pk_q(q - q^{\star})$$

$$+M(q)k_q(M^{-1}(q)p - \dot{q}^{\star}).$$

It is now possible to obtain a perfect global asymptotic tracking in absence of disturbances ($\delta(t) = 0$): this can be done by designing the control torque ν in order to delete the "bad" term $\Pi(\cdot)$, to shape the energy of the error system to have a minimum in the origin[2] and to add some damping action in order to have this minimum globally attractive. Keeping this in mind, the control action is defined to be

$$\nu = \Pi(q, p, q^{\star}(t), \dot{q}^{\star}(t), \ddot{q}^{\star}(t)) + D(q)M^{-1}(q)p + \frac{\partial P(q)}{\partial q} - k_p M^{-1}(\bar{q})\bar{p} + \tau$$

$$-[M^{-1}(q)M(\bar{q}) + k_q\frac{1}{2}\bar{p}\frac{\partial M(\bar{q})^{-1}}{\partial\bar{q}}M(\bar{q})]^{\mathrm{T}}\left(\frac{1}{2}\bar{p}^{\mathrm{T}}\frac{\partial M(\bar{q})^{-1}}{\partial\bar{q}}\bar{p} + \bar{q}\right)$$

(21)

where k_p is a symmetric positive definite design matrix ($-k_p$ is Hurwitz) and τ is an additional control torque that will be used in the following to compensate the presence of additional torque disturbances.

The error system with the controller (21) writes as

$$\dot{\bar{q}} = -k_q\frac{\partial H'}{\partial\bar{q}} + J_1(\bar{q}, \bar{p}, q^{\star}(t))\frac{\partial H'}{\partial\bar{p}}$$

$$\dot{\bar{p}} = -J_1(\bar{q}, \bar{p}, q^{\star}(t))^{\mathrm{T}}\frac{\partial H'}{\partial\bar{q}} - k_p\frac{\partial H'}{\partial\bar{p}} + \tau + v(t)$$

(22)

---

[2] Note that $\bar{q} = 0$ means that the tracking is achieved as $q \to q^{\star}(t)$.

with

$$J_1(\bar{q}, \bar{p}, q^\star(t)) = \left[M^{-1}(q)M(\bar{q}) + k_q \frac{1}{2}\bar{p}\frac{\partial M(\bar{q})^{-1}}{\partial \bar{q}} M(\bar{q}) \right]$$

and the new Hamiltonian function defined as

$$H' = \frac{1}{2}\bar{p}^{\mathrm{T}} M^{-1}(\bar{q})\bar{p} + \frac{1}{2}\bar{q}^{\mathrm{T}}\bar{q}\,. \tag{23}$$

It is easy to realize that the tracking objective is globally asymptotically achieved in absence of external disturbances ($\delta = 0$ and hence $\tau = 0$): it is, in fact, straightforward to choose the Hamiltonian function H' as a Lyapunov function and to state, thanks to Barbalat Lemma, that

$$\lim_{t\to\infty} q(t) = q^\star(t)\,, \quad \lim_{t\to\infty} p(t) = p^\star(t)\,.$$

3.2   Problem Statement and Internal Model Design

It is now possible to state the input disturbance suppression problem considering the torque disturbance $\delta(t) = \Gamma w$ as generated by a neutrally stable autonomous exosystem like the one defined by (2), (3) and (4): the problem fits now in the framework presented in section 2 and it is possible to design an adaptive internal model controller following the procedure stated by Theorem 1.

Remark 4. It is important to point out the instrumental role played by the change of coordinates (19) and in particular by the additional term $M(q)k_q\bar{q}$ that makes the presented solution different from the one in [6] and [2]. The effect of this term is to introduce a new damping action in the generalized error coordinates $\bar{q}$. The error system (22) satisfies all conditions imposed by Theorem 1 and, as $J_1(\bar{q}, \bar{p}, q^\star(t))$ depends on time, by remark 2. In particular, (22) is now characterized by a dissipation matrix R' defined as

$$R' = \begin{pmatrix} k_q & 0 \\ 0 & k_p \end{pmatrix}$$

and hence matrix Q in Theorem 1 is equal to the identity ($Q = I$); moreover the origin of (22) is globally asymptotically attractive in force of the Barbalat Lemma applied to the auxiliary system

$$\dot{\bar{q}} = -k_q \frac{\partial H'}{\partial \bar{q}} + J_1 \frac{\partial H'}{\partial \bar{p}}$$

$$\dot{\bar{p}} = -J_1^{\mathrm{T}} \frac{\partial H'}{\partial \bar{q}} - k_p \frac{\partial H'}{\partial \bar{p}} + \sigma$$

$$\dot{\sigma} = -\frac{\partial H'}{\partial \bar{p}}\,.$$

The resulting control law generated by the adaptive internal model unit (17) is defined by

96   L. Gentili, A. Paoli, and C. Bonivento

$$
\begin{cases}
\dot{\xi} = (F + G\hat{\Psi})\xi - FG\bar{p} - GJ_1^{\mathrm{T}} \left(\frac{1}{2}\bar{p}^{\mathrm{T}} \frac{\partial M(q)}{\partial q}^{-1} \bar{p} + \bar{q} \right) - Gk_p M^{-1}(q)\bar{p} \\
\qquad -G\hat{\Psi}G\bar{p} \\[2mm]
\dot{\hat{\Phi}} = - \begin{pmatrix} (\xi - G\bar{p}) & 0 & \cdots & 0 \\ 0 & (\xi - G\bar{p}) & \cdots & 0 \\ \vdots & \vdots & \vdots & \vdots \\ 0 & 0 & \cdots & (\xi - G\bar{p}) \end{pmatrix} (M(\bar{q})^{-1}\bar{p}) = -J_2(\xi, \bar{p})(M(\bar{q})^{-1}\bar{p}) \\[2mm]
\tau = \hat{\Psi}\xi - \hat{\Psi}G\bar{p} \, .
\end{cases}
$$

$$(24)$$

where $A = G$ and $\hat{\Phi}$ is defined by (10).

To conclude the discussion of this example it is worth to define in this particular case the whole error system (15) and to proceed to the effective design of the characteristic gain matrices to point out the constructive part of the proof of Theorem 1.

To this aim, apply again the changes of coordinates (8). The new error system identifies an interconnection described by:

$$
\dot{\tilde{x}} = [\tilde{J}(\tilde{x}) - \tilde{R}(\tilde{x})]\frac{\partial H_x(\tilde{x})}{\partial \tilde{x}} + \Lambda(\tilde{x})
\tag{25}
$$

with state

$$
\tilde{x} = \left(\bar{q} \; \bar{p} \; \chi \; \bar{\Phi} \; w \right)^{\mathrm{T}} ,
$$

the Hamiltonian function $H_x(\tilde{x})$ defined by

$$
H_x(\tilde{x}) = \frac{1}{2}\bar{p}^{\mathrm{T}} M(\bar{q})^{-1}\bar{p} + \frac{1}{2}\bar{q}^{\mathrm{T}}\bar{q} + \frac{1}{2}\chi^{\mathrm{T}}\chi + \frac{1}{2}\bar{\Phi}^{\mathrm{T}}\bar{\Phi} + \frac{1}{2}w^{\mathrm{T}}w \, ,
$$

the skew-symmetric interconnection matrix $\tilde{J}(x)$ defined by

$$
\tilde{J}(\tilde{x}) = \begin{pmatrix}
0 & J_1 & 0 & 0 & 0 \\
-J_1^{\mathrm{T}} & 0 & \Psi & J_2^{\mathrm{T}} & 0 \\
0 & -\Psi^{\mathrm{T}} & 0 & 0 & 0 \\
0 & -J_2 & 0 & 0 & 0 \\
0 & 0 & 0 & 0 & S
\end{pmatrix} ,
$$

the damping matrix $\tilde{R}$ defined by

$$
\tilde{R} = \begin{pmatrix}
k_q & 0 & 0 & 0 & 0 \\
0 & k_p & 0 & 0 & 0 \\
0 & 0 & -F & 0 & 0 \\
0 & 0 & 0 & 0 & 0 \\
0 & 0 & 0 & 0 & 0
\end{pmatrix} ,
$$

and finally $\Lambda(\tilde{x})$ defined by

$$
\Lambda(\tilde{x}) = \left(0 \; 0 \; \Psi^{\mathrm{T}}\bar{p} \; 0 \; 0 \right)^{\mathrm{T}} .
$$

Specializing for system (25) the proof of Theorem 1, choose the Hamiltonian function as Lyapunov function $V = H_x(\tilde{x})$ and compute its time derivative:

$$\dot{V} = -\frac{\partial^{\mathrm{T}} H_{\tilde{x}}(\tilde{x})}{\partial \tilde{x}} \tilde{R} \frac{\partial H_{\tilde{x}}(\tilde{x})}{\partial \tilde{x}} + \frac{\partial^{\mathrm{T}} H_{\tilde{x}}(\tilde{x})}{\partial \tilde{x}} \Lambda(\tilde{x}) =$$

$$= -\left(\frac{1}{2}\bar{p}^{\mathrm{T}}\frac{\partial M^{-1}(\bar{q})}{\partial \bar{q}}\bar{p}\right)^{\mathrm{T}} k_q \left(\frac{1}{2}\bar{p}^{\mathrm{T}}\frac{\partial M^{-1}(\bar{q})}{\partial \bar{q}}\bar{p}\right) - \bar{q}^{\mathrm{T}} k_q \bar{q}$$

$$-(M^{-1}(\bar{q})\bar{p})^{\mathrm{T}} k_p (M^{-1}(\bar{q})\bar{p}) + \chi^{\mathrm{T}} F \chi + \chi^{\mathrm{T}} \Psi^{\mathrm{T}} \bar{p}.$$

There exist real numbers $\eta_{k_p} \in \mathbb{R}^-$, $\eta_F \in \mathbb{R}^-$, $\eta_{k_q} \in \mathbb{R}^-$ (depending on design matrices k_p, F, k_q) and $\eta_\Psi \in \mathbb{R}$, such that

$$\dot{V} \le \eta_{k_q}\|\bar{q}\|^2 + \eta_{k_p}\|\bar{p}\|^2 + \eta_F\|\chi\|^2 + \eta_\Psi\|\bar{p}\|\|\chi\| .$$

Using a Young's inequality argumentation we can write:

$$\dot{V} \le \eta_{k_q}\|\bar{q}\|^2 + \eta_{k_p}\|\bar{p}\|^2 + \eta_F\|\chi\|^2 + \frac{\eta_\Psi}{2}\varepsilon\|\bar{p}\|^2 + \frac{\eta_\Psi}{2\varepsilon}\|\chi\|^2 ,$$

for a certain value of ε. Now choosing $\varepsilon = -\eta_{k_p}/\eta_\Psi$, we obtain

$$\dot{V} \le \eta_{k_q}\|\bar{q}\|^2 + \frac{\eta_{k_p}}{2}\|\bar{p}\|^2 + \left(\eta_F - \frac{\eta_\Psi^2}{2\eta_{k_p}}\right)\|\chi\|^2 .$$

Hence choosing matrix F such that

$$\eta_F < \frac{\eta_\Psi^2}{2\eta_{k_p}}$$

we have that $\dot{V} \le 0$.

The asymptotic behavior of this error system is then defined by Barbalat Lemma that assures the asymptotic convergence of $\bar{q}$, $\bar{p}$ and χ to the origin. The original tracking problem is then asymptotically solved and the exogenous disturbances are perfectly compensated by the adaptive internal model unit.

Remark 5. As pointed out in [2] and [3], the disturbance estimation phase can be performed by comparing the state of the internal compensation unit which automatically offsets the disturbance effect with a suitably tuned threshold. In fact, notice that $\xi(t)$ asymptotically converge to $Yz(t)$ which is zero in the nominal case and different from zero when a disturbance is acting.

4   Conclusions

In this paper an internal model approach to input disturbance suppression for pHs is presented. The main contribution of the paper is the introduction of an adaptive internal model design procedure able to solve a regulation problem in presence of periodic input disturbances for a generic pHs, exploiting the energy-based characteristic properties of this formalism in order to prove the global asymptotical stability of the solution.

To point out the effectiveness of the design procedure, a tracking control problem is discussed for a robotic manipulator affected by torque ripples.

References

1. Astolfi A, Isidori A, Marconi L (2003) A note on disturbance suppression for hamiltonian systems by state feedback. 2nd IFAC Workshop LHMNLC, Seville, Spain
2. Bonivento C, Gentili L, Paoli A (2004a) Internal model based fault tolerant control of a robot manipulator. 43rd Conference on Decision and Control, Paradise Island, Bahamas
3. Bonivento C, Isidori A, Marconi L, Paoli A (2004b) Implicit fault tolerant control: Application to induction motors. Automatica 40(3):355–371
4. Byrnes C, Delli Priscoli F, Isidori A (1997a) Output regulation of uncertain nonlinear systems. Birkhäuser, Boston
5. Canudas de Wit C, Praly L (2000) Adaptive eccentricity compensation. IEEE Transactions on Control Systems Technology 8(5):757–766
6. Fujimoto K, Sakurama K, Sugie T (2003) Trajectory tracking control of port-controlled hamiltonian systems via generalized canonical transformations. Automatica 39(12):2059–2069
7. Gentili L, van der Schaft A (2003) Regulation and input disturbance suppression for port-controlled Hamiltonian systems. 2nd IFAC Workshop LHMNLC, Seville, Spain
8. Isidori A (1995) Nonlinear Control Systems. Springer-Verlag, London
9. Isidori A, Marconi L, Serrani A (2003) Robust Autonomous Guidance: An Internal Model-based Approach. Limited series Advances in Industrial Control, Springer Verlag, London
10. Maschke B, van der Schaft A (1992) Port-controlled hamiltonian system: modelling origins and system theoretic approach. 2nd IFAC NOLCOS, Bordeaux, France
11. Nikiforov V (1998) Adaptive non-linear tracking with complete compensation of unknown disturbances. European Journal of Control 4:132–139
12. Ortega R (2003) Some applications and recent results on passivity based control. 2nd IFAC Workshop on Lagrangian and Hamiltonian Methods for Nonlinear Control, Seville, Spain
13. Serrani A, Isidori A, Marconi L (2001) Semiglobal output regulation with adaptive internal model. IEEE Transaction On Automatic Control 46(8):1178–1194
14. van der Schaft A (1999) L_2-gain and Passivity Techniques in Nonlinear Control. Springer-Verlag, London
15. Bonivento C, Gentili L, Marconi L (2005) Balanced Robust Regulation of a Magnetic Levitation System. IEEE Tran. Control System Technology 13(6):1036–1044
16. Alleyne A (2000) Control of a class of nonlinear systems subject to periodic exogenous signals. IEEE Trans. on Control Systems Technology 8(2):279–287
17. Marino R, Santosuosso G L, Tomei P (2003) Robust adaptive compensation of biased sinusoidal disturbances with unknown frequency. Automatica 19(10):1755–1761
18. Bodson M, Douglas S C (1997) Adaptive algorithms for the rejection of periodic disturbances with unknown frequencies. Automatica 33(12):2213-2221
19. Ortega R, van der Schaft A, Maschke B, Escobar G (1999) Interconnection and damping assignment passivity-based control of port-controlled Hamiltonian systems. Automatica 38(4):585–596
20. Bonivento C, Gentili L, Paoli A (2005) Internal model based framework for tracking and fault tolerant control of a permanent magnet synchronous motor. IFAC World Congress, Praha
21. Canudas de Wit C, Praly L (2000) Adaptive eccentricity compensation. IEEE Transactions on Control Systems Technology 8(5):757–766
22. Khalil H K (2002) Nonlinear Systems 3rd ed. Prentice Hall

A Systems Theory View of Petri Nets

Alessandro Giua and Carla Seatzu

Dip. Ingegneria Elettrica ed Elettronica, Università di Cagliari, Italy
{giua,seatzu}@diee.unica.it

Summary. Petri nets are a family of powerful discrete event models whose interest has grown, within the automatic control community, in parallel with the development of the theory of discrete event systems. In this tutorial paper our goal is that of giving a flavor, by means of simple examples, of the features that make Petri nets a good model for systems theory and of pointing out at a few open areas for research. We focus on Place/Transitions nets, the simplest Petri net model. In particular we compare Petri nets with automata, and show that the former model has several advantages over the latter, not only because it is more general but also because it offers a better structure that has been used for developing computationally efficient algorithms for analysis and synthesis.

Keywords: Discrete Event Systems, Petri Nets, Models of Concurrency, Controllability.

1   Introduction

The object of the study of traditional control theory have been *time-driven systems*, i.e., systems of continuous and synchronous discrete variables, modeled by differential or difference equations. However, as the scope of control theory is being extended into the domains of manufacturing, robotics, computer and communication networks, and so on, there is an increasing need for different models, capable of describing systems that evolve in accordance with the abrupt occurrence, at possibly unknown irregular intervals, of physical events. Such systems, whose states have logical or symbolic, rather than numerical, values that change in response to events which may also be described in nonnumerical terms, are called *discrete event systems* and the corresponding models are called *discrete event models* [5].

These systems require control and coordination to ensure the orderly flow of events. As controlled (or potentially controllable) dynamic systems, discrete event systems qualify as a proper subject for control theory. Hence a fundamental issue arises: we need classes of formal models that are capable of capturing the essential features of discrete, asynchronous and possibly nondeterministic systems and that are endowed with efficient mathematical tools for analysis and control.

Petri nets are a family of models developed from the original model presented in 1962 by Carl Adam Petri in his doctoral dissertation: "Kommunikation mit Automaten" (Communication with Automata). The theory of Petri nets is now well established and many different Petri net models have been defined, capable of describing: logical (i.e., untimed) systems; timed systems, both deterministic and stochastic; hybrid systems.

We claim that Petri nets are a powerful discrete event model and, in fact, the interest for this model has grown, within the automatic control community, in parallel with the

C. Bonivento et al. (Eds.): Adv. in Control Theory and Applications, LNCIS 353, pp. 99–127, 2007.
springerlink.com

development of the theory of discrete event systems. In this tutorial paper the goal is not that of providing a comprehensive survey of the research in this area, but rather that of giving a flavor, by means of simple examples, of the features that make Petri nets a good model for systems theory and of pointing out at a few open areas for research.

We compare Petri nets with automata, and show that the former model has several advantages over the latter, not only because it is more general but also because it offers a better structure that has been used for developing computationally efficient algorithms for analysis and synthesis. This gives credit to our belief that the study of automata — that is an integral part of the introductory courses on discrete event systems — should always be complemented with the presentation of Petri nets.

The paper is structured as follows. In Section 2 the definition of Place/Transition net (the most well-known Petri net model) is given and its dynamic behavior is described. Section 3 deals with the modeling of physical systems with Petri nets, with an example taken from the manufacturing domain. In Section 4 the main analysis techniques pertaining to this model are discussed, with a particular focus on the techniques based on the state equation and on the reachability graph. In Section 5 we look at Petri nets as language generators and characterize the classes of languages accepted and generated by this model. In Section 6 we show that Petri nets are a generalization of automata and point out some of advantages the first model has with respect to the latter. In Section 7 we discuss how many classical control properties may be extended to the context of discrete event systems and, as an example, discuss controllability in the framework of Petri nets. Finally, in Section 8 a few areas of research that are still opened in the Petri net domain are presented.

2   Petri Nets: Main Definitions

In this paper we consider the basic Petri net model called *Place/Transition net* (*P/T net* for short). It is a purely *logic* model that takes into account the order of occurrence of events, without associating time to them. For a comprehensive introduction to Petri nets see also the paper by Murata [27], and the books by Peterson [32] and by David and Alla [9].

2.1   Net Structure

Definition 1. *A* Place/Transition net *is a structure* $N = (P, T, Pre, Post)$ *where:*

- $P = \{p_1, p_2, \ldots, p_m\}$ *is a set of* places *represented by circles;*
- $T = \{t_1, t_2, \ldots, t_n\}$ *is a set of* transitions *represented by bars;*
- $Pre : P \times T \to \mathbb{N}$ *is the pre-incidence function that specifies the weight of the arcs directed from places to transitions;*
- $Post : P \times T \to \mathbb{N}$ *is the* post-incidence function *that specifies the weight of the arcs directed from transitions to places.*   ▲

Example 1. Fig. 1 shows a net $N = (P, T, Pre, Post)$ with set of places $P = \{p_1, p_2, p_3\}$, and set of transitions $T = \{t_1, t_2, t_3, t_4\}$. Here

$$Pre = \begin{bmatrix} 1 & 0 & 0 & 0 \\ 0 & 1 & 1 & 0 \\ 0 & 0 & 0 & 1 \end{bmatrix} \begin{matrix} p_1 \\ p_2 \\ p_3 \end{matrix} \qquad Post = \begin{bmatrix} 0 & 1 & 0 & 0 \\ 1 & 0 & 0 & 0 \\ 0 & 0 & 2 & 1 \end{bmatrix} \begin{matrix} p_1 \\ p_2 \\ p_3 \end{matrix}$$
$$\quad\; t_1 \;\; t_2 \;\; t_3 \;\; t_4 \qquad\qquad\qquad\quad t_1 \;\; t_2 \;\; t_3 \;\; t_4$$

$\blacksquare$

The information contained in the two matrices Pre and $Post$ is often summarized in a single matrix, defined as

$$C = Post - Pre : P \times T \to \mathbb{Z} \tag{1}$$

and called *incidence matrix*. Note however that the incidence matrix does not contain the same information of Pre and $Post$, namely the structure of the net cannot be univocally determined starting from C. This is clearly illustrated in the following example.

Example 2. The incidence matrix of the net in Fig. 1 is

$$C = \begin{bmatrix} -1 & 1 & 0 & 0 \\ 1 & -1 & -1 & 0 \\ 0 & 0 & 2 & 0 \end{bmatrix} \begin{matrix} p_1 \\ p_2 \\ p_3 \end{matrix}$$
$$t_1 \;\; t_2 \;\; t_3 \;\; t_4$$

In this matrix a negative element corresponds to a pre arc, and a positive element to a post arc. Note, however, that when a transition and a place form a loop, the weight of the pre and post arc may cancel out. In this net such is the case for the loop formed by p_3 and t_4: since $C(p_3, t_4) = 0$ no information on this loop is contained in C. $\blacksquare$

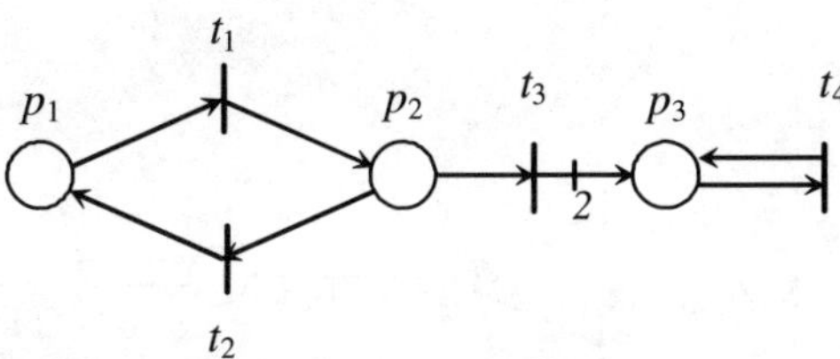

Fig. 1. A Place/Transition net

In the following we denote as $^\bullet t$ the set of *input places* of transition t, namely the set of places $p \in P$ that have an arc going from p to t, and $t^\bullet$ the set of *output places* of transition t, namely the set of places $p \in P$ that have an arc going from t to p.

Analogously, $^\bullet p$ and $p^\bullet$ denote respectively, the set of *input* and *output transitions* of place p, namely the set of transitions $t \in T$ that have an arc going from t to p, and from p to t, respectively.

Example 3. Let consider the net in Fig. 1. It holds $^\bullet t_1 = \{p_1\}$, $t_1^\bullet = \{p_2\}$, $^\bullet p_3 = \{t_3, t_4\}$ and $p_3^\bullet = \{t_4\}$. $\blacksquare$

2.2   Dynamic Behavior

Definition 1 only refers to the structure of the net. To associate a dynamic behavior to it, we need to introduce the notion of *state* and to definite the *rules* that govern the occurrence of the discrete events. In particular, in the P/T framework, the state corresponds to the *marking* of the net, and the evolution corresponds to the *firing* of transitions that may occur provided that appropriate enabling conditions are verified.

Definition 2. *A* marking *is a function* $M : P \rightarrow \mathbb{N}$ *that associates to each place a non negative number of tokens. The initial marking is denoted* M_0. ▲

Definition 3. *A net* N *with initial marking* M_0 *is a dynamical system. It is called* net system *and is denoted as* $\langle N, M_0 \rangle$. ▲

Graphically, tokens are represented as black dots within places.

Example 4. Let us consider the net in Fig. 1. A possible initial marking is

$$M_0 = [M_0(p_1)\ M_0(p_2)\ M_0(p_3)]^T = [1\ 0\ 0]^T$$

that is shown in Fig. 2.(a). Here the only marked place is p_1, that contains one token. Another possible initial marking is $M_0 = [0\ 1\ 0]^T$ that is shown in Fig. 2.(b). Here the only marked place is p_2 that contains one token. ∎

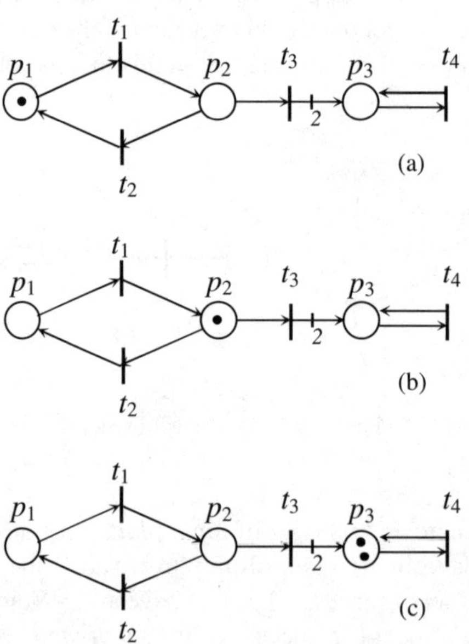

Fig. 2. Place/Transition net systems

Definition 4. *A transition t is* enabled *at marking M if*

$$M \geq Pre(\cdot, t)$$

where $Pre(\cdot, t)$ denotes the column of matrix Pre relative to transition t. We write $M[t\rangle$ to denote this condition . ▲

In simple words, the enabling condition of a transition only depends on the marking of its input places. In particular, t is enabled at M if each place $p \in {}^\bullet t$ contains at least $Pre(p, t)$ tokens, i.e., place p contains a number of tokens greater or equal to the weight of the arc going from p to t.

Example 5. Let us consider the net system in Fig. 2.(a). The only enabled transition is t_1. ■

Definition 5. *A transition t that is enabled at M may* fire. *The firing of t removes $Pre(p, t)$ tokens from each place $p \in P$ and adds $Post(p, t)$ tokens to each place $p \in P$. Thus the firing of t at M determines a new marking*

$$M' = M - Pre(\cdot, t) + Post(\cdot, t) = M + C(\cdot, t). \tag{2}$$

To denote this we write $M[t\rangle M'$. ▲

Note that, since $Pre(p, t) \neq 0$ only if $p \in {}^\bullet t$, and $Post(p, t) \neq 0$ only if $p \in t^\bullet$, then the firing of t at M removes $Pre(p, t)$ tokens from each input place p to t, and adds $Post(p, t)$ tokens to each output place p to t.

Moreover, by looking at Definition 5 it is immediate to observe that the enabling condition given by Definition 4 guarantees the non-negativity of the marking.

Example 6. Let us consider the net system $\langle N, M_0 \rangle$ in Fig. 2.(a). If transition t_1 fires the net reaches the new marking in Fig. 2.(b) because one token is removed from p_1 and added to p_2.

Now, both transitions t_2 and t_3 are enabled. If t_3 fires, the net reaches the new marking in Fig. 2.(c) because one token is removed from p_2 and two tokens are added to p_3, being 2 the weight of the arc going from t_3 to p_3.

Now, the only enabled transition is t_4, but its firing does not change the marking being $C(p_3, t_4) = 0$. ■

Definition 6. *A sequence $\sigma = t_{j_1} t_{j_2} \ldots t_{j_k} \in T^*$ is* enabled *at M if: t_{j_1} is enabled at M and its firing brings to a new marking M_1 that enables t_{j_2}; the firing of t_{j_2} at M_1 brings to a new marking M_2 that enables t_{j_3}, and so on.*

In such a case we write

$$M[t_{j_1}\rangle M_1[t_{j_2}\rangle \ldots M_{k-1}[t_{j_k}\rangle M_k$$

or simply $M[\sigma\rangle M_k$. An enabled sequence σ is called a firing sequence. ▲

Definition 7. *A marking M is* reachable *in $\langle N, M_0 \rangle$ if there exists a firing sequence σ such that $M_0[\sigma\rangle M$.*

The reachability set *of* $\langle N, M_0 \rangle$, *denoted as* $R(N, M_0)$, *is the set of markings that are reachable from* M_0, *i.e.,*

$$R(N, M_0) = \{M \in \mathbb{N}^m \mid \exists \sigma \in T^* : M_0[\sigma\rangle M\}.$$

▲

The reachability set may never be an empty set because it always includes at least the initial marking. Moreover, it may either be finite or infinite.

Example 7. In the case of the P/T net system in Fig. 2.(a) it is easy to verify that

$$R(N, M_0) = \{[1\ 0\ 0]^T, [0\ 1\ 0]^T, [0\ 0\ 2]^T\}.$$

Consider now the P/T net system in Fig. 3. In this case the initial marking is $M_0 = [0]$ but transition t_1 has no input arcs (it is a *source* transition) hence it is always enabled and can fire as many times as desired, adding each time a token to place p. On the contrary transition t_2 is only enabled if place p is marked: its firing removes one token from p. This simple net thus describes an unbounded queueing system: the initial marking in the figure corresponds to a queue initially empty. The reachability set is thus $R(N, M_0) = \mathbb{N}$. ■

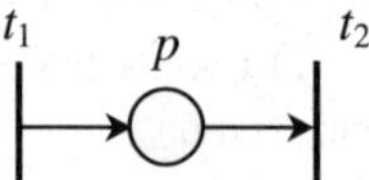

Fig. 3. The P/T net of an unbounded queueing system

The fact that the reachability set of a P/T system may be infinite is one of the main advantages of Petri nets with respect to other discrete event models, such as automata. In fact, using Petri nets we are able to represent with a finite structure a discrete event system with an infinite number of states.

3   Modeling with Petri Nets

Petri nets have been applied in a large variety of application domains, such as operational research, manufacturing systems, flexible production systems, transportation systems, and so on. The book by DiCesare *et al.* [10] provides a nice survey of Petri net approaches for the modeling and control of manufacturing systems.

In this section we first discuss the main primitives of concurrent systems that can be modeled using Petri nets. If one is interested in the order of event occurrences, the basic structures are *sequency, choice,* and *concurrency.* On the contrary, if one is interested in describing the use of available resources, the three most common structures are *disassembly, assembly, mutual exclusion.* Finally, we present in detail an example taken from the manufacturing domain, representing an assembly system.

3.1   Main Structures

Let us consider the Petri net systems in Fig. 4. Figure 4.(a) models *sequency*. Given the initial marking, only event e_1 may occur. Then, event e_2 may only occur after the occurrence of event e_1, and event e_3 may only occur after the firing of e_2. Note that here we are talking indifferently of events and transition firings.

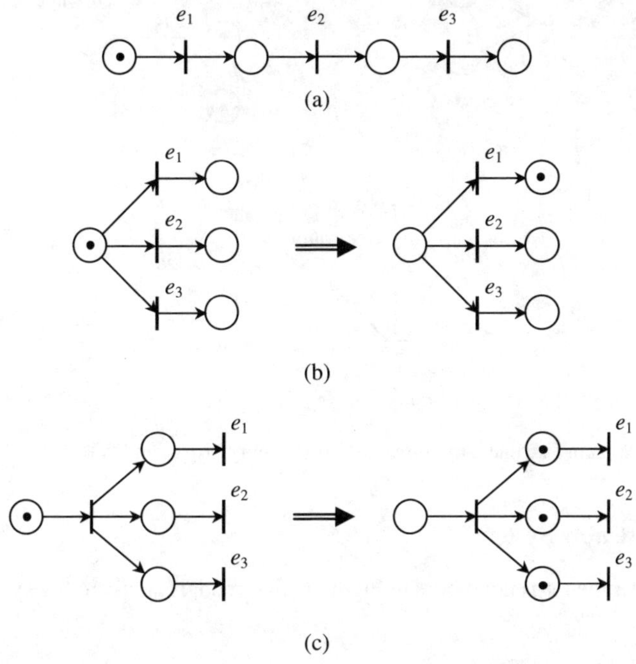

Fig. 4. Three main Petri net structures: (a) sequency, (b) choice, (c) concurrency

Fig. 4.(b) models the *choice* among events. Given the actual marking, all the events e_1, e_2, and e_3 are enabled. However, if any of such events occurs, then the others are disabled. We also say that these events are in *conflict* among them.

Finally, Fig. 4.(c) models *concurrency*. After the firing of the only enabled transition at the initial marking, all the events e_1, e_2 and e_3 are independently enabled and may occur in any order, even simultaneously.

If tokens represent available resources, three other main structures can be defined, as summarized in Fig. 5.

Fig. 5.(a) provides an example of a *disassembly* operation. If a vehicle is disassembled, then we get 4 wheels and one chassis.

Fig. 5.(b) provides an example of an *assembly* operation. If milk, espresso and cocoa are appropriately combined, then a cappuccino is obtained.

Fig. 5.(c) models *mutual exclusion*. Assume that two machines, M_1 and M_2, share a resource, namely a robot, whose task is that of loading them. At the initial marking the robot may either load M_1 or M_2. However, if it starts loading M_1, then it is not available for M_2. It is ready to load M_2 only after it has finished to process M_1. Analogously, if it is working on M_2, it cannot load M_1 until the loading of M_2 is finished.

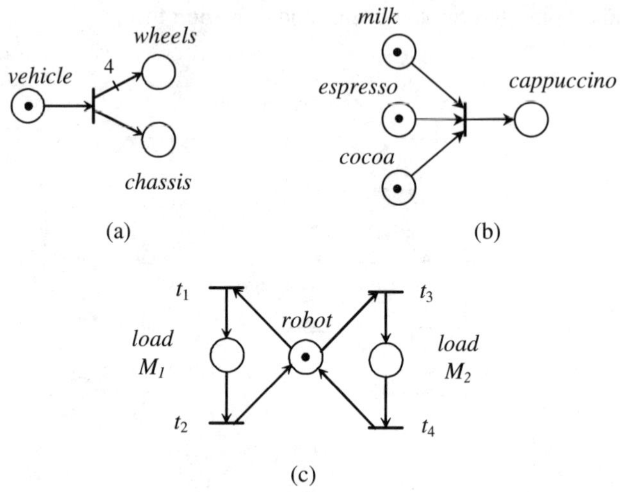

Fig. 5. Three main Petri net structures: (a) disassembly, (b) assembly, (c) mutual exclusion

3.2   An Assembly System

Let us consider the Petri net model in Figure 6, that models an assembly system [15, 10].

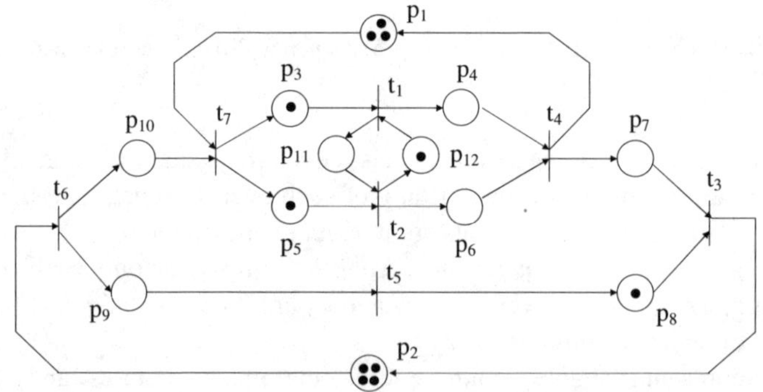

Fig. 6. The Petri net model of the assembly system in Subsection 3.2

It consists of five machines, $\mathcal{M}_1$, $\mathcal{M}_2$, $\mathcal{M}_3$, $\mathcal{M}_4$ and $\mathcal{M}_5$ whose operational process is modeled by the firing of transitions t_1, t_2, t_3, t_4 and t_5, respectively. Two principal types of operations are involved in this manufacturing system: *regular operations* and *assembly operations*. Regular operations (modeled by transitions t_1, t_2 and t_5) just transform a component of the intermediate product. Assembly operations (modeled by transitions t_3 and t_4) put components together to obtain a more complex component of a final product or the final product itself.

Note that this model uses transitions (t_6 and t_7) which do not represent operations but the beginning of the manufacturing of components which are required to assemble a more complex component or the final product.

In this example there are two manufacturing levels, the primary one, performed by $\mathcal{M}_3$, leads to finite product, the secondary one, performed by $\mathcal{M}_4$, leads to semi–finished (in–working) product.

The markings of places p_1 and p_2 represent the number of assembly servers for t_4 and t_3 respectively. The marking of places p_3, p_5, and p_9 represent the availability of parts to be processed (raw materials), while the marking of places p_4, p_6, p_7 and p_8 represent the availability of semi–finished products. Places p_{11} and p_{12} ensure that machines $\mathcal{M}_1$ and $\mathcal{M}_2$ work alternatively.

4   Analysis Techniques

As discussed in the previous section, P/T nets are a formal model that allows one to describe many interesting features of concurrent systems. Once a physical system has been modeled by a P/T net, the properties of interest of the system map fairly well into properties of the corresponding model. The formal definition of these properties, such as *reachability*, *boundedness*, *reversibility*, *liveness*, *deadlock-freeness*, *fairness*, etc., goes beyond the scope of this paper, but we address to [5] for a comprehensive discussion of this topic.

Many algorithms, with a well developed mathematical and practical foundation, have been developed to study these properties. The analysis techniques for Petri nets may be divided into the following groups.

- *Structural analysis*. It permits the demonstration of several properties almost independently of the initial marking. Structural analysis may be based on the study of the state equation of the net or on the study of the net graph.
- *Analysis by enumeration*. It requires the construction of the *reachability graph* representing the set of reachable markings and transition firings. If this set is not finite, a finite *coverability graph* may be constructed.
- *Analysis by transformation*. A net N_1 is transformed, according to particular rules, into a net N_2 while maintaining the properties of interest. The analysis of the net N_2 is assumed to be simpler than the analysis of the net N_1. Examples of this analysis technique are *reduction methods*, that permit the simplification of the structure of a net.
- *Simulation analysis*. It is useful to study the behavior of nets that interact with an external environment.

An extensive literature on these topics has appeared in last decades. In particular, we address to [9, 32, 34, 36] for more details. In the rest of this section, only the first two techniques will be partially described. Furthermore, we will limit our analysis to the basic *reachability problem*, that consists in establishing if a given marking is reachable starting from the initial marking.

4.1   State Equation

A linear algebraic equation can be written to describe the evolution of the net system after the firing of a sequence $\sigma \in T^*$. Such equation is based on Definition 5 and on the definition of *firing vector*.

Definition 8. *Given a net N with set of transitions $T = \{t_1, t_2, \ldots, t_n\}$ and a firing sequence $\sigma \in T^*$, we call* firing vector *relative to σ, the vector $\boldsymbol{\sigma} \in \mathbb{N}^n$ whose i-th component is equal to the number of times t_i appears in σ.*   ▲

Next result follows immediately from Definition 5.

Proposition 1. *Let us consider a net system $\langle N, M_0 \rangle$ with incidence matrix C. If M is reachable from M_0 firing σ, then*

$$M = M_0 + C \cdot \boldsymbol{\sigma}. \tag{3}$$

▲

Eq. (2), or sometimes its transitive closure given by Eq. (3), is called the *state equation* of $\langle N, M_0 \rangle$.

Example 8. Let us consider the net system in Fig. 2.(a) and the firing sequence $\sigma = t_1 t_2 t_1 t_2 t_1 t_3$. The firing vector associated to σ is $\boldsymbol{\sigma} = [3\ 2\ 1\ 0]^T$ and we can easily verify that the marking $M = [0\ 2\ 0]^T$ obtained from M_0 firing σ satisfies eq. (3) where C is given in Example 2.   ■

It is important to stress that the state equation only provides a *necessary* (but *not sufficient*) condition for reachability. Indeed, the existence of a vector $\boldsymbol{\sigma} \in \mathbb{N}^n$ such that $M = M_0 + C \cdot \sigma \in \mathbb{N}^m$ does not imply the existence of a firing sequence σ whose firing vector is $\boldsymbol{\sigma}$, and that is enabled at M_0.

Example 9. Let us consider the net system in Fig. 7 where $M_0 = [1\ 0\ 0\ 0]^T$, and $\boldsymbol{\sigma} = [1\ 1]^T$. The marking $M = M_0 + C \cdot \boldsymbol{\sigma} = [0\ 0\ 0\ 1]^T$ is a non-negative marking however it is not reachable from M_0. In fact, no transition is enabled at the initial marking. Hence neither $\sigma' = t_1 t_2$ nor $\sigma'' = t_2 t_1$, i.e., no sequence whose firing sequence is $\boldsymbol{\sigma}$, may fire from M_0.   ■

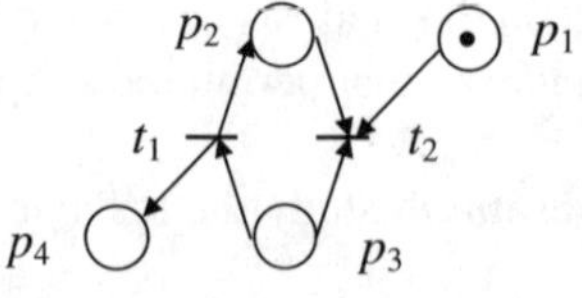

Fig. 7. The P/T system in Example 9

4.2   Reachability Graph

In this section we focus on a particular class of P/T nets, namely *bounded Petri nets*, for which the reachability problem can be solved constructing the so-called *reachability graph*.

Definition 9. *A Petri net system* $\langle N, M_0 \rangle$ *is* bounded *if and only if there exists a* finite *constant K such that* $\forall p \in P$ *and* $\forall M \in R(N, M_0)$, $M(p) \leq K$.   ▲

Thus a Petri net system is bounded if and only if the marking of each place is bounded for any reachable marking. An obvious result is the following.

Proposition 2. *A Petri net system is bounded if and only if its reachability set is* finite. ▲

For bounded Petri net systems, it is possible to enumerate in a systematic way the reachability set by means of the *reachability graph*. Here each node corresponds to a reachable marking, and each arc corresponds to a transition. The reachability graph may be constructed using the following algorithm that terminates in a finite number of steps if the reachability set is finite.

Algorithm 1 (Reachability graph). *Let* $\langle N, M_0 \rangle$ *be a marked net with incidence matrix C.*

1. *The root node is M_0. This node has initially no label.*
2. *Let us consider a node M with no label.*
   (a) For each transition t enabled at M:
      i.   Let $M' = M + C(\cdot, t)$.
      ii. If there does not exists a node M' in the graph, add it.
      iii. Add an arc t from M to M'.
   (b) Label the node M "old".
3. *It there are nodes with no label, goto step 2.*
4. *Remove all labels from nodes.*   ▲

Example 10. Let us consider the P/T system in Fig. 2.(a). Using Algorithm 1 we obtain the reachability graph in Fig. 8.   ■

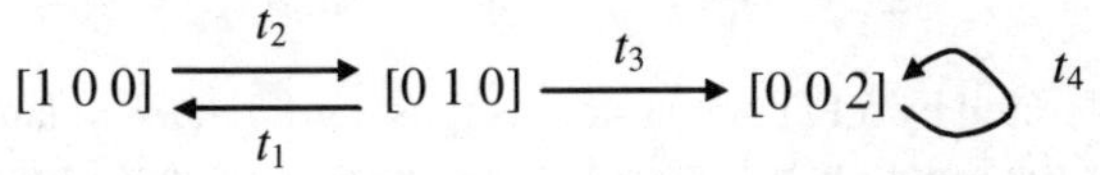

$$[1\ 0\ 0] \underset{t_1}{\overset{t_2}{\rightleftarrows}} [0\ 1\ 0] \xrightarrow{t_3} [0\ 0\ 2] \circlearrowright\ t_4$$

Fig. 8. The reachability graph of the P/T net system in Fig. 2.(a)

Looking at the reachability graph of a P/T system $\langle N, M_0 \rangle$, one can immediately determine which markings are reachable, because a node M is reachable from M_0 if and only if it belongs to the graph. Furthermore:

(1) a marking M' is reachable from a reachable marking M *iff* there exist two nodes M and M' in the graph and there exists an oriented path that goes from M to M';

(2) a sequence σ is firable from a reachable marking M *iff* there exists an oriented path that starts from M whose sequence of arc labels is σ.

If the reachability set is infinite, then obviously the reachability graph is infinite as well. In such a case a different algorithm can be used to compute a finite graph, called the *coverability graph*, where each arc still corresponds to a transition, while each node either corresponds to a single reachable marking, or it represents an infinite set of reachable markings. Note, however, that in such a case there is a price to pay for representing with a finite graph an infinite set: the coverability graph usually provides only necessary (but not sufficient) conditions for determining if a marking is reachable or if a sequence is firable. See [9, 32] for details.

5 Petri Net Languages

In the previous section we introduced the notion of reachability and highlighted the importance of characterizing the reachability set of a net system. However, the modeling power of a discrete event system is also strictly related to the sequences of events it can generate, i.e., in the Petri net framework, to the sequences of transitions that can fire. A sequence of transitions is a string, and a set of strings is a *language*. In this section we focus on the classes of languages defined by Petri nets. In particular, we first recall the notion of *generated* and *accepted* languages, and define *labeled* Petri nets. Then, we provide the definition of L-type, G-type and P-type Petri net languages. Finally, we provide some important relationships among these classes and the class of *regular languages*.

A good introduction to Petri net languages can be found in the classic book of Peterson [32], while some generalizations and more recent results can be found in the paper by Gaubert and Giua [12]. All the material presented in this section is taken from these two references.

5.1 Generated and Accepted Languages

Definition 10. *The* language generated *by* $\langle N, M_0 \rangle$ *is the set of sequences that are enabled at the initial marking* M_0, *i.e.,*

$$L(N, M_0) = \{\sigma \in T^* \mid M_0[\sigma\rangle\}.$$ ▲

The language generated by a P/T net system is thus a *prefix-closed* language. Note that it always includes the empty word (usually denoted as ε) because for any $M \in \mathbb{N}^m$, it holds $M[\varepsilon\rangle M$.

Example 11. Let us consider the net system in Fig. 2.(a). The language of this net can be easily described with a regular expression as

$$L(N, M_0) = (t_1 t_2)^* [\varepsilon + t_1 + t_1 t_3 t_4^*].$$

This means that the sequence $t_1 t_2$ may fire indefinitely from the initial marking. Then, either no other sequence fires, or it fires t_1, or it fires the sequence $t_1 t_3$: at this point the only enabled transition is t_4 that can fire indefinitely. ■

Definition 11. *Let us consider a P/T system $\langle N, M_0 \rangle$. Let F be a set of* final *(or ac-cepting)* markings. *The* language accepted *by $\langle N, M_0 \rangle$ is the set of sequences that are enabled at the initial marking M_0 and that lead to a marking $M \in F$, i.e.,*

$$L_F(N, M_0) = \{\sigma \in T^* \mid (\exists M \in F) \, M_0[\sigma\rangle M\}. \qquad \blacktriangle$$

Depending on the final set F, the language accepted by a P/T net system may not be prefix-closed. Moreover, it includes the empty word if and only if $M_0 \in F$.

Example 12. Let us consider the net system in Fig. 2.(a). Assume $F = \{[0\ 1\ 0]^T\}$. The language accepted by $\langle N, M_0 \rangle$ is $L_F(N, M_0) = (t_1 t_2)^* t_1$. $\qquad \blacksquare$

5.2   Labeled P/T Nets

When observing the evolution of a net, it is common to assume that each transition t is assigned a label $\ell(t)$ and that the occurrence of t generates an observable output $\ell(t)$. This leads to the definition of labeled nets.

Definition 12. *Given a Petri net N with set of transitions T, a* labeling function $\ell :$ $T \to \Sigma$ *assigns to each transition $t \in T$ a symbol from a given set of labels Σ, that may also include the empty string ε.*

A Σ-labeled Petri net system *is a 3-tuple $G = \langle N, M_0, \ell \rangle$ where $N = (P, T, Pre, Post)$, M_0 is the initial marking, and $\ell : T \to \Sigma$ is the labeling function.* $\qquad \blacktriangle$

Also in the case of labeled P/T nets we can distinguish among generated and accepted language. In particular, the following definitions hold.

Definition 13. *The* language generated *by a Σ-labeled P/T net system $\langle N, M_0, \ell \rangle$ is the ℓ-image of the set of firing sequences that are enabled at M_0, i.e.,*

$$L(N, M_0, \ell) = \{\ell(\sigma) \mid \sigma \in T^*, \, M_0[\sigma\rangle\}. \qquad \blacktriangle$$

Definition 14. *Let us consider a Σ-labeled P/T net system $\langle N, M_0, \ell \rangle$. Let F be a set of final markings. The* language accepted *by $\langle N, M_0, \ell \rangle$ is the ℓ-image of the set of firing sequences leading to a final marking, i.e.,*

$$L_F(N, M_0, \ell) = \{\ell(\sigma) \mid \sigma \in T^*, (\exists M \in F) \, M_0[\sigma\rangle M\}. \qquad \blacktriangle$$

Example 13. Let us consider again the net system in Fig. 2.(a). Assume $\ell(t_1) = \ell(t_4) = a$, $\ell(t_2) = \ell(t_3) = b$. Then $L(N, M_0, \ell) = (ab)^*[\varepsilon + a + aba^*]$. Moreover, if $F = \{[0\ 1\ 0]^T$, the accepted language is $L_F(N, M_0, \ell) = (ab)^* a$. $\qquad \blacksquare$

5.3   Classes of Languages

Different classes of *accepted* Petri net languages may be defined depending on the set of final markings F and on the labeling function ℓ [32].

Definition 15. *The accepted language of a Petri net system $\langle N, M_0 \rangle$ with set of accepting markings F, can be classified as follows.*

- *L*-type: $L_F(N, M_0)$ *is an* L-type Petri net language *if the set of final markings F is* finite.
- *G*-type: $L_F(N, M_0)$ *is a* G-type Petri net language *if the set of final markings F is the covering set of a given* finite *set $\bar{F}$. This means that a marking M is final if and only if $M \geq \bar{M}$ for a given $\bar{M} \in \bar{F}$. Languages in this class are usually called* weak *languages.*
- *P*-type: $L_F(N, M_0)$ *is a* P-type Petri net language *if the set of final markings F coincides with the reachability set $R(N, M_0)$. In such a case the accepted language is equal to the generated language and it is obviously prefix-closed.*   ▲

Moreover, four classes of labeling functions may be defined.

Definition 16. *The labeling function of a labeled Petri net system $\langle N, M_0, \ell \rangle$ can be classified as follows.*

- free: *if all transitions are labeled distinctly, namely a different label is associated to each transition, and no transition is labeled with the empty string.*
- deterministic: *if no transition is labeled with the empty string, and the following condition[1] holds: for all $t, t' \in T$, with $t \neq t'$, and for all $M \in R(N, M_0)$: $M[t\rangle \wedge M[t'\rangle \Rightarrow [\ell(t) \neq \ell(t')]$ i.e., two transitions simultaneously enabled may not share the same label. This ensures that the knowledge of the firing labels $\ell(\sigma)$ is sufficient to reconstruct the marking M that the firing of σ yields.*
- λ-free: *if no transition is labeled with the empty string[2].*
- arbitrary: *if no restriction is posed on the labeling function ℓ.*   ▲

Each of these type of labeling is a generalization of the previous one. Furthermore all types of labeling only depend on the structure of the net, but for the deterministic labeling, that depends both on the structure and on the behavior of the net.

Example 14. Let us consider the nets in Fig. 9. If we only look at the net structure — that is the same in both nets — we can say that the labeling is λ-free. However, in the first net the labeling is also deterministic because the two transitions labeled a can never be simultaneously enabled from any reachable marking. The second net is nondeterministic, because the two transitions labeled a can be simultaneously enabled.

Assume that the string aa is observed in the second net. The first a is certainly due to the occurrence of transition t_1, the only one enabled at M_0, whose firing yields the new marking $M = [1\ 1\ 0]^T$. From this marking, however, both t_1 and t_2 are enabled and one cannot determine if the second a yield $M = [0\ 2\ 0]^T$ or $M = [1\ 0\ 1]^T$.   ■

---

[1] A looser condition is sometimes given: for all $t, t' \in T$, with $t \neq t'$, and for all $M \in R(N, M_0)$: $M[t\rangle \wedge M[t'\rangle \Rightarrow [\ell(t) \neq \ell(t')] \vee [Post(\cdot, t) - Pre(\cdot, t) = Post(\cdot, t') - Pre(\cdot, t')]$. Thus two transitions with the same label may be simultaneously enabled at a marking M, if the two markings reached from M by firing t and t' are the same.

[2] In the Petri net literature the empty string is denoted λ, while in the formal language literature it is denoted ε. In this paper we denote the empty string ε but, for consistency with the Petri net literature, we still use the term λ-*free* for a non erasing labeling function $\ell : T \to \Sigma$.

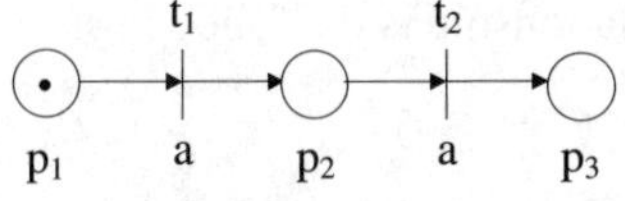

deterministic

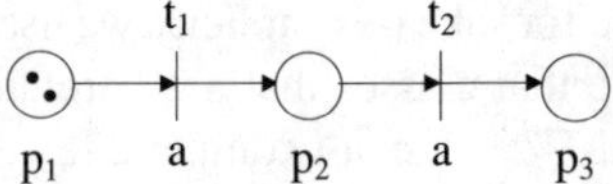

λ-free (nondeterministic)

Fig. 9. A deterministic labeled net (a) and a nondeterministic one

Twelve different classes of Petri net languages result from the cross product of the three types of final marking sets in Definition 15 and the four types of labeling in Definition 16, as summarized in Table 1.

Here the classes of L-type, G-type, and P-type λ-free languages are denoted, respectively, $\mathcal{L}$, $\mathcal{G}$, and $\mathcal{P}$. An additional superscript f, det or λ denotes, respectively, the corresponding classes of free, deterministic, and arbitrary languages.

Table 1. The 12 classes of Petri net languages

	free	deterministic	λ-free	arbitrary
$\mathcal{L}$-type	$\mathcal{L}^f$	$\mathcal{L}^{det}$	$\mathcal{L}$	$\mathcal{L}^\lambda$
$\mathcal{G}$-type	$\mathcal{G}^f$	$\mathcal{G}^{det}$	$\mathcal{G}$	$\mathcal{G}^\lambda$
$\mathcal{P}$-type	$\mathcal{P}^f$	$\mathcal{P}^{det}$	$\mathcal{P}$	$\mathcal{P}^\lambda$

5.4   Relationships Among Classes of Petri Net Languages

The above classes of Petri net languages are closely related. In particular, some intuitive relationships hold:

$$\begin{aligned}
\mathcal{L}^f \subsetneqq \mathcal{L}^{det} \subsetneqq \mathcal{L} \subsetneqq \mathcal{L}^\lambda, \\
\mathcal{G}^f \subsetneqq \mathcal{G}^{det} \subsetneqq \mathcal{G} \subsetneqq \mathcal{G}^\lambda, \\
\mathcal{P}^f \subsetneqq \mathcal{P}^{det} \subsetneqq \mathcal{P} \subsetneqq \mathcal{P}^\lambda,
\end{aligned} \tag{4}$$

where the symbol $\subsetneqq$ denotes strict inclusion.

Note that as a consequence of the strict inclusions (4), it is not possible to provide determinization procedures to convert a nondeterministic Petri net (namely a Petri net with an arbitrary labeling function) into an equivalent deterministic Petri net. On the contrary, this is possible with finite state automata where a systematic approach exists to convert a nondeterministic finite state automaton into an equivalent deterministic one [5].

Another quite intuitive relationship is the following

$$\mathcal{P}^f \subsetneq \mathcal{G}^f, \qquad \mathcal{P}^{det} \subsetneq \mathcal{G}^{det}, \qquad \mathcal{P} \subsetneq \mathcal{G}, \qquad \mathcal{P}^\lambda \subsetneq \mathcal{G}^\lambda. \tag{5}$$

In fact, every P-type language is a G-type language if F is a singleton containing the null marking.

Other less intuitive relationships have also been proved and can be summarized graphically as in Fig. 10. Here for sake of simplicity we use $\to$ to denote $\subsetneq$, i.e., $A \to B$ is equivalent to $A \subsetneq B$. Note that classes that are unrelated in the table (such as $\mathcal{L}^{det}$ and $\mathcal{G}^{det}$, or such as $\mathcal{L}^{det}$ and $\mathcal{P}^{det}$) are not comparable.

$$
\begin{array}{ccccccc}
\mathcal{L}^f & \to & \mathcal{L}^{det} & \to & \mathcal{L} & \to & \mathcal{L}^\lambda \\
 & & & & & & \uparrow \\
\mathcal{G}^f & \to & \mathcal{G}^{det} & \to & \mathcal{G} & \to & \mathcal{G}^\lambda \\
\uparrow & & \uparrow & & \uparrow & & \uparrow \\
\mathcal{P}^f & \to & \mathcal{P}^{det} & \to & \mathcal{P} & \to & \mathcal{P}^\lambda
\end{array}
$$

Fig. 10. Relationships among classes of Petri net languages

This plethora of Petri net languages may generate some confusion, even more considering the fact that additional classes can be defined as mentioned in [12]. Note, however, that not all these classes are useful in practice. In fact, the classes of free languages are very restricted, in the sense they do not contain all regular languages. On the contrary, for the largest classes of λ-free or arbitrary languages the problems of language equivalence or inclusion is not *decidable*. Thus we may conclude that the only interesting classes of Petri net languages are the deterministic ones [12], and we will consider them as representative of Petri net languages.

5.5   Relationships Among Petri Net Languages and Regular Languages

One of the classes of formal languages that has received most attention in the literature, is the class of *regular languages* [23] that we denote as $\mathcal{R}$. Regular languages are characterized by regular expressions and are generated by regular grammars. Moreover, it has been proved that the class of regular languages is coincident with the class of languages accepted by finite state automata.

The following important result expresses the most important relationship among Petri net languages and regular languages.

Theorem 2. [12] *The intersection of the classes of L-type and G-type regular Petri net is the class of regular languages, i.e.,* $\mathcal{R} = \mathcal{G}^{det} \cap \mathcal{L}^{det}$. ▲

Therefore, $\mathcal{L}^{det}$ and $\mathcal{G}^{det}$ provide proper and distinct extensions of regular languages.

Other interesting relationships among Petri net languages and other classes of languages, such as *contex-free* languages, *bounded contex-free* languages, *context-sensitive* languages, have been proved and are reported in Fig. 11. Here we can see that Petri net languages are a subclass of context-sensitive languages, and a superclass of regular languages. Petri net languages are not comparable with context-free languages.

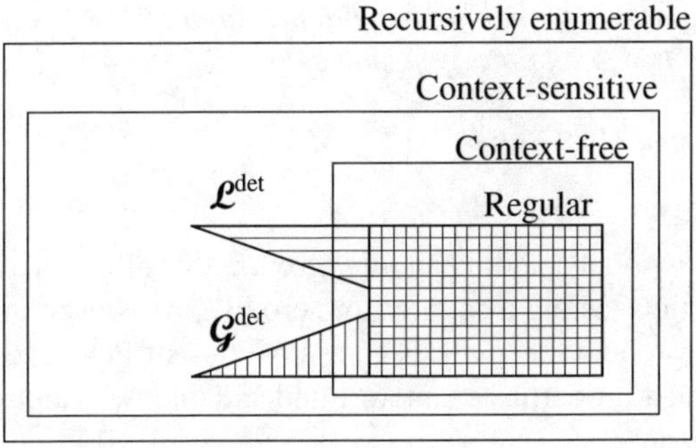

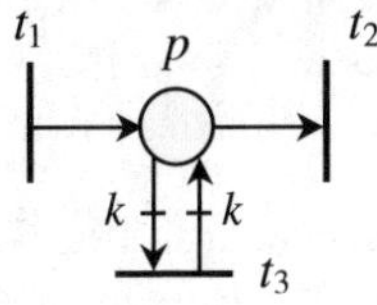

Fig. 11. Relationships among classes of formal languages

6   Comparison with Automata

The language analysis in the previous session shows that Petri nets are a generalization of automata. In this section we want to focus on the relationship between P/T nets and automata and show what are the main advantages the former model offers with respect to the latter. Five different aspects will be considered: the *state representation power*, the *language power*, the *modularity*, the *structural representation of primitives*, and the *linear algebraic structure*.

6.1   State Representation Power

A Petri net is a finite state automaton additionally equipped with *weak counters*, i.e., with the possibility of testing if a counter has reached a fixed value:

$$M(p) \geq k?$$

Example 15. Let us consider the net in Fig. 12. Place p is the counter, whose value is increased by the firing of t_1, and decreased by the firing of t_2. If there are k or more tokens in p, transition t_3 is also enabled and may fire (test of the counter) without changing the value of the counter.   ∎

It is important to stress that places in a P/T net are weak counters, i.e., may be tested only for inequalities of the type $\geq$ while a test for $\leq$ is not allowed. In fact, from the enabling rule in Definition 4 follows this obvious result.

Fig. 12. A weak counter

Monotonicity property. *If something can happen from M it can also happen from any marking greater than M, i.e., for any sequence $\sigma \in T^*$:*

$$M[\sigma\rangle \quad and \quad M' \geq M \quad \implies \quad M'[\sigma\rangle.$$

This property can be violated adding an *inhibitor arc* that allows a transition to fire only if a place is empty, thus testing a counter for zero [37]. However, this feature increases the modeling power — and the analysis complexity — of Petri nets to that of a Turing machine, making most properties of interest undecidable: we cannot properly consider these models as P/T nets.

6.2   Language Power

A Petri net is a generator of regular languages with the additional feature of generating *one-sided Dyck languages*, i.e., of testing if a string of parenthesis

$$((()))()((\cdots$$

is well formed [31].

Example 16. Let us consider the net in Fig. 3. Here, the firing of t_1 corresponds to the opening of a parenthesis "(", while the firing of t_2 corresponds to the closing of a parenthesis ")". All firing sequences generated by this net correspond to well formed strings of parenthesis. ∎

6.3   Modularity

With modular synthesis, complex systems may be constructed by aggregation of simpler modules. The most common operator that allows to automatically construct the model of a complex system from the models of the subsystems that compose it, is the *concurrent composition operator*, that can be defined both for automata and Petri nets.

There are, however, two main advantages in using Petri nets rather than automata.

- When applying the concurrent composition operator to Petri nets, the structure of the modules is kept in the composed net.
- The composition of k automata, each with a state space Q_i of cardinality n, yields a composed model with state space $Q \subseteq Q_1 \times \cdots \times Q_k$, i.e., the composed automata has a state space of cardinality up to n^k (exponential growth). On the contrary, the composition of k Petri nets, each with set of place P_i of cardinality m yields a composed model with set of places $P = P_1 \cup \cdots \cup P_k$, i.e., the composed net has a set of places of cardinality $k \cdot m$ (linear growth).

Example 17. Let us consider the automata in Fig. 13.(a) that represent two machines with state space Q' and Q'' respectively. Here event t_2 is shared between the two modules and their concurrent composition is shown in Fig. 13.(b). Note that each state of the new automaton is a pair $(q', q'') \in Q' \times Q''$. The structure of the two modules is lost in the composed system in Fig. 13.(b), in the sense that it is not possible to partition its structure into two parts, each corresponding to one of the two modules.

In Fig. 13.(c) we have represented the P/T net models of the two machines, whose concurrent composition is given by the net in Fig. 13.(d). Note that the composed model is obtained by the modules simply fusing the transitions with the same label. The set of places of the composed net is the union of the set of places of the modules, whose structure can still be clearly identified in Fig. 13.(d). ∎

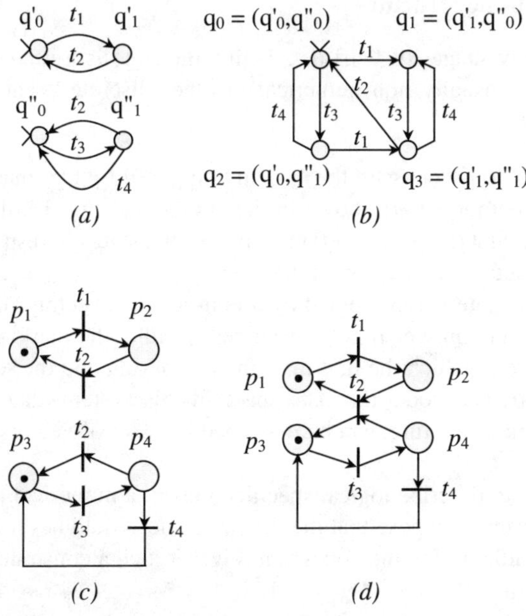

Fig. 13. The automata models of two machines (a) and their concurrent composition (b); the Petri net models of two machines (c) and their concurrent composition (d)

6.4 Structural Representation of Primitives

In Section 3.1 we have discussed several primitives that can be represented by Petri nets, such as "sequence", "choice", "concurrency". Each of these primitives correspond to a clear Petri net structure: sequence corresponds to a path in the graph, choice to a place inputting to more than one transition, concurrency to parallel transitions.

In the case of automata concurrency may not be represented, because an automaton can only describe the interleaving of events and not their simultaneous occurrence. However, one may think that the other primitives can be well described by structures similar to those described in Section 3.1. Here we point out that this is not always true with a simple example.

Example 18. Let us consider again the system composed by two machines whose automaton and Petri net model are shown in Fig. 13.(b) and Fig. 13.(d).

In the automaton structure we identify the path $q_0 - t_1 - q_1 - t_3 - q_3$. *Can we conclude that events t_1 and t_3 are in a sequency relation?*

In the automaton structure, from state q_0 both event t_1 and t_3 are enabled. *Can we conclude that events t_1 and t_3 are in a choice relation?*

The answer to both questions is no: transitions t_1 and t_3 are concurrent as can be seen from the Petri net model. In fact, the two transitions belong to different subsystems and can fire concurrently when both are enabled. ■

6.5 Linear Algebraic Structure

One of the main advantages of Petri nets is that the state is a *vector of non-negative integers*, while it is usually non numerical in other discrete event models, such as automata.

Example 19. Let us consider again the system composed by two machines whose automaton and Petri net model are shown in Fig. 13.(b) and Fig. 13.(d). State q_0' (resp., q_0'') denotes that the first (resp., second) machine is idle; state q_1' (resp., q_1'') denotes that the first (resp., second) machine is working.

In the Petri net a state is represented by a non-negative vector. Marking $[1\ 0\ 1\ 0]^T$ corresponds to the state in which both machines are idle; the marking $[0\ 1\ 1\ 0]^T$ corresponds to the state in which the first machine is working and the second is idle, and so on. Using a Petri net model the state space of this system, that is a series of labels with no algebraic structure, can be described by a set of vectors, i.e., by a highly structured set.

This also allows to describe logical specifications in a numerical form. Assume for instance, that we want to impose that the first machine should never be working if the second machine is idle. Using the notation in Fig. 13 such a constraint can be imposed forcing the constraint $M(p_2) + M(p_3) \leq 1$. ■

The possibility offered by Petri nets to describe the state space of a discrete event system that may have absolutely no algebraic structure, with a set of integers vectors has an important implication. In fact, it is possible to apply algebraic formalisms such as integer programming for the analysis and control of these systems. Within this area of research that, as we mentioned before, is called *structural analysis*, several well-founded formal approaches have been developed. Unfortunately a survey of this area is still missing, and we cannot provide comprehensive references; see however [22, 36] for a few interesting examples.

7 Mapping Classical Properties into Discrete Event Systems

Classical control theory deals with *time-driven* systems modeled by difference or differential equations. However, many properties of dynamical systems have been defined in very general terms that are *model independent*.

It seems natural to study these properties in the context of *discrete event systems*, and more specifically in the context of Petri nets. This rather standard approach has been used by many researchers, and it has proved to be extremely fruitful, inspiring many of the current research areas in Petri nets.

It is important, however, to point out that the extension from time-driven to discrete event systems must be taken with care. To give a flavor of the problems one may face, in this section we discuss the classical property of *controllability* and the way in which it has been handled in the framework of Petri nets.

Note that much of this discussion is essentially due to Murata [26]. As far as we know, his 1977 work was the first paper dealing with Petri nets published in an IEEE journal. The fact that this paper was published on an Automatic Control journal is emblematic of the appeal that the algebraic structure of Petri nets has to control engineers.

7.1 Controllability

In Subsection 4.1 we introduced the state equation of a Petri net system, that can be rewritten as

$$M_{k+1} = M_k + C \cdot \sigma_k \tag{6}$$

where $\sigma_k \in \{0, 1\}^n$ is the firing vector relative to the transition that has fired at the marking M_k, thus leading to the new marking M_{k+1}.

This clearly reminds one of the state equation of a *discrete-time linear stationary time-driven* system, namely

$$\mathbf{x}_{k+1} = A\mathbf{x}_k + B\mathbf{u}_k \tag{7}$$

where $\mathbf{x}_k \in \mathbb{R}^m$ (resp., $\mathbf{x}_{k+1} \in \mathbb{R}^m$) is the state vector at the sampling time k (resp., $k + 1$), and $\mathbf{u}_k \in \mathbb{R}^n$ is the control input vector at sampling time k. More precisely, equation (6) is a particular case of (7) with $A = I$ and $B = C$, where I denotes the identity matrix.

As well known, the following definition of controllability holds.

Definition 17. *A discrete-time linear stationary system is* controllable *if and only if it is possible, by appropriately acting on the input, to transfer the state of the system from any initial state $\mathbf{x}_0$ to any other state x_f, called the* target state, *in a finite number of sampling steps $f \geq 0$.* ▲

Theorem 3. *Given the discrete-time linear stationary system (7), we call* controllability matrix *the $(m \times n \cdot m)$ matrix*

$$\Gamma = \left[\, B \,|\, AB \,|\, A^2 B \,|\, \ldots \,|\, A^{m-1} B \,\right].$$

A necessary and sufficient condition for the controllability of (7) is that

$$n_c \triangleq rank\ \Gamma = m.$$ ▲

Therefore, using Theorem 3, the controllability matrix of a Petri net is

$$\Gamma = \left[\, C \,|\, C \,|\, C \,|\, \ldots \,|\, C \,\right],$$

thus

$$\mathrm{rank}\ \Gamma = \mathrm{rank}\ C,$$

i.e., the rank of the controllability matrix always coincides with the rank of the incidence matrix.

We now observe that:

- the condition rank $\Gamma = m$ is only a *necessary* condition for controllability if we restrict the control input to $\mathbf{u}_k \in \{0, 1\}^m$ and to $M_k + C \cdot \boldsymbol{\sigma}_k \geq \mathbf{0}$;
- moreover, as already discussed above (see Example 9), the state equation of a Petri net system only provides a *necessary* condition for reachability.

Therefore, as a result of this analysis, the following conclusion may be drawn: *in the Petri net framework, rank $\Gamma = m$ only provides a* necessary *(but not sufficient) condition for controllability.*

This conclusion is not surprising — discrete event systems are much more difficult to study than linear systems — but, as Murata observes, does not address the real issue. What in fact is totally missing from this analysis is a discussion of how significant for a discrete event system is the property of controllability that derives from Definition 17. In fact, this classical notion does not fit well with discrete event systems, and it is hardly meaningful. As an example, consider a Petri net model of a manufacturing system where the marking of a place denotes the availability of a resource. It is not meaningful to investigate if the marking of such a place may reach *any* value starting from *any* other marking. As a trivial example, in an assembly system described by a Petri net starting from a state in which there are only *two* wheels available, it may be possible to reach state in which *one* bicycle has been assembled, but not a state in which *ten* bicycles have been assembled.

It seems thus natural to introduce different notions of controllability, more suited to describe the desired properties of discrete event systems. Here are some possible examples.

- Given a Petri net with incidence matrix C of dimension $m \times n$, we say that $\mathbf{x} \in \mathbb{Z}^m$ is a *P-flow* if $\mathbf{x}^T \cdot C = \mathbf{0}$.

  A P-flow imposes an invariant law on the reachability set of a net: in fact if marking M is reachable from the initial marking M_0 it must hold

$$\mathbf{x}^T \cdot M = \mathbf{x}^T \cdot M_0$$

  as can be seen multiplying the state equation (3) by $\mathbf{x}^T$ from the left.

  This condition, however, is necessary but not sufficient for reachability. Given a net system $\langle N, M_0 \rangle$ with incidence matrix C, let X be a matrix whose columns are P-flows forming a basis of the left-null space of matrix C. It holds

$$R(N, M_0) \subseteq I_X(N, M_0) \stackrel{\text{def}}{=} \{M \in \mathbb{N}^m \mid X^T \cdot M = X^T \cdot M_0\}.$$

  Thus a meaningful definition for a Petri net system may be the following: *a Petri net system $\langle N, M_0 \rangle$ is controllable if $R(N, M_0) = I_X(N, M_0)$.*
- Yet a different definition of controllability may be given for *timed* Petri nets, namely P/T nets in which a *time interval* is associated to transitions: an enabled transition may fire provided that it has been enabled for a time that belongs to its time interval. Such a model is particularly useful when making performance analysis. Clearly imposing a timing structure over a logical model influences its reachability set. In fact, since a timed model can be seen as a logical model with additional timing

constraints, the reachability set of the timed net is usually a subset of that of the underlying untimed one. One may define a timed Petri net system controllable if its reachability set coincides with that of the underlying untimed model.

- Finally, in Supervisory Control — one of the most interesting approaches to the control of discrete event systems — controllability is not defined as a property of a system alone, but is defined with respect to a given *specification*, i.e., with respect to a set of legal states or to a set of legal words. This definition has often been used to define a Petri net system controllable if its evolution can be restricted to a given set of legal markings or to a given set of legal words.

8   Current Research Areas in Petri Nets

In the last two decades a large number of researchers from the automatic control community have devoted their effort to the study of Petri nets. There are, however, a certain number of basic problems that are still open. Here, we mention the following four significant areas of on-going research.

- *Control*: as in classical control, the control problem consists in finding a control law that constraints the controlled system behavior to satisfy a given specification.
- *Deadlock*: a deadlock represents an anomalous state from which no further evolution is possible. This is an issue that appears in most practical applications, and appropriate strategies should be adopted in order to prevent it.
- *Observability*: the problem is that of determining efficient ways of reconstructing the state of a net based on observed events occurrence and/or on partial marking observation.
- *Identification*: this problem consists in determining a Petri net system starting from examples/counterexamples of its language, or from the structure of its reachability (or coverability) graph.

In the following we recall the main results that have been proposed in the above areas.

8.1   Control

The most interesting and original approach to the control of discrete event systems, that has directly or indirectly shaped much of the research in this area, is *Supervisory Control Theory* (SCT), originated by the work of P.J. Ramadge and W.M. Wonham [33]. According to the paradigm of SCT, a discrete event system G is a language generator whose behavior, i.e., language, is denoted $L(G)$. Given a legal language K, the basic control problem is to design a supervisor that restricts the closed loop behavior of the plant to $K \cap L(G)$, disabling *controllable* events; the events whose occurrence cannot be disabled are called *uncontrollable*. It is also usually required that the closed loop system satisfies additional qualitative specifications, such as absence of blocking, reversibility, etc. Since Petri nets can be seen as language generators, it is also possible to use them as discrete event models for SCT; in this case it is assumed that some transitions, that we call controllable, can be disabled by an external agent. See [5, 20] for a review of this topic.

A similar approach can also be taken when considering the state evolution of a discrete event system, rather than the traces of events it generates. This approach, that we call *state-based*, is particularly attractive when Petri nets are used to represent the plant and was used by several authors, as reviewed in [20]. Let us consider a Petri net system $\langle N, M_0 \rangle$ with m places, whose set of reachable markings is $R(N, M_0)$. Assume we are given a set of legal markings $\mathcal{M} \subseteq \mathbb{N}^m$: the basic control problem consisting in designing a supervisor that restricts the reachability set of plant in closed loop to $\mathcal{M} \cap R(N, M_0)$, while satisfying some qualitative properties of interest.

Of particular interest are those Petri net state-based control problems where the set of legal markings $\mathcal{M}$ is expressed by one — or more — linear inequality constraints called Generalized Mutual Exclusion Constraints (GMEC) [13]. In this case we write $\mathcal{M}(\mathbf{w}, k) = \{M \in \mathbb{N}^m \mid \mathbf{w}^T M \leq k\}$ to denote that $\mathcal{M}$ is expressed by the GMEC $(\mathbf{w}, k)$ with $\mathbf{w} \in \mathbb{Z}^m, k \in \mathbb{Z}$. Problems of this kind have been considered by several authors and this special structure of the legal set has the advantage that the supervisor for this class of problems takes the form of a place, called *monitor*, which has arcs going to and coming from some transitions of the plant net. The plant and the controller are both described by a net in order to have a useful linear algebraic model for control analysis and synthesis. Moreover the synthesis is not computationally demanding since it involves only a matrix multiplication. See [22] for a recent survey.

The use of Petri nets for the control of discrete event systems is still an active research area and we believe that it will continue to remain central during the next decade. In fact, there exist many interesting Petri net analysis tools — as an example, the partial order based techniques, such as *unfolding* — whose applicability to control is still largely unexplored.

8.2   Deadlock

Deadlock is a major issue to be addressed when designing a supervisory controller. A Petri net is said to be *deadlocked* if no transition is enabled. Clearly, this is an undesirable condition in quite all real applications. This is the reason why this problem has been largely investigated in the literature, particularly within the area of flexible manufacturing systems.

Deadlock problems may be seen from two different prospectives: deadlock *prevention* refers to static policies — usually coded in the net structure — for eliminating deadlocks, whereas deadlock *avoidance* refers to dynamic policies applied on-line.

The first significant contribution in this area dates back to 1990 and is due to Viswanadham *et. al* [39]. Here the authors used the reachability graph of the Petri net model to arrive at static resource allocation policies. For deadlock avoidance, they proposed an on-line monitoring and control system.

Many other significant contributions on deadlock prevention are based on a linear algebraic characterization of deadlock in *ordinary*[3] net. In fact, it is well know that a necessary and sufficient condition for an ordinary net to be deadlocked is the following:

---

[3] A net is ordinary if all the arc weight are unitary.

the set of empty places of the net forms a *siphon*[4] and each transitions has at least one empty input place. An interesting deadlock prevention procedure has been proposed by Iordache, Moody, and Antsaklis in [21]: the approach consists in adding to the Petri net that models the plant a number of additional places that prevent reaching empty siphons, thus ensuring deadlock freeness. Other significant contributions on deadlock prevention based on a linear algebraic characterization of empty siphons are due to Ezpeleta *et al.* [11], to Chu and Xie [6], to Li and Zhou [24], and more recently to Reveliotis [35]. In particular, in [11] the authors consider a particular class of Petri nets and proved that for such a class deadlock prevention also ensures liveness. Finally, in [35] Reveliotis develop a general theory that provides a unifying framework for all the relevant existing results based on siphon analysis, and reveals the key structures and mechanisms that connect the resource allocation systems (RAS) non-liveness to the concept of deadly marked and empty siphon.

The most important contributions in the development of deadlock avoidance strategies are due to Park and Reveliotis. In [28] the authors shown that a significant class of deadlock avoidance policies, known as algebraic polynomial kernel–deadlock avoidance policies, originally developed in the finite-state automata paradigm, can be analyzed using results from Petri net structural analysis. Other interesting results in this framework have been given by Park and Reveliotis in [29, 30].

Despite the above important contributions, deadlock prevention and deadlock avoidance are still open research areas because in the case of very large scale problems the computational complexity of most of the existing approaches may be prohibitive and, in the case of deadlock prevention, the number of places that should be added is too large. Moreover, using the above approaches it is not always possible to deal with the case of uncontrollable and unobservable transitions.

8.3   Observability

If the marking of a Petri net system is not measurable, different information can be used to reconstruct it, or at least to estimate it.

Benasser in [1] has studied the possibility of defining the set of markings reached firing a "partially specified" set of transitions using logical formulas, without having to enumerate this set.

Meda *et al.* in [25] have discussed the problem of estimating the marking of a Petri net using a mix of transition firings and place observations. Finally, Zhang and Holloway [40] used a Controlled Petri Net model for forbidden state avoidance under partial *event* observation assuming that the initial marking is known.

We have also worked in this area and studied the following cases.

- The initial marking of the net is not known (or only a partial information of it is available) but all events are observable.
- The initial marking of the net is known but the events occurrence is observed though a labeling function, i.e., a mask, that makes some events *undistinguishable* or *silent*.

---

[4] A *siphon* is a set of places $S \subseteq P$ such that $\bigcup_{p \in S} {}^{\bullet}p \subseteq \bigcup_{p \in S} p^{\bullet}$, i.e., all transitions outputting to one place of the set are also inputting from one place of the set.

While the first case can be studied using *unlabeled* Petri nets, the second case requires *labeled* models. In particular, undistinguishable events are modeled with transitions that share the same label; silent events are modeled with transitions whose label is equal to the empty string.

In all cases the goal is that of characterizing the set of markings in which the system may be, given the actual observation, and the information, if any, on the initial marking and on the structure of the net. We denote this set as $\mathcal{C}(w)$, where w is the word of observed events, and we call it the set of *markings consistent with w*.

Some important contributions in this topic have been given in [7, 14, 18] where it has been proved that in all the above cases the set $\mathcal{C}(w)$ may be characterized by a *finite* set of *linear algebraic* constraints whose structure keeps the same regardless of the length of the observed word w, and depends on some parameters that can be easily computed using appropriate recursive algorithms. The main advantage of such an approach is that it does not require the enumeration of the set of consistent markings.

In [15, 16] it has also been shown how such characterizations can be efficiently used when the observer is included in a control loop, and when designing a diagnoser for fault detection.

8.4   Identification

The first partial but interesting approach to identification is due Hiraishi [19] on the synthesis of safe Petri nets. Bourdeaud'huy and Yim [2] have presented an approach based on logic constraints that can deal with positive examples of firing sequences but not with counterexamples.

A different approach is based on the *theory of regions* whose objective is that of deciding whether a given graph is isomorphic to the reachability graph of some free labeled net and then constructing it. An excellent survey of this approach, that also presents some efficient algorithms for net synthesis based on linear algebra, can be found in the paper by Badouel and Darondeau [8].

Recently, in a series of paper [3, 4, 17] we have presented a general approach to identification based on integer programming. In particular, in [17] the problem of identifying a Petri net system, given a finite language that it generates, has been considered. Note that this approach allows one to specify not only examples of the systems behavior (i.e., strings that belong to the language) but also counterexamples (i.e., strings that do not belong to the language). It has been shown that the identification problem can be solved via an integer programming problem, and additional structural constraints can also be easily imposed to the net. The above results have been extended in [3] to the case of *labeled* Petri nets.

In [4] the following identification problem has been dealt with: given an automaton that represents the coverability graph of a net, determine a net system whose coverability graph is isomorph to the automaton. Again the proposed approach requires solving an integer programming problem whose set of unknowns contains the elements of the pre and post incidence matrices and the initial marking of the net.

Finally, in a recent paper Sreenivas [38] dealt with a related topic: the minimization of Petri net models. Given a λ-free labeled Petri net generator and a measure function — that associates to it, say, a non negative integer — the objective is that of finding a

Petri net that generates the same language of the original net while minimizing the given measure. Unfortunately, these minimization procedures only exist for restricted families of Petri net languages where language-containment is decidable, and for a restricted class of measures.

9 Conclusions

In this paper we considered Petri nets, an efficient and powerful formalism for the simulation, analysis, and control of discrete event systems. The purpose of this paper is that of presenting the main features of such a model to the automatic control community, whose interest in the area of discrete event systems has been constantly increasing during the last years. In particular, we focus on the basic Petri net model, namely on Place/Transition nets, and discuss via some simple but intuitive examples, its modeling power, and its main advantages with respect to other formalisms, such as automata. A discussion on the main current research area is also presented, together with a survey of the classical results in this framework.

References

1. Benasser A (2000) Reachability in Petri nets: an approach based on constraint programming, Ph.D. Thesis, Université de Lille, France (in French)
2. Bourdeaud'huy T, Yim P (2004) Synthse de rseaux de Petri partir d'exigences, Actes de la 5me conf. francophone de Modlisation et Simulation, (Nantes, France), 413–420, (in French)
3. Cabasino M P , Giua A, Seatzu C (2006) Identification of deterministic Petri nets, Proceedings of 8th International Workshop on Discrete Event Systems, Ann Arbor, Michigan, USA
4. Cabasino M P, Giua A, Seatzu C (2006) Identification of unbounded Petri nets from their coverability graph, Proceedings of 45th IEEE Conference on Decision and Control
5. Cassandras C G, Lafortune S (1999) Introduction to Discrete Event Systems. Kluwer Academic Publishers
6. Chu F, Xie X (1997) Deadlock analysis of Petri nets using siphons and mathematical programming, IEEE Transactions on Robotics and Automation, 13: 793-804
7. Corona D, Giua A, Seatzu C (2004) Marking estimation of Petri nets with silent transitions, Proceedings of 43rd IEEE Conference on Decision and Control
8. Badouel E, Darondeau P (1998) Theory of regions. Lecture Notes in Computer Science: Lectures on Petri Nets I: Basic Models, Springer-Verlag, (eds. Reisig, W. and Rozenberg, G.), 1491: 529–586
9. David R, Alla H (2005) Discrete, continous and hybrid Petri nets. Springer Verlag, Heidelberg
10. DiCesare F, Harhalakis G, Proth J M, Silva M, Vernadat F B (1993) Practice of Petri Nets in Manufacturing. Chapman and Hall
11. Ezpeleta J, Colom J M, Martinez J (1995) A PN based deadlock prevention policy for flexible manufacturing systems, IEEE Transactions on Robotics and Automation, 11: 173–184
12. Gaubert S, Giua A (1999) Petri net languages and infinite subsets of $\mathbb{N}^m$, Journal of Computer and System Sciences, 59: 373-391
13. Giua A, DiCesare F, Silva M (1992) Generalized mutual exclusion constraints on nets with uncontrollable transitions, Prooceedings of IEEE International Conference on Systems, Man and Cybernetics

14. Giua A, Seatzu C (2002) Observability of Place/Transition nets, IEEE Transactions on Automatic Control, 47: 1424-1437
15. Giua A, Seatzu C, Basile F (2004) Observer based state-feedback control of timed Petri nets with deadlock recovery, IEEE Transactions on Automatic Control, 49: 17-29
16. Giua A, Seatzu C (2005) Fault detection for discrete event systems using labeled Petri nets," Proocedings of the 44th IEEE Conference on Decision and Control and European Control Conference
17. Giua A, Seatzu C (2005) Identification of free-labeled Petri nets via integer programming, Proocedings of the 44th IEEE Conference on Decision and Control and European Control Conference
18. Giua A, Corona D, Seatzu C (2005) State estimation of λ-free labeled Petri nets with contact-free nondeterministic transitions, Journal of Discrete Event Dynamic Systems, 15:85-108
19. Hiraishi K (1992) Construction of a class of safe Petri nets by presenting firing sequences. Lecture Notes in Computer Science; 13th International Conference on Application and Theory of Petri Nets 1992, Sheffield, UK, Springer-Verlag, (ed. K. Jensen), 616: 244–262
20. Holloway L E, Krogh B H, Giua A (1997) A survey of Petri nets methods for controlled discrete event systems, Discrete Event Dynamic Systems, 7:151–190
21. Iordache M V , Moody J O , Antsaklis P J (2002) Synthesis of deadlock prevention supervisors using Petri nets, IEEE Transactions on Robotics and Automation, 18: 59–68
22. Iordache M V , Antsaklis P J (2006) Supervisory Control of Concurrent Systems. A Petri Net Structural Approach. Series: Systems and Control: Foundations and Applications. Birkhuser
23. Lewis H R, Papadimitriou C H (1981) Elements of the Theory of Computation. Prentice-Hall
24. Li Z, Zhou M (2004) Elementary siphons of Petri nets and their application to deadlock prevention in flexible manufacturing systems, IEEE Transactions on Systems, Man, Cybernetics, Part A, 34: 38–51
25. Meda M E , Ramírez A, Malo A (1998) Identification in discrete event systems, Proceedings of 1998 IEEE International Conference on Systems, Man and Cybernetics, 740–745
26. Murata T (1977) State equation, controllability, and maximal matching of Petri nets, IEEE Transactions on Automatic Control, 22:412–416
27. Murata T (1989) Petri nets: properties, analysis and applications, Proceedings IEEE 77: 541–580
28. Park J, Reveliotis S A (2000) Algebraic synthesis of efficient deadlock avoidance policies for sequential resource allocation systems, IEEE Transactions on Automatic Control, 16: 190–195
29. Park J, Reveliotis S A (2002) Policy mixtures: a novel approach for enhancing the operational flexibility of resource allocation systems with alternate routings, IEEE Transactions on Robotics and Automation, 18:616–620
30. Park J, Reveliotis S A (2002) Liveness-enforcing supervision for resource allocation systems with uncontrollable behavior and forbidden states, IEEE Transactions on Robotics and Automation, 18: 234–238
31. Parigot M, Peltz E (1985) A logical formalism for the study of the finite behavior of Petri nets, in Advances in Petri Nets 1985, Lecture Notes in Computer Science 222, G. Rozenberg (ed.), Springer Verlag, 346–361
32. Peterson J L (1981) Petri Net Theory and the Modeling of Systems. Prentice-Hall
33. Ramadge P J, Wonham W M (1989) The control of discrete event systems, Proceedings IEEE,77: 81–98
34. Recalde L (1998) Structural methods for the design and analysis of concurrent systems modeled with Place/Transition nets, PhD Thesis, DIIS. Univ. Zaragoza
35. Reveliotis S A (2005) Siphon-based characterization of liveness and liveness-enforcing supervision for sequential resource allocation systems, In Deadlock Resolution in Computer-Integrated Systems, M. Zhou and M. P. Fanti, (Eds), Marcel Dekker, Inc., 283–307

36. Silva M, Colom J M, Campos J (1992) Linear algebraic techniques for the analysis of Petri nets, Proceedings of International Symposium on Mathematical Theory of Networks and Systems
37. Sreenivas R S , Krogh B H (1992) On Petri net models of infinite state supervisors, IEEE Transactions on Automatic Control, 37: 274–277
38. Sreenivas R S (2002) On minimal representations of Petri net languages, 6th Workshop on Discrete Event Dynamic Systems, Zaragoza, Spain, 237–242
39. Viswanadham N, Narahari Y, Johnson T L (1990) Deadlock prevention and deadlock avoidance in flexible manufacturing systems using Petri net models, IEEE Transactions on Robotics and Automation, 6: 713–723
40. Zhang L, Holloway L E (1995) Forbidden state avoidance, Proceedings of 33rd Allerton Conference, Monticello, Illinois, 146–155

Wireless Sensing with Power Constraints

Orhan C. Imer[1] and Tamer Başar[2]

[1] General Electric Global Research Center, 1 Research Circle, Niskayuna, NY
`imer@research.ge.com`
[2] University of Illinois at Urbana-Champaign, 1308 W. Main Street, Urbana, IL
`tbasar@control.csl.uiuc.edu`

Summary. We introduce two conceptual models for wireless sensing and control with power-limited sensors and controllers. The limited battery power of the wireless device is captured in the models by imposing hard constraints on either the number of available transmissions the device can make, or on the number of cycles it can stay awake. Such hard constraints can be viewed as a measurement budget, under which estimation or control policies will have to be developed over a given decision horizon. Among the two representative models studied here, the first one is one of optimal scheduling of a finite measurement budget for a Gauss-Markov process over an observation horizon. The second one is an optimal estimation problem where the number of transmissions the wireless sensor can make is limited to a number, M, which is less than the observation horizon, N. It is shown that both problems can be solved by employing dynamic-programming type arguments, and their solutions have a threshold characterization.

Keywords: Wireless Sensing and Control, Optimal Scheduling, Power-Limited Estimation, Dynamic Programming, Threshold Policies.

1 Introduction

Recent advances in wireless technology and standards, such as ZigBee and IEEE 802.15.4, have made wireless sensing solutions feasible for industrial applications [1, 2, 3, 4]. Most of these applications use battery-powered integrated wireless sensing/communication devices, also called motes, for data logging and monitoring purposes [2]. Often times, data collected from sensors is relayed back to a central processing unit where it is analyzed for trends. In most monitoring applications, the data is collected on a near real-time basis. Early adapters of this wireless technology in industry have been combating several design and performance challenges for a reliably operating system. First and foremost is the issue of data reliability which is intricately linked to the reliability of the communication channel. Interference from other RF sources, such as IEEE 802.11b/g devices or microwave ovens, and multipath effects can severely degrade the performance of the wireless monitoring system. A careful study of all these effects is essential [4]. Another very important issue that needs to be addressed in designing these systems is the power-limited nature of the wireless devices, which is the focus of this paper. In most industrial applications a battery lifespan in the

C. Bonivento et al. (Eds.): Adv. in Control Theory and Applications, LNCIS 353, pp. 129–160, 2007.
springerlink.com

order of several years is required for feasible commercial operation [1]. This requirement imposes severe restrictions on the duration of time the wireless device can be on/awake and the number of transmissions it can make. This is because the radio frequency (RF) communication consumes a significant portion of the battery power when the wireless unit is awake. Therefore, life of the wireless device can be lengthened by optimizing the duty cycle (or reporting frequency) of the unit as well as by transmitting data only when it is necessary.

In this paper, we introduce two conceptual models for wireless sensing with power-limited sensors. The focus is on wireless systems where the sensor can only make a *limited* number of transmissions [5, 6, 7]. The models we consider here are idealized for ease of presentation and mathematical tractability. However, the basic thinking behind these models can easily be adopted to some real-world applications. When doing so, one needs to consider several other requirements imposed on the system, such as communication requirements to keep connectivity and time synchronization, which we ignore in this paper.

In both conceptual models considered, we start with a mathematical description of the process that is under observation. In most applications a model for the process is available, or can be developed from historic data using some regression analysis. In this paper, the process model is assumed to be discrete-time and Markovian [8]. The limited battery power of the wireless device is modeled by imposing a hard constraint on either the number of available transmissions it can make, or on the number of cycles it can stay awake [5, 6, 7]. We think of this hard constraint as a measurement budget, and determine as to how to best spend this budget by scheduling the measurements over a decision horizon.

The rest of the paper is organized as follows. In Section 2, we introduce the problem of optimal scheduling of a finite measurement budget over an observation horizon. Section 3 discusses the optimal estimation problem where the number of transmissions the wireless sensor can make is limited to a number M, which less than the observation horizon, $N > M$. The paper ends with the concluding remarks of Section 4.

2 Optimal Measurement Scheduling with Limited Measurements

In this section, we introduce the problem of estimating a process with limited measurement resources. Under different performance criteria, we show how to best spend a finite measurement budget by scheduling the measurement times over a time horizon.

2.1 Problem Definition

Let $\{X_n, n \geq 0\}$ be a Markov process, where $X_0 = x_0$ is known *a priori*. We would like to measure X_n over a measurement horizon of length N, i.e., $1 \leq n \leq N$, but measurements are expensive. We are given a measurement budget which allows us to make $M < N$ observations of the process. We assume that there is no measurement noise, i.e. when we decide to measure the process we can do it with infinite precision[1].

Let $\{\hat{X}_n, n \geq 0\}$ be the sequence of estimates of the process $\{X_n\}$. Since X_0 is known *a priori*, we have $\hat{X}_0 = x_0$, and $\hat{X}_n = X_n$ for $n \in \mathbf{M}$, where $\mathbf{M} \subset [1, N]$ denotes the set of times a measurement is made. The estimates at other times $n \notin \mathbf{M}$ are determined through an optimization process whereby some estimation error criteria is minimized.

In this section, we will consider two types of estimation criteria. The first one is the standard mean-square error criterion where the performance index is the average cumulative estimation error over the decision horizon $[1, N]$:

$$\sum_{n=1}^{N} E\{(X_n - \hat{X}_n)^2\}$$

where the expectation is taken over the statistics of $\{X_n\}$.

We will also consider a "threshold-error" criterion, which is defined as follows[2]. If $\{X_n\}$ is a discrete-state process taking values on $\mathbf{Z}$, we let T_L be the *stopping time*

$$T_L := \inf\{k \geq 1 : X_k \in \mathbf{L}\}$$

where $\mathbf{L} \subset \mathbf{Z}^3$.

The objective of the threshold-error estimation criterion is to have an accurate estimate of the process as it crosses into the threshold set $\mathbf{L}$. Thus, we would like to pick the measurement instances such that the following probability is maximized:

$$P[X_{T_L} = \hat{X}_{T_L} | X_0 = x_0]$$

The continuous-time counterpart of this error criterion is defined when $\{X_n\}$ is a continuous-state process taking values on $\mathbf{R}$. In this case, we define the stopping time

$$N_\tau := \inf\{k \geq 1 : |X_k| \geq \tau\}$$

and determine the time instances we should observe X_n so that the estimation error

$$E\{(X_{N_\tau} - \hat{X}_{N_\tau})^2 | X_0 = x_0\}$$

is minimized.

Finally, we would like to draw attention to the difference between *open-loop* and *closed-loop* measurement system designs. In an open-loop design, the measurement times are determined *a priori* before any value of the process is observed (except $X_0 = x_0$). Since there is no penalty in waiting to decide what time the next measurement should be taken, in a closed-loop design, we wait until a measurement is made to decide on the next measurement time. The advantage of closed-loop design is that it is more robust to process noise, and it can lead

---

[1] Results of this paper can be extended to the case when there is measurement noise.
[2] The threshold-error criterion is defined over an infinite-time horizon, i.e., $N \to \infty$.
[3] We assume that the Markov process $\{X_n\}$ is irreducible.

to lower values for the performance metrics. Note that the available information about the process increases only when it is observed. Hence, the decision as to when to observe the process next can be made with maximum information at the end of the current observation period. In the context of wireless sensing, this corresponds to the sensor deciding on when to wake-up next before it goes into the sleep-mode to conserve energy.

More precisely, we assume that the information I_n available at time n to decide on the estimate $\hat{X}_n$ is limited to the observed process values up until time n. So, if no measurement is made between times n and $m > n$, there is no additional information gained, i.e., $I_m = I_n$. Thus, if we measure the process say at time n_k and again at time n_{k+1}, at time n_k we can decide on the sequence of estimates $\hat{X}_n$ between the times n_k and $n_{k+1} - 1$. In the case of threshold-error estimation criterion, an additional information structure may be considered where the event $\{X_k \notin \mathbf{L}\}^4$ at time k is observable (measurable) by the decision maker. In this case, at a measurement instance we cannot decide on the sequence of estimates until the next measurement instance, since as the process evolves in time our information about it increases. The time of the next measurement cannot be determined at the time of the current measurement for the same reason. This type of information structure may be well-suited for certain applications, but in the case of wireless sensing, there is no opportunity for the wireless sensors to make intermediate observations about the process between the measurements. Hence, the former information structure, where the event $\{X_k \notin \mathbf{L}\}$ is *not* observable, is more applicable in this case.

In what follows we describe and provide solutions to three class of measurement design problems representative of the more general problems in their respective classes.

2.2 Measurement Schedule Optimization Problems

Problem I

Consider the Gauss-Markov process defined by

$$X_{n+1} = AX_n + W_n, \ n = 0, 1, \ldots \tag{1}$$

where $X_n, A, W_n \in \mathbf{R}$, $X_0 = x_0$ is known, and $\{W_n\}$ is an i.i.d. Gaussian sequence with zero mean and variance σ_w^2. The first problem we consider is estimating the process $\{X_n\}$ over a decision horizon of length N with $M < N$ measurement opportunities. The objective is to minimize the mean-square error

$$e = \sum_{n=1}^{N} E\{(X_n - \hat{X}_n)^2\}$$

As we will see next, this is one of the problems where due to symmetry, the open-loop measurement design will coincide with the closed-loop one. To see

---

^{4} Or $\{|X_k| < \tau\}$ for a continuous-state process.

this, we first note that if n is a time of measurement, then the estimation error component

$$e_n = E\{(X_n - \hat{X}_n)^2]\}$$

is zero, as $\hat{X}_n = X_n$ for $n \in \mathbf{M}$.

The optimal estimator that minimizes the estimation error e for those times when no measurement is made is given by the conditional expectation

$$\hat{X}_n = E\{X_n | I_n\}, \ n \notin \mathbf{M} \tag{2}$$

where I_n denotes the information available at time n. Since information is obtained only at measurement times, I_n will include the measured process values X_n up to time n. However, due to the Markov nature of the process, knowing the most recent measurement prior to time n is sufficient in determining the conditional expectation in (2). Therefore, we have

$$\hat{X}_n = A^{n-m_n} x_{m_n}, \ n \notin \mathbf{M}$$

where m_n denotes the time of the last measurement prior to time n. With this estimator structure, the estimation-error component e_n becomes

$$e_n = \sum_{k=1}^{n-m_n} A^{2(k-1)} \sigma_w^2, \ n \notin \mathbf{M}$$

Note that e_n is *not* a function of the absolute time n, but only of the difference $n - m_n$, i.e., the time since the last measurement. Also, e_n does not depend on any of the measurements made up until time n. Since $e_n = e_{n-m_n}$ increases with the difference $n - m_n$, to minimize the error we must keep this difference as small as possible for all $n \notin \mathbf{M}$. Thus, minimizing the mean-square error e is equivalent to minimizing

$$\mathcal{M} = \sum_{n \notin \mathbf{M}} n - m_n$$

over the M-element subsets of $[1, N]$, i.e., $\mathbf{M}$.

Now, it can be seen that the minimum value of $\mathcal{M}$ is attained when $\mathbf{M}^*$ is the set with its M elements evenly distributed over the measurement interval $[1, N]$. Note that the solution may not be unique, as there may be more than one way to achieve a uniform distribution of M measurement times over the interval $[1, N]$. For example, when $N = 4$ and $M = 2$, $\mathbf{M}^* = \{1, 3\} = \{2, 3\} = \{2, 4\}$ all achieve the minimum $\mathcal{M}^* = 2$.

In summary, the optimal measurement schedule for the Gauss-Markov process (1) can be determined *offline*[5], and is given by a uniform distribution of the measurement opportunities over the measurement horizon.

---

[5] Therefore, the open-loop and closed-loop schedules are identical.

Problem II

Let $\{X_n, n \geq 0\}$ be a simple walk[6] on integers defined by

$$X_{n+1} = \begin{cases} X_n + 1, \text{ w.p. } p \\ X_n, \quad\quad \text{ w.p. } 1-p \end{cases}$$

where $p \in (0, 1)$ is the probability of an up-move, and $X_0 = x_0$ is given.
Let T_L be the stopping time

$$T_L := \inf\{k \geq 1 : X_k = L\}$$

where $L > x_0$ is a given integer.

The objective is to detect the process as it crosses the threshold L. Therefore, we want to maximize the probability

$$\mathcal{P} = P[X_{T_L} = \hat{X}_{T_L}|X_0 = x_0\}$$

If we were given an infinite number of observation opportunities, we could make this probability as close to 1 as possible by continuously observing the process. However, measurements are expensive, and therefore we are only allowed to make M of them. In this paper, we only consider the case when $M = 1$, but the results can be extended to an arbitrary $M > 1$. We also assume that the information available at time k to decide on the estimate $\hat{X}_k$ is limited to the observed process values up until time k, i.e., $\{X_k \neq L\}$ is not measurable at time k.

Let $m \geq 1$ be the time of the measurement. The probability $\mathcal{P}$ can be written as

$$\mathcal{P} = P_{x_0}[X_{T_L} = \hat{X}_{T_L}] = \sum_{k=L-x_0}^{\infty} P_{x_0}[X_k = \hat{X}_k]P_{x_0}[T_L = k]$$

We would like to minimize this expression over $m \geq 1$ and $\{\hat{X}_n, n \geq 1\}$. Note that for a given $m \geq 1$, the estimate of X_n that maximizes $\mathcal{P}$ is its maximum likelihood (ML) estimate. For $n < m$, $X_n - x_0$ is Binomial with (n, p), for $n = m$, $X_n = x_m$, and for $n > m$, $X_n - x_m$ is Binomial with $(n - m, p)$. Since the maximum likelihood estimate of a Binomial random variable with parameters (n, p) is given by $\lfloor (n+1)p \rfloor$, we have

$$\hat{X}_n = \begin{cases} x_m + \lfloor (n - m + 1)p \rfloor, n \geq m \\ x_0 + \lfloor (n + 1)p \rfloor, \quad\quad n < m \end{cases}$$

Now, the probabilities $P_{x_0}[X_k = \hat{X}_k]$ can be calculated as follows: for $k < m$

$$P_{x_0}[X_k = \hat{X}_k] = \binom{k}{\lfloor (k+1)p \rfloor} p^{\lfloor (k+1)p \rfloor}(1 - p)^{k - \lfloor (k+1)p \rfloor}$$

---

[6] One may also consider the symmetric random walk version of this problem.

for $k = m$

$$P_{x_0}[X_k = \hat{X}_k] = 1$$

and for $k > m$:

$$P_{x_0}[X_k = \hat{X}_k] = \binom{k-m}{\lfloor(k-m+1)p\rfloor} p^{\lfloor(k-m+1)p\rfloor} (1-p)^{k-m-\lfloor(k-m+1)p\rfloor}$$

We next calculate $P[T_L = k]$:

$$\begin{aligned}
P_{x_0}[T_L = k] &= P_{x_0}[X_{k-1} = L-1, X_k = L] \\
&= P_{x_0}[X_k = L | X_{k-1} = L-1] P_{x_0}[X_{k-1} = L-1] \\
&= p P_{x_0}[X_{k-1} = L-1] \\
&= \binom{k-1}{L-1-x_0} p^{L-x_0} (1-p)^{k-L+x_0}
\end{aligned}$$

Now, m can be found by solving the optimization problem

$$\max_{m \geq 1} \sum_{k=L-x_0}^{\infty} P_{x_0}[X_k = \hat{X}_k] \binom{k-1}{L-1-x_0} p^{L-x_0} (1-p)^{k-L+x_0}$$

The solution to this optimization problem depends on the difference $L - x_0$, and a numerical solution can be obtained using Matlab.

In Figure 1, we plot the successful estimation probability $\mathcal{P}$ at time T_L as a function of the measurement time, $m \geq 1$. The threshold is set at $L = 10$, the

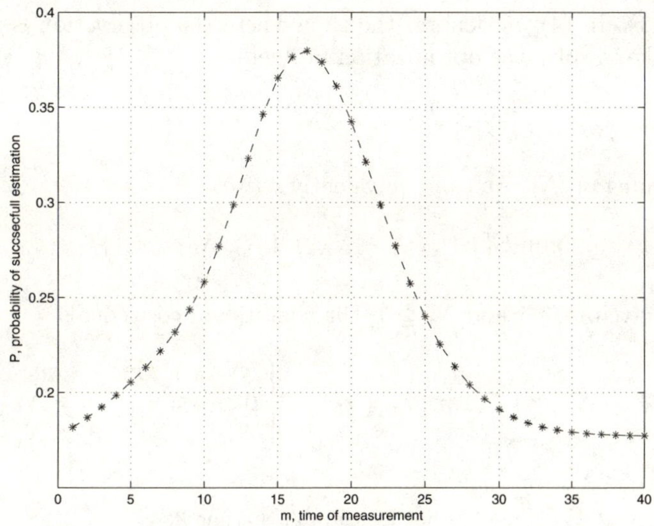

Fig. 1. m vs. $\mathcal{P}$. $p = 0.5, L = 10$.

walk starts at $x_0 = 0$, and the probability of an up-move is $p = 0.5$. Note that, the probability is maximized when $m^* = 17$. Therefore, if we make a measurement at time $m = 17$, we have approximately 38% chance of being able to catch the process crossing the threshold $L = 10$.

Problem III

Problem III is the continuous-time counterpart of the Problem II. Let X_n be the Gauss-Markov process defined by

$$X_{n+1} = X_n + W_n, \ n = 0, 1, \ldots$$

where $X_0 = x_0$ is known, and $\{W_n\}$ is an i.i.d. Gaussian sequence with zero mean and variance σ_w^2.

Let N_τ be the stopping time

$$N_\tau := \inf\{k \geq 1 : |X_k| \geq \tau\}$$

where $\tau > |x_0|$ is a given threshold. We would like to estimate X_n but again the observations are costly. Say we are allowed to observe the process only once. What time instance should we observe X_n so that the estimation error

$$e(\tau, x_0) = E\{(X_{N_\tau} - \hat{X}_{N_\tau})^2 | X_0 = x_0\} \tag{3}$$

is minimized. In (3), $\hat{X}_n$ denotes the estimate of X_n at time n, and for $n \geq 1$ it is given by[7]

$$\hat{X}_n = \begin{cases} \hat{X}_{n-1}, & m \neq k \\ X_n, & m = k \end{cases} \tag{4}$$

with $\hat{X}_0 = x_0$. In (4), m denotes the time where the observation is made, and we would like to solve the optimization problem:

$$\min_{m \geq 1} E\{(X_{N_\tau} - \hat{X}_{N_\tau})^2 \mid X_0 = x_0\}$$

Conditioning on N_τ, we can equivalently write

$$\min_{m \geq 1} E\{E\{(X_{N_\tau} - \hat{X}_{N_\tau})^2 \mid N_\tau, X_0 = x_0\}\}$$

Now, for a given $m \geq 1$ and $N_\tau \geq 1$, the conditional cost equals

$$E\{(X_{N_\tau} - \hat{X}_{N_\tau})^2 \mid N_\tau, m, X_0 = x_0\} = \begin{cases} (N_\tau - m)\sigma_w^2, & 1 \leq m < N_\tau \\ 0, & m = N_\tau \\ N_\tau \sigma_w^2, & m > N_\tau \end{cases}$$

---

[7] Since the event $\{|X_k| < \tau\}$ is not measurable at time k.

Hence, the average cost for $m \geq 2$ can be written as

$$e_m(\tau, x_0) = \sum_{k=1}^{m-1} k P[N_\tau = k | X_0 = x_0] \sigma_w^2$$

$$+ \sum_{k=m+1}^{\infty} (k - m) P[N_\tau = k | X_0 = x_0] \sigma_w^2$$

and for $m = 1$

$$e_1(\tau, x_0) = \sum_{k=m+1}^{\infty} (k - m) P[N_\tau = k | X_0 = x_0] \sigma_w^2$$

Note that for $m \geq 2$, we have

$$e_m(\tau, x_0) = E\{N_\tau | X_0 = x_0\} \sigma_w^2 - m \sum_{k=m}^{\infty} P[N_\tau = k | X_0 = x_0] \sigma_w^2$$

$$= E\{N_\tau | X_0 = x_0\} \sigma_w^2 - m \left(1 - \sum_{k=1}^{m-1} P[N_\tau = k | X_0 = x_0]\right) \sigma_w^2$$

and for $m = 1$

$$e_1(\tau, x_0) = E\{N_\tau | X_0 = x_0\} \sigma_w^2 - \sigma_w^2$$

Since $\{X_k\}$ is a Markov chain, we write $P[N_\tau = k | X_0 = x_0]$ as

$$P[N_\tau = k | X_0 = x_0] = P_{x_0}[|X_k| \geq \tau, |X_{[1,k]}| < \tau]$$

$$= P_{x_0}[|X_k| \geq \tau, |X_1| < \tau, \ldots, |X_{k-1}| < \tau]$$

$$= P_{x_0}[|X_k| \geq \tau | |X_{k-1}| < \tau]$$

$$\cdots P_{x_0}[|X_2| < \tau | |X_1| < \tau] P_{x_0}[|X_1| < \tau]$$

Let $p_k(\tau, x_0)$ denote the conditional probability

$$p_k(\tau, x_0) = P[|X + W| < \tau | |X| < \tau]$$

where $W \sim N(0, \sigma_w^2)$, and $X \sim N(x_0, k\sigma_w^2)$, $k \geq 1$, and (X, W) are independent. By definition

$$p_k(\tau, x_0) = \frac{P[|X + W| < \tau, |X| < \tau]}{P[|X| < \tau]}$$

Now,

$$P[|X| < \tau] = \Phi\left(\frac{\tau - x_0}{\sqrt{k}\sigma_w}\right) - \Phi\left(\frac{-\tau - x_0}{\sqrt{k}\sigma_w}\right)$$

where $\Phi(\cdot)$ is the CDF of the standard Gaussian random variable. Also, note that

$$P[|X + W| < \tau, |X| < \tau]$$

$$= \int_{-\tau}^{\tau} \frac{1}{\sqrt{2\pi k \sigma_w^2}} e^{-\frac{(x - x_0)^2}{2k\sigma_w^2}} \left[\Phi\left(\frac{\tau - x}{\sigma_w}\right) - \Phi\left(\frac{-\tau - x}{\sigma_w}\right)\right] dx$$

Thus,

$$p_k(\tau, x_0) = \frac{\int_{-\tau}^{\tau} \frac{1}{\sqrt{2\pi k \sigma_w^2}} e^{-\frac{(x-x_0)^2}{2k\sigma_w^2}} \left[\Phi\left(\frac{\tau-x}{\sigma_w}\right) - \Phi\left(\frac{-\tau-x}{\sigma_w}\right) \right] dx}{\Phi\left(\frac{\tau-x_0}{\sqrt{k}\sigma_w}\right) - \Phi\left(\frac{-\tau-x_0}{\sqrt{k}\sigma_w}\right)}$$

Let $p_0(\tau, x_0)$ denote

$$p_0(\tau, x_0) = P[|X_1| < \tau | X_0 = x_0] = \Phi\left(\frac{\tau - x_0}{\sigma_w}\right) - \Phi\left(\frac{-\tau - x_0}{\sigma_w}\right)$$

Using $p_k(\tau, x_0)$'s, we write the probability distribution of N_τ as

$$P[N_\tau = k | X_0 = x_0] = \begin{cases} 1 - p_0(\tau, x_0), & k = 1 \\ (1 - p_{k-1}(\tau, x_0)) \prod_{n=0}^{k-2} p_n(\tau, x_0), & k \geq 2 \end{cases}$$

Substituting the expression for $P[N_\tau = k | X_0 = x_0]$ into the error expression $e_m(\tau, x_0)$ for $m \geq 3$ yields

$$e_m(\tau, x_0) = E\{N_\tau | X_0 = x_0\}\sigma_w^2$$
$$-m \left(1 - \sum_{k=1}^{m-1}(1 - p_{k-1}(\tau, x_0)) \prod_{n=0}^{k-2} p_n(\tau, x_0)\right) \sigma_w^2$$

and for $m = 1, 2$, we have

$$e_1(\tau, x_0) = E\{N_\tau | X_0 = x_0\}\sigma_w^2 - \sigma_w^2$$
$$e_2(\tau, x_0) = E\{N_\tau | X_0 = x_0\}\sigma_w^2 - 2p_0(\tau, x_0)\sigma_w^2$$

We next look at the normalized difference

$$\delta_m(\tau, x_0) := \frac{e_{m+1}(\tau, x_0) - e_m(\tau, x_0)}{\sigma_w^2}$$

For $m \geq 2$, calculating this difference yields

$$\delta_m(\tau, x_0) = mP[N_\tau = m | X_0 = x_0] + \sum_{k=1}^{m} P[N_\tau = k | X_0 = x_0] - 1$$

and for $m = 1$, we have

$$\delta_1(\tau) = 1 - 2p_0(\tau, x_0)$$

Note that, by telescoping the last term, $\delta_m(\tau, x_0), m \geq 2$ can be written as

$$\delta_m(\tau, x_0) = m(1 - p_{m-1}(\tau, x_0)) \prod_{n=0}^{m-2} p_n(\tau, x_0) - 1$$
$$+ \sum_{k=1}^{m}(1 - p_{k-1}(\tau, x_0)) \prod_{n=0}^{k-2} p_n(\tau, x_0)$$

$$= m(1 - p_{m-1}(\tau, x_0)) \prod_{n=0}^{m-2} p_n(\tau, x_0) - 1 + 1 - \prod_{n=0}^{m-1} p_n(\tau, x_0)$$

$$= m \prod_{n=0}^{m-2} p_n(\tau, x_0) - (m+1) \prod_{n=0}^{m-1} p_n(\tau, x_0)$$

The error sequence $\{e_m(\tau, x_0)\}$ for $m \geq 1$ is decreasing if and only if

$$\delta_m(\tau, x_0) < 0$$

Therefore, the estimation error is minimum for $m^*(\tau, x_0)$ such that

$$m^*(\tau, x_0) = \inf\{m \geq 1 : \delta_m(\tau, x_0) > 0\}$$

The comparison $\delta_m(\tau, x_0) > 0$ is equivalent to, for $m = 1$

$$\delta_1(\tau, x_0) > 0 \Leftrightarrow 1 - 2p_0(\tau, x_0) > 0 \Leftrightarrow p_0(\tau, x_0) < \frac{1}{2}$$

and for $m \geq 2$

$$\delta_m(\tau, x_0) > 0 \Leftrightarrow m \prod_{n=0}^{m-2} p_n(\tau, x_0) - (m+1) \prod_{n=0}^{m-1} p_n(\tau, x_0) > 0$$

$$\Leftrightarrow p_{m-1}(\tau, x_0) < \frac{m}{m+1}$$

Hence, to minimize the estimation error we pick $m^*(\tau, x_0)$ such that

$$m^*(\tau, x_0) = \inf \left\{ m \geq 1 : p_{m-1}(\tau, x_0) < \frac{m}{m+1} \right\}$$

Note that a solution always exists, since $p_m(\tau, x_0) \downarrow$ as $m \uparrow$, and $\frac{m}{m+1} \rightarrow$ 1, as $m \rightarrow \infty$. This feature of the solution is illustrated in Figure 2, where both $p_{m-1}(\tau, x_0)$ (red-square) and the function $\frac{m}{m+1}$ (blue-diamond) are plotted against m. In Figure 2, the threshold is set as $\tau = 3$, $x_0 = 0$, and $\sigma_w^2 = 1$. For these parameters, the optimal measurement time is given by $m^* = 9$.

3   Optimal Estimation with Limited Measurements

In this section, we turn our attention into a sequential estimation problem with two decision makers who work as members of a team [6]. One of the decision makers is the wireless sensor and it makes sequential measurements about the state of an underlying stochastic process for a fixed period of time. Note that this is different than the setup considered in Section 2 where the wireless sensor schedules its measurements across time each time before it goes to sleep. The sensor (or observer) upon measuring the process makes a decision as to whether to transmit some information about the process to the estimator. The estimator

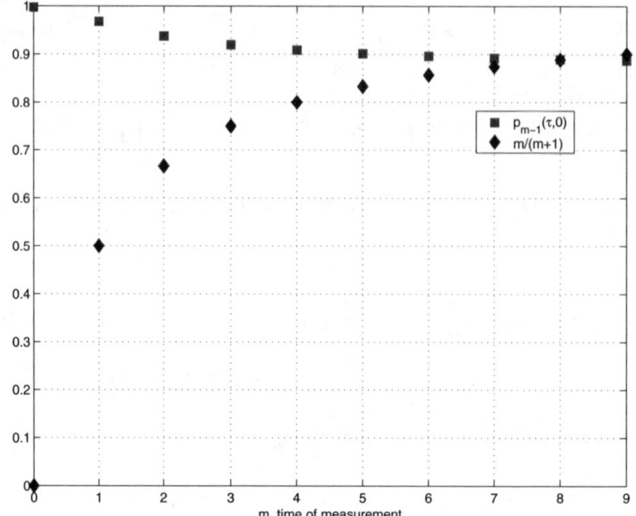

Fig. 2. m vs. $p_{m-1}(\tau, x_0)$ (red-square) and $\frac{m}{m+1}$ (blue-diamond). $x_0 = 0, \tau = 3$.

sequentially estimates the state of the process. The objective is to minimize a performance criterion with the constraint that the sensor may only transmit a limited number of measurements.

More specifically, we consider estimating a stochastic process over a decision horizon of length N using only $M \leq N$ measurements. Both the measurement and estimation of the process is carried out sequentially by two different decision makers called the *observer* and the *estimator*[8], respectively. Over the decision horizon of length N, the observer agent has exactly M opportunities to disclose some information about the process to the estimator. These information disclosures, or transmissions, are assumed to be error and noise free, and the problem is to jointly determine the best observation and estimation policies that minimize the average estimation error between the process and its estimate.

3.1   Problem Statement

Problem Definition

The problem of optimal estimation with limited measurements can be treated in the more general framework of a communication system with limited channel uses. For this purpose, consider the generic communication system whose block diagram is given in Figure 3 [9]. The source outputs some data b_k for $0 \leq k \leq N - 1$, that needs to be communicated to the user over a channel. The data b_k are generated according to some *a priori* known stochastic process, $\{b_k\}$, which

______

[8] As we show next, in a communication-theoretic setting we may call them an *encoder* and a *decoder*, respectively.

may be i.i.d., or correlated as in a Markov process. An encoder (or an observer) and a decoder (or an estimator) is placed after the source output and the channel output, respectively, to communicate the data to the user efficiently. In the most general case, the encoder/observer may have access to a noise-corrupted version of the source output:

$$z_k = b_k + v_k, \ 0 \le k \le N - 1$$

where $\{v_k\}$ is an independent[9] noise process.

The main constraint is that the encoder/observer can access the channel only a *limited*, $M \le N$, number of times. The goal is to design an observer-estimator pair[10], $(\mathcal{O}, \mathcal{E})$, that will "causally" (or sequentially) observe/encode the data measurements, z_k, and estimate/decode the channel output, y_k, so as to minimize the *average distortion* or *error* between the observed data, b_k, and estimated data, $\hat{b}_k$.

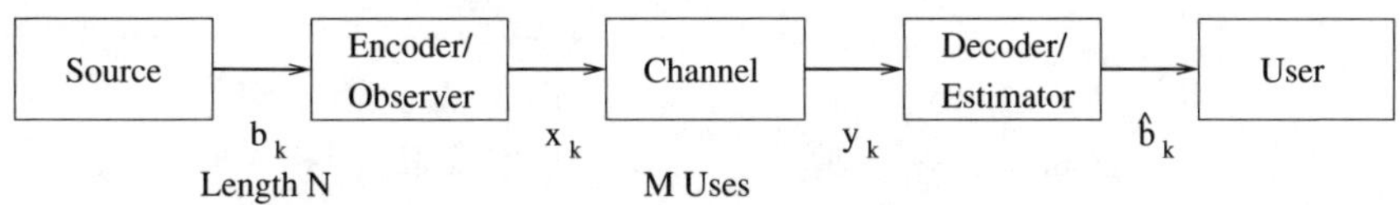

Fig. 3. Communication with limited channel use

The channel is assumed to be memoryless, and is completely characterized by the conditional probability distribution $P_c(y|x)$ on $y \in \mathcal{Y}$ for each $x \in \mathcal{X}$, where $\mathcal{X}$ and $\mathcal{Y}$ are the set of allowable channel inputs, and the set of possible channel outputs, respectively.

The average distortion $D_{(M,N)}$ depends on the distortion measure and may vary depending on the underlying application. Some examples are the average mean-square error

$$D_{(M,N)} = E\left\{ \frac{1}{N} \sum_{k=0}^{N-1} (b_k - \hat{b}_k)^2 \right\} \tag{5}$$

or the Hamming (probability of error) distortion measure

$$D_{(M,N)} = E\left\{ \frac{1}{N} \sum_{k=0}^{N-1} \mathcal{I}_{b_k \ne \hat{b}_k} \right\} \tag{6}$$

where $\mathcal{I}_S$ denotes the indicator function of the set S.

From a communication-theoretic standpoint, with the channel, source, and the distortion measure defined, we can formally state our main problem: Given a source and a memoryless channel, for a given decision-horizon N, and number of

---

[9] Independent across time and from the source output process b_k.

[10] Or depending on the application, an encoder-decoder pair $(\mathcal{E}, \mathcal{D})$.

channel uses M, what is the minimum attainable value of the average distortion $D_{(M,N)}$? This minimization is carried out over the choice of possible encoder-decoder (observer-estimator) pairs which are *causal.*

In this paper, we present a solution to this problem when the source process is i.i.d. with a continuous or discrete probability density function, and the encoder/observer has access to the noiseless or a noisy version of the source output. We assume that the channel is noiseless, and hence, it is completely characterized by the probability distribution $P_c(y|x) = \delta(y - x)$. We also present the solution to the case when the source process is Gauss-Markov.

Note that, in wireless sensing applications the desired length of time the wireless device will be in operation can be related to the decision horizon N in some appropriate time unit, and the size of the battery installed in the sensor can be related to the possible number of transmissions or channel uses M (see Figure 4).

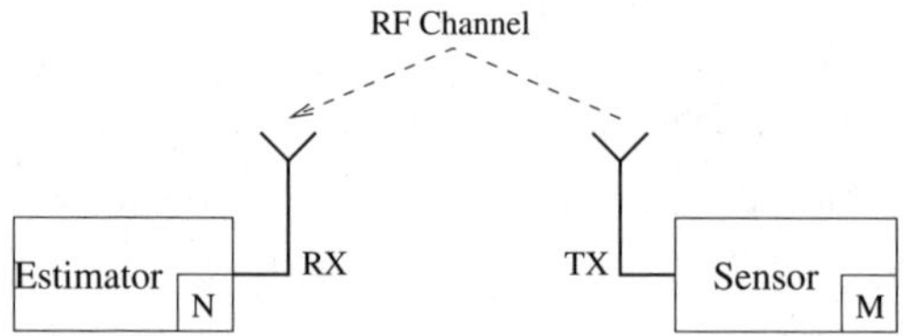

Fig. 4. Optimal transmission scheduling with limited channel access

Hence, given an underlying performance criterion $D_{(M,N)}$, the problem is to design the best transmission schedule, and estimation policies for the wireless device and the remote monitoring station, respectively.

3.2   Estimating an i.i.d. Random Sequence with Limited Measurements

Problem Definition

Consider the special case of the general problem defined in Section 3.1, where the source outputs a zero-mean[11] i.i.d. random sequence b_k, $0 \leq k \leq N - 1$. Let $\mathcal{B}$ denote the range of the random variable b_k. We assume that b_k's have a finite second moment, $\sigma_b^2 < \infty$, but their probability distribution remains unspecified for now. At time k, the encoder/observer makes a sequential measurement of b_k, and determines whether to access the channel for transmission, which it can only do a limited, $M \leq N$, number of times. The channel is noiseless and thus has a capacity to transmit the source output error-free when it is used to transmit. Note that, even when it decides not to use the channel for transmission, the observer/encoder may still convey a 1-bit information to the estimator/decoder. In view of this, the channel input x_k belongs to the set $\mathcal{X} := \mathcal{B} \cup \{\text{NT}\}$, where NT stands for "no transmission."

---

[11] This is not restrictive, as the known mean can be subtracted out by the estimator.

More precisely, we let s_k denote the number of channel uses (or transmissions) left at time k. Now if $s_k \geq 1$, we have $y_k = x_k$ for $x_k \in \mathcal{B} \cup \{\mathrm{NT}\}$. If $s_k = 0$, on the other hand, the channel is useless, since we have exhausted the allocated number of channel uses. Note that, when the channel is noiseless, both the encoder and the decoder can keep track of s_k by initializing $s_0 = M$ and decrementing it by 1 every time a transmission decision is taken.

We want to design an estimator/decoder

$$\hat{b}_k = \hat{\mu}_k(I_k^d) \text{ for } 0 \leq k \leq N - 1$$

based on the available information I_k^d at time k. Clearly, the information available to the estimator is controlled by the observer. The average distortion between the observed and estimated processes can be taken to be the average mean square error as given by (5), or the probability of error distortion measure which is given by (6).

The information I_k^d available to the estimator at time k is a result of an outcome of decisions taken by the observer up until time k. Let the observer's decision at time k be

$$x_k = \mu_k(I_k^e)$$

where I_k^e is the information available to the observer at time k. Assuming perfect recall, we have

$$I_0^e = \{(s_0, t_0); b_0\}$$
$$I_k^e = \{(s_k, t_k); b_0^k; x_0^{k-1}\}, \ 1 \leq k \leq N - 1$$

where t_k denotes the number of time, or decision slots left at time k. We have

$$t_{k+1} = t_k - 1, \ 0 \leq k \leq N - 2$$

with $t_0 = N$.

The range of $\mu_k(\cdot)$ is the space $\mathcal{X} = \mathcal{B} \cup \{\mathrm{NT}\}$. Let σ_k denote the decision whether the observer has decided to transmit or not. Assume $s_k \geq 1$, and let $\sigma_k = 1$ if a transmission takes place; i.e., $x_k \in \mathcal{B}$, and $\sigma_k = 0$ if no transmission takes place. We have

$$s_{k+1} = s_k - \sigma_k, \ 0 \leq k \leq N - 2$$

with $s_0 = M$.

The observer's decision at time k is a function of its k past measurements, and $k - 1$ past decisions, i.e.,

$$\mu_k(I_k^e) : \mathcal{B}^k \times \mathcal{X}^{k-1} \to \mathcal{X}, \ 0 \leq k \leq N - 1$$

Now, the information I_k^d available to the estimator at time k can be written as

$$I_k^d = \{(s_k, t_k); y_0^k\}, \ 0 \leq k \leq N - 1$$

By definition, the channel output y_k satisfies $y_k = x_k$ if $s_k \geq 1$, and $y_k \in \emptyset$ (i.e., no information) if $s_k = 0$.

Consider the class of observer-estimator (encoder-decoder) policies consisting of a sequence of functions

$$\Pi = \{\mu_0, \hat{\mu}_0, \ldots, \mu_{N-1}, \hat{\mu}_{N-1}\}$$

where each function μ_k maps I_k^e into $\mathcal{X}$, and $\hat{\mu}_k$ maps I_k^d into $\mathcal{B}$[12], with the additional restriction that μ_k can map to $\mathcal{B}$ at most M times. Such policies are called *admissible*.

We want to find an admissible policy $\pi^* \in \Pi$ that minimizes the average N-stage distortion, or estimation error:

$$e_{(M,N)}^{\pi} = E\left\{ \sum_{k=0}^{N-1} (b_k - \hat{\mu}_k(I_k^d))^2 \right\} \tag{7}$$

or for source processes, b_k, with discrete probability densities:

$$e_{(M,N)}^{\pi} = E\left\{ \sum_{k=0}^{N-1} \mathcal{I}_{b_k \neq \hat{\mu}_k(I_k^d)} \right\} \tag{8}$$

That is

$$e_{(M,N)}^{*} = \min_{\pi \in \Pi} e_{(M,N)}^{\pi}$$

Note that, we omitted the factor of $\frac{1}{N}$ from the average error expressions for convenience.

If $M \geq N$, this problem has the trivial solution where the observer writes the source output b_k directly into the channel at each time k (i.e., $\mu_k^*(b_k) = b_k$), and since the channel is noiseless, the estimator can use an identity mapping (i.e., $\hat{\mu}_k^*(I_k^d) = b_k$), resulting in zero distortion. Therefore, we only consider the case when $M < N$.

Before closing our account on this section, we would like to note the nonclassical nature of the information in this problem. Clearly, the observer's action affects the information available to the estimator, and there is no way in which the estimator can infer the information available to the observer. Also note the order of actions between the decision makers in the problem. At time k, first the random variable b_k becomes available, then the observer acts by transmitting some data or not, and finally, the estimator acts by estimating the state with $\hat{\mu}_k$, the cost is incurred, and we move to the next time $k+1$.

Structure of the Solution

We first consider the problem of finding the optimal estimator $\hat{\mu}_k^*$ at time k. Note that the estimator $\hat{\mu}_k$ appears only in a single term in the error expressions (7)-(8). Thus, for the mean-square error criterion, the optimal estimator is simply the solution of the quadratic minimization problem

---

[12] Note that we do not distinguish between the source and user sets.

$$\min_{\hat{\mu}_k(I_k^d)} E\left\{(b_k - \hat{\mu}_k(I_k^d))^2 | I_k^d\right\}$$

which is given by the conditional expectation of b_k given the available information at time k:

$$\hat{\mu}_k^*(I_k^d) = E\{b_k | I_k^d\} = E\{b_k | (s_k, t_k); y_0^k\} \tag{9}$$

Similarly, for the probability of error distortion criterion, the optimal estimator is the solution of the minimization problem

$$\min_{\hat{\mu}_k(I_k^d)} E\left\{\mathcal{I}_{b_k \neq \hat{\mu}_k(I_k^d)} | I_k^d\right\}$$

If at time k the channel can still be used ($s_k \geq 1$), the solution to this problem is given by the maximum *a posteriori* probability (MAP) estimate of the random variable b_k given the available information at time k:

$$\hat{\mu}_k^*(I_k^d) = \arg \max_{m_i \in \mathcal{B}_k(I_k^d)} \delta(y_k - i)p_i = \arg \max_{m_i \in \mathcal{B}_k((s_k, t_k); y_0^k)} p_i \tag{10}$$

where $\mathcal{B}_k(I_k^d) \subset \mathcal{B}$ is some subset of the range of the random variable b_k, which we assume is countable. Let m_i denote the values the random variable b_k takes. Then, p_i's denote the probability mass function of the random variable b_k, i.e., $p_i = P[b_k = m_i]$.

Note that, for the probability of error distortion criterion, if the channel is useless at time k (i.e., $s_k = 0$), the best estimate of b_k is simply given by

$$\hat{\mu}_k^*(I_k^d) = \arg \max_{m_i \in \mathcal{B}} p_i \tag{11}$$

since the past channel outputs, y_0^{k-1}, are independent of b_k.

Similarly, for the mean-square error criterion, the channel output y_k has no information on b_k if $s_k = 0$. Thus, in this case, the conditional expectation in (9) equals

$$\hat{\mu}_k^*(I_k^d) = E\{b_k | (0, t_k); y_0^{k-1}, y_k\} = E\{b_k\} = 0 \tag{12}$$

since again the past channel outputs, y_0^{k-1}, are generated by the σ-algebra of random variables b_0^{k-1}, and hence are independent from b_k.

If $s_k \geq 1$, the channel output $y_k = x_k$, but since $y_0^{k-1} = x_0^{k-1}$ is the outcome of a Borel-measurable function defined on the σ-algebra generated by b_0^{k-1}, the conditional expectation in (9) is equivalent to

$$\hat{\mu}_k^*(I_k^d) = E\{b_k | (s_k, t_k); x_k\} \tag{13}$$

By a similar argument we can write (10) as

$$\hat{\mu}_k^*(I_k^d) = \arg \max_{m_i \in \mathcal{B}_k((s_k, t_k); x_k)} p_i \tag{14}$$

Now, substituting the optimal estimators (13)-(14) back into the estimation error expressions (7)-(8) yields

$$e^{\pi}_{(M,N)} = E\left\{\sum_{k=0}^{N-1}(b_k - E\{b_k|(s_k,t_k);x_k\})^2\right\} \tag{15}$$

and

$$e^{\pi}_{(M,N)} = E\left\{\sum_{k=0}^{N-1}\mathcal{I}_{b_k \neq arg\max_{m_i \in \mathcal{B}_k((s_k,t_k);x_k)}p_i}\right\} \tag{16}$$

which we seek to minimize over the observer/encoder policies $\mu_k(I^e_k), 0 \leq k \leq N-1$. Since $x_k = \mu_k(I^e_k)$, we see that the choice of an observer policy affects the cost only through the information made available to the estimator.

In general, the observer's decision μ_k at time k depends on (s_k, t_k), all past measurements b_0^{k-1}, the present measurement b_k, and its past actions x_0^{k-1}. However, as we show next, there is nothing the observer can gain by having access to its past measurements b_0^{k-1} and its past actions x_0^{k-1} as far as the optimization of the criteria (15)-(16) are concerned. Thus, a *sufficient statistics* for the observer are the current measurement b_k and the remaining number of channel uses (transmission opportunities) and decision instances, i.e. (s_k, t_k).

Proposition 1. *The set $S^e_k = \{(s_k, t_k); b_k\}$ constitutes sufficient statistics $S^e_k(I^e_k)$ for the optimal policy μ^*_k of the observer. In other words,*

$$\mu^*_k(I^e_k) = \bar{\mu}(S^e_k(I^e_k))$$

for some function $\bar{\mu}$.

Proof. Suppose we would like to determine the optimal observer policy $\mu^*_k(I^e_k)$ at time k, where $0 \leq k \leq N-1$ is arbitrary. Due to the sequential nature of the decision problem, any observer policy we decide on at time k will only affect the error e_k incurred after time k, i.e.[13],

$$e_k = E\left\{\sum_{n=k}^{N-1}(b_n - E\{b_n|(s_n,t_n);x_n\})^2\right\}$$

Taking the conditional expectation given the available information I^e_k, under any observer policy $\mu_k(I^e_k)$ we have

$$E\{e_k|(s_k,t_k);b_0^k;x_0^{k-1}\} = E\{e_k|(s_k,t_k);b_k\}$$

because b_k^{N-1} is independent of b_0^{k-1}, and x_0^{k-1} is the outcome of a Borel-measurable function defined on the σ-algebra generated by b_0^{k-1}. Hence, at time k, the knowledge of b_0^{k-1} and x_0^{k-1} is redundant.

A consequence of Proposition 1 is that the observer's decision to use the channel to transmit a source measurement or not is based purely on the current observation b_k and its past actions only through (s_k, t_k).

---

[13] Here, we give the proof only for the error criterion (15). An identical proof can be constructed for the probability of error distortion criterion (16).

Since μ_k depends explicitly only on the current source output b_k, the search for an optimal observer policy can be narrowed down to the class of policies of the form[14]

$$\mu_k(I_k^e) = \bar{\mu}((s_k, t_k); b_k) = \begin{cases} b_k & \text{if } b_k \in \mathcal{T}_{(s_k, t_k)} \\ \text{NT} & \text{if } b_k \in \mathcal{T}_{(s_k, t_k)}^c \end{cases} \tag{17}$$

where $\mathcal{T}_{(s_k, t_k)}$ is a measurable set on $\mathcal{B}$ and is a function of (s_k, t_k). The complement of the set $\mathcal{T}_{(s_k, t_k)}$ is taken with respect to $\mathcal{B}$, i.e., $\mathcal{T}_{(s_k, t_k)}^c = \mathcal{B} \backslash \mathcal{T}_{(s_k, t_k)}$. When probability of error distortion criterion is used, Proposition 1 implies that $\mathcal{B}_k((s_k, t_k); \text{NT}) = \mathcal{T}_{(s_k, t_k)}^c$, and $\mathcal{B}_k((s_k, t_k); m_i) = m_i$.

Note that the optimal estimators (13) and (14) have access to (s_k, t_k) as well. Thus, even when the observer chooses not to transmit b_k, it can still pass a 1-bit information about b_k to the estimator provided that $s_k \geq 1$. If k is such that all M transmissions are concluded prior to time k (i.e., $s_k = 0$), the estimators are given by (11)-(12), irrespective of b_k.

Now, observe that the optimization over the observer policies is equivalent to optimization over the sets $\mathcal{T}_{(s_k, t_k)}$ for all k such that

$$\max\{0, M - k\} \leq s_k \leq \min\{t_k, M\}$$

and $t_k = N - k$. The nonnegativity of s_k is a result of the limited channel use constraint. Note that if $s_{k_0} = 0$ for some k_0, then $s_k = 0$ for all k such that $k_0 \leq k \leq N - 1$. At the other extreme, we must have $s_k \leq N - k$, since if $s_k = N - k$, this means there are as many channel uses left as there are decision instances, and the optimal observer and estimator policies in this case are obvious.

The Solution with the Mean-Square Error Criterion

Let $(s_k, t_k) = (s, t)$, and $e_{(s,t)}^*$ denote the optimal value of the estimation error (or distortion) (15) when the decision horizon is of length t, and the observer is limited to s channel uses, where $s \leq t$. We know that at time k, the optimal observation policy will be of the form (17).

Now, at time $k + 1$, depending on the realization of the random variable b_k, the remaining $(t-1)$-stage estimation error is either $e_{(s-1, t-1)}^*$, or $e_{(s, t-1)}^*$. Thus, inductively by the DP equation [10], we can write[15]

$$e_{(s,t)}^* = \min_{\mathcal{T}_{(s,t)}} \left\{ e_{(s-1,t-1)}^* \int_{b \in \mathcal{T}_{(s,t)}} f(b)db + e_{(s,t-1)}^* \int_{b \in \mathcal{T}_{(s,t)}^c} f(b)db \right.$$

$$\left. + \int_{b \in \mathcal{T}_{(s,t)}^c} \left[b - E\{b | b \in \mathcal{T}_{(s,t)}^c\} \right]^2 f(b)db \right\}$$

---

[14] As long as k is such that all M measurements are not exhausted, i.e., $s_k \geq 1$.

[15] Assuming that the random variables $\{b_k\}$ are continuous with a well-defined probability density function (pdf) $f(b)$.

where $f(b)$ is the pdf of the random variables b_k. If b_k's are discrete random variables with a probability mass function (pmf), one has to replace the integrals in the above expression with sums. Expanding out the expectation yields

$$e^*_{(s,t)} = \min_{\mathcal{T}_{(s,t)}} \left\{ e^*_{(s-1,t-1)} \int_{b \in \mathcal{T}_{(s,t)}} f(b)db + e^*_{(s,t-1)} \int_{b \in \mathcal{T}^c_{(s,t)}} f(b)db \right.$$

$$\left. + \int_{b \in \mathcal{T}^c_{(s,t)}} \left[b - \frac{\int_{b \in \mathcal{T}^c_{(s,t)}} bf(b)db}{\int_{b \in \mathcal{T}^c_{(s,t)}} f(b)db} \right]^2 f(b)db \right\} \tag{18}$$

To solve for $e^*_{(s,t)}$, we first note the boundary conditions $e^*_{(t,t)} = 0$, and $e^*_{(0,t)} = t\sigma_b^2$, $\forall t \geq 0$, where σ_b^2 is the variance of b_k. The term $e^*_{(s,t)}$ remains undefined for $s > t$. The optimal sets satisfy the boundary conditions $\mathcal{T}^*_{(t,t)} = \mathcal{B}$, and $\mathcal{T}^*_{(0,t)} = \emptyset$, $\forall t \geq 0$. The recursion of (18) needs to be solved offline and the optimal sets $\mathcal{T}^*_{(s,t)}$ must be tabulated starting with smaller values of (s,t)[16]. The solution to the original problem can then be determined as follows:

Initialize $s_0 = M$, $t_0 = N$. For each k in $0 \leq k \leq N-1$ do the following:

1. Look up the optimal set $\mathcal{T}^*_{(s_k,t_k)}$ from the table that was determined offline.
2. Observe b_k, and apply the observation policy

$$\bar{\mu}^*((s_k, t_k); b_k) = \begin{cases} b_k & \text{if } b_k \in \mathcal{T}^*_{(s_k,t_k)} \\ \text{NT} & \text{if } b_k \in \mathcal{T}^{*c}_{(s_k,t_k)} \end{cases}$$

3. Apply the estimation policy

$$\hat{\mu}^*_k(\mathcal{T}^*_{(s_k,t_k)}) = E\{b_k | b_k \in \mathcal{T}^{*c}_{(s_k,t_k)}\} = \frac{\int_{b \in \mathcal{T}^{*c}_{(s_k,t_k)}} bf(b)db}{\int_{b \in \mathcal{T}^{*c}_{(s_k,t_k)}} f(b)db}$$

4. Update

$$s_{k+1} = s_k - \sigma_k, \quad t_{k+1} = t_k - 1$$

In tabulating $\mathcal{T}^*_{(s,t)}$ one should start with solving for $\mathcal{T}^*_{(1,2)}$, and the corresponding estimation error $e^*_{(1,2)}$. To determine the optimal set at (s,t), we need to know the optimal costs at $(s, t-1)$, and $(s-1, t-1)$. Hence, we can propagate our calculations as shown in Figure 5 starting with $(s,t) = (1,2)$.

Now, we come back to the problem of minimizing (18) over $\mathcal{T}_{(s,t)}$. Expanding out the expression inside the minimization we get

$$e^*_{(s,t)} = e^*_{(s-1,t-1)} + \min_{\mathcal{T}^c_{(s,t)}} \left\{ - \left(e^*_{(s-1,t-1)} - e^*_{(s,t-1)} \right) \int_{b \in \mathcal{T}^c_{(s,t)}} f(b)db \right.$$

$$\left. + \int_{b \in \mathcal{T}^c_{(s,t)}} b^2 f(b)db - \frac{\left[\int_{b \in \mathcal{T}^c_{(s,t)}} bf(b)db \right]^2}{\int_{b \in \mathcal{T}^c_{(s,t)}} f(b)db} \right\} \tag{19}$$

---

[16] Note that $(1,2)$ is the smallest possible nontrivial value.

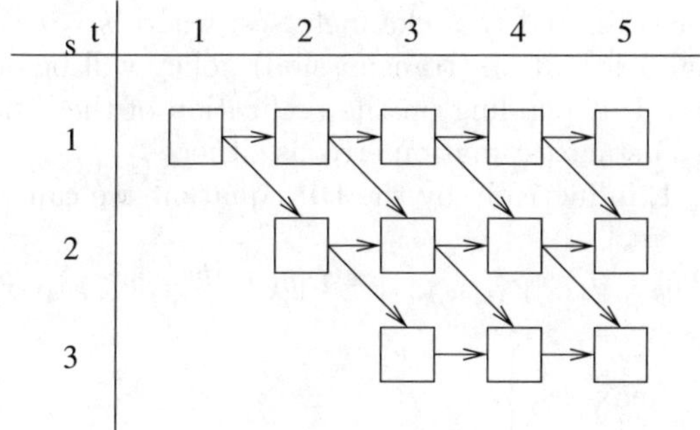

Fig. 5. Recursive calculation of $e^*_{(s,t)}$

where we used the fact that $\int_{b \in T_{(s,t)}} f(b)db = 1 - \int_{b \in T^c_{(s,t)}} f(b)db$.

This an optimization problem over measurable sets $T^c_{(s,t)}$ on the real line, and since these sets are not countable, there is no known method for carrying out this minimization in a systematic manner. Therefore, we restrict our search to the sets that are in the form of simple symmetric intervals, i.e., $T^c_{(s,t)} = [-\beta_{(s,t)}, \beta_{(s,t)}]$, where $0 \le \beta_{(s,t)} \le \infty$.

Now, because of symmetry, the last term on the right-hand side of (19) disappears from the minimization. Differentiating the remaining terms inside the curly brackets, we obtain the first-order necessary condition:

$$- \left(e^*_{(s-1,t-1)} - e^*_{(s,t-1)} \right) f(\beta_{(s,t)}) + \beta^2_{(s,t)} f(\beta_{(s,t)}) = 0$$

From which the critical point $\beta^*_{(s,t)}$ can be determined as[17]

$$\beta^*_{(s,t)} = \sqrt{e^*_{(s-1,t-1)} - e^*_{(s,t-1)}} \tag{20}$$

Note that, we always have $e^*_{(s,t-1)} \le e^*_{(s-1,t-1)}$, since for the same decision horizon, $t - 1$, the minimum average distortion achieved by s channel uses, is always less than that achieved by $s - 1$ channel uses. So, $\beta^*_{(s,t)}$ always exists.

From the first-order condition, we observe that the objective function is strictly decreasing on the interval $[0, \beta^*_{(s,t)})$, and it is strictly increasing on the interval $(\beta^*_{(s,t)}, \infty)$. Thus, $\beta^*_{(s,t)}$ must be a strict global minimizer. Thus, in the class of symmetric intervals, the best set $T^c_{(s,t)}$ is given by the interval

$$T^{*c}_{(s,t)} = [-\sqrt{e^*_{(s-1,t-1)} - e^*_{(s,t-1)}}, \sqrt{e^*_{(s-1,t-1)} - e^*_{(s,t-1)}}] \tag{21}$$

The Solution with the Probability of Error Criterion

As in Section 3.2, let $(s_k, t_k) = (s, t)$, and let $e^*_{(s,t)}$ denote the optimal value of the estimation error (or distortion) (16) when the decision horizon is of length

---

[17] The other critical point, namely $\beta_{(s,t)} = +\infty$, yields a larger cost.

t, and the observer is limited to s channel uses, where $s \leq t$. We know that at time k, the optimal observation (transmission) policy will be of the form (17).

Now, at time $k + 1$, depending on the realization of the random variable b_k, the remaining $(t-1)$-stage estimation error is either $e^*_{(s-1,t-1)}$, or $e^*_{(s,t-1)}$. Thus, assuming that $s \geq 1$, inductively by the DP equation, we can write

$$e^*_{(s,t)} = \min_{\mathcal{T}_{(s,t)}} \left\{ P[b_k \in \mathcal{T}_{(s,t)}]e^*_{(s-1,t-1)} + P[b_k \in \mathcal{T}^c_{(s,t)}]e^*_{(s,t-1)} + P[b_k \in \mathcal{T}^c_{(s,t)}] \right.$$
$$\left. - \max_{m_j \in \mathcal{T}^c_{(s,t)}} p_j \right\}$$

or equivalently

$$e^*_{(s,t)} = \min_{\mathcal{T}^c_{(s,t)}} \left\{ \left(1 - P[b_k \in \mathcal{T}^c_{(s,t)}]\right) e^*_{(s-1,t-1)} \right.$$
$$\left. + P[b_k \in \mathcal{T}^c_{(s,t)}]e^*_{(s,t-1)} + P[b_k \in \mathcal{T}^c_{(s,t)}] - \max_{m_j \in \mathcal{T}^c_{(s,t)}} p_j \right\}$$

Plugging in $P[b_k \in \mathcal{T}^c_{(s,t)}] = \sum_{m_i \in \mathcal{T}^c_{(s,t)}} p_i$, and rearranging the terms, we obtain the following error recursion:

$$e^*_{(s,t)} = e^*_{(s-1,t-1)} + \min_{\mathcal{T}^c_{(s,t)}} \left\{ -(e^*_{(s-1,t-1)} - e^*_{(s,t-1)}) \sum_{m_i \in \mathcal{T}^c_{(s,t)}} p_i + \sum_{m_i \in \mathcal{T}^c_{(s,t)}} p_i \right.$$
$$\left. - \max_{m_j \in \mathcal{T}^c_{(s,t)}} p_j \right\} \tag{22}$$

We next show that the error difference, $e^*_{(s-1,t-1)} - e^*_{(s,t-1)}$, can be bounded from below and above.

Proposition 2. *Suppose* $1 \leq s \leq t$. *Then, the error difference* $e^*_{(s-1,t-1)} - e^*_{(s,t-1)}$ *satisfies:*
$$0 \leq e^*_{(s-1,t-1)} - e^*_{(s,t-1)} \leq 1$$

Proof. The lower bound can be established by observing that for the same decision horizon, $t - 1$, the minimum average distortion achieved by s channel uses, is always at least as small as the one that can be achieved by $s - 1$ channel uses. For the upper bound, one needs to observe that the maximum stage-wise estimation error is bounded by 1.

Using Proposition 2, we will next show that the optimum choice for the sets $\mathcal{T}^c_{(s,t)}$ is the singleton $\mathcal{T}^{c*}_{(s,t)} = \{m_{i^*}\}$, where $i^* = \arg\max_{m_i \in \mathcal{B}} p_i$.

In other words, the optimal solution is not to transmit the *most likely* outcome, and transmit all the other outcomes of the source process b_k. Moreover, this policy is independent of the number of decision instances left, t_k, and the number of transmission opportunities left, s_k, provided that $s_k \geq 1$. Recall that, the optimum estimator is the MAP estimator and is given by (14).

In order to show that this is indeed the optimal observer (or transmission) policy, we first set the cardinality of the set $\mathcal{T}_{(s,t)}$ to $|\mathcal{T}^c_{(s,t)}| = 0$, and determine that the expression inside the curly brackets in (22) is just 0.

We next set $|\mathcal{T}^c_{(s,t)}| = 1$, and note that the minimization of the function inside the curly brackets in (22) is equivalent to the following minimization:

$$\min_i \left\{ (1 - p_i)e^*_{(s-1,t-1)} + p_i e^*_{(s,t-1)} + p_i - p_i \right\}$$

Here i is such that $m_i \in \mathcal{B}$. Canceling p_i's and rearranging, we obtain an equivalent minimization problem:

$$\min_i -p_i(e^*_{(s-1,t-1)} - e^*_{(s,t-1)})$$

By Proposition 2, the error difference, $e^*_{(s-1,t-1)} - e^*_{(s,t-1)}$, is nonnegative; thus, the minimum is achieved by picking i as

$$i^* = \arg \max_{m_i \in \mathcal{B}} p_i$$

This choice yields a minimum value of $-(e^*_{(s-1,t-1)} - e^*_{(s,t-1)})p_{i*}$. Note that this value is at least as good as the value we obtained when we set the cardinality of the set $|\mathcal{T}^c_{(s,t)}| = 0$. Thus, we never pick $\mathcal{T}^c_{(s,t)}$ such that it has zero cardinality.

Finally, we let $|\mathcal{T}^c_{(s,t)}| \geq 2$, and let $p_{\max}$ denote the element of $\mathcal{T}^c_{(s,t)}$ with the maximal probability. That is,

$$p_{\max} = \max_{m_j \in \mathcal{T}^c_{(s,t)}} p_j$$

Since the number of elements of $\mathcal{T}^c_{(s,t)}$ is at least 2, the minimization problem inside the curly brackets in (22) can be written as

$$\min_{\mathcal{T}^c_{(s,t)}} \left\{ -(e^*_{(s-1,t-1)} - e^*_{(s,t-1)})p_{\max} + (1 - (e^*_{(s-1,t-1)} - e^*_{(s,t-1)})) \sum_{m_i \in \mathcal{T}^c_{(s,t)} \setminus m_{j*}} p_i \right\}$$

where

$$m_{j*} = \arg \max_{m_i \in \mathcal{T}^c_{(s,t)}} p_i$$

Now, by Proposition 2, the term multiplying the sum $\sum_{m_i \in \mathcal{T}^c_{(s,t)} \setminus m_{j*}} p_i$ is always nonnegative; hence, we can conclude that the above minimum is bounded from below by

$$-(e^*_{(s-1,t-1)} - e^*_{(s,t-1)})p_{\max}$$

for any choice of the set $\mathcal{T}^c_{(s,t)}$ with cardinality $|\mathcal{T}^c_{(s,t)}| \geq 2$. However the expression $-(e^*_{(s-1,t-1)} - e^*_{(s,t-1)})p_{\max}$ satisfies

$$-(e^*_{(s-1,t-1)} - e^*_{(s,t-1)})p_{\max} \geq -(e^*_{(s-1,t-1)} - e^*_{(s,t-1)})p_{i^*}$$

since $p_{i^*} \geq p_{\max}$. Therefore, the minimum value of the function inside the curly brackets in (22) is achieved when $\mathcal{T}^c_{(s,t)} = \{m_{i^*}\}$, as claimed.

In summary, when the distortion criterion is the probability of error, at time k, the optimal observer first observes the source output b_k. Then, it checks to see if $s_k \geq 1$; if so, it transmits b_k unless $b_k = m_{i^*}$, i.e., the most likely outcome. The estimator (or decoder), on the other hand, employs the MAP estimation rule given the output of the channel.

Gaussian Case

Suppose b_k's are zero-mean, i.i.d. Gaussian. Let $\Phi(\cdot)$ denote the cumulative density function (CDF) of the standard Gaussian random variable with zero mean and unit variance. In the Gaussian case, we can generalize our search for an optimum in (19) to more general intervals of the form $\mathcal{T}^c_{(s,t)} = [\alpha_{(s,t)}, \beta_{(s,t)}]$, where $-\infty \leq \alpha_{(s,t)} \leq \beta_{(s,t)} + \infty$.

Figure 6 shows the plot of the objective function on the right-hand side of (19) for the case when $\mathcal{T}_{(s,t)} = [a, b]$, $\sigma_b^2 = 1$, $e^*_{(s-1,t-1)} = 3$, and $e^*_{(s,t-1)} = 1$. Note that the minimum occurs at $b^* = -a^* = \sqrt{3-1} = \sqrt{2} = 1.4142$. Thus, even though we did not restrict ourselves to symmetric intervals, the solution is still a symmetric interval around zero. To show that this is indeed the case in general, one needs to differentiate the objective function inside the curly brackets in (19) with respect to both $\alpha_{(s,t)}$ and $\beta_{(s,t)}$, and show that the minimum occurs at $\beta^*_{(s,t)} = -\alpha^*_{(s,t)}$ when $f(b)$ is the Gaussian pdf [5].

To evaluate the optimum estimation error $e^*_{(s,t)}$ in terms of $e^*_{(s-1,t-1)}$ and $e^*_{(s,t-1)}$, we substitute the optimum interval solution (21) into the right-hand side of (19), and use the standard properties of the Gaussian density that we listed above to obtain

$$e^*_{(s,t)} = e^*_{(s-1,t-1)} - \left[e^*_{(s-1,t-1)} - e^*_{(s,t-1)} - \sigma_b^2\right]$$

$$\times \left[2\Phi\left(\sqrt{\frac{e^*_{(s-1,t-1)} - e^*_{(s,t-1)}}{\sigma_b^2}}\right) - 1\right]$$

$$-\frac{2\sigma_b^2}{\sqrt{2\pi\sigma_b^2}}\sqrt{e^*_{(s-1,t-1)} - e^*_{(s,t-1)}}\,e^{-\frac{e^*_{(s-1,t-1)} - e^*_{(s,t-1)}}{2\sigma_b^2}} \qquad (23)$$

We can normalize the optimal estimation error by letting

$$\epsilon_{(s,t)} = \frac{e^*_{(s,t)}}{\sigma_b^2} \qquad (24)$$

and rewrite the recursion (23) in a simpler form:

$$\epsilon_{(s,t)} = \epsilon_{(s-1,t-1)} - \left[\epsilon_{(s-1,t-1)} - \epsilon_{(s,t-1)} - 1\right]\left[2\Phi\left(\sqrt{\epsilon_{(s-1,t-1)} - \epsilon_{(s,t-1)}}\right) - 1\right]$$

$$-\frac{2}{\sqrt{2\pi}}\sqrt{\epsilon_{(s-1,t-1)} - \epsilon_{(s,t-1)}}\,e^{-\frac{\epsilon_{(s-1,t-1)} - \epsilon_{(s,t-1)}}{2}} \qquad (25)$$

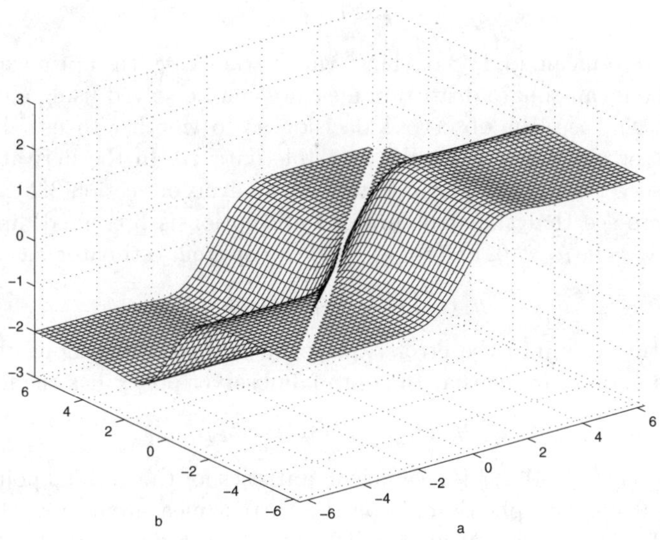

Fig. 6. Plot of the objective function in the Gaussian case with $\mathcal{T}_{(s,t)} = [a, b]$ when $\sigma_b^2 = 1$, $e^*_{(s-1,t-1)} = 3$, and $e^*_{(s,t-1)} = 1$

with the initial conditions

$$\epsilon(t,t) = 0, \quad \epsilon(0,t) = t, \quad \forall t \geq 0$$

and $\epsilon(s,t)$ is undefined for $s > t$.

Hence, we can provide a solution to the problem of optimal sequential estimation of an i.i.d. Gaussian process of finite length over a noiseless channel that can only be used a limited number of times. First, a table has to be formed by an offline numerical computation of the recursion (25). Then, the table can be scaled, if needed, to the actual variance of the process through (24). Next, the transmission intervals for the observer are determined via (20), and tabulated for all feasible pairs (s, t). For the online computation, as illustrated in Section 3.2, the observer has to keep two states, (s_k, t_k). Each time unit k, after observing the realization of the random variable b_k, the observer compares the realized value of the random variable to the optimum decision interval corresponding to the current state (s_k, t_k), and makes a transmission decision. The estimator, on the other hand, has access to the same tabulated values of the transmission intervals, $\mathcal{T}^*_{(s,t)}$, and it keeps track of the states (s_k, t_k) in the same way the observer does. Upon receiving the transmitted data, y_k, from the channel, the estimator simply applies the estimation policy given in Section 3.2.

Gaussian Case with Noisy Measurements

Let the source process b_k be i.i.d. Gaussian. If the observer has access to a noisy version of the source output, i.e.,

$$z_k = b_k + v_k$$

where v_k is zero-mean, i.i.d. Gaussian[18] with variance σ_v^2, the optimization problem with the mean-square distortion measure can be solved using a similar approach. In this case, the observer's decision as to whether to use the channel to transmit or not depends on the available data z_k. In the derivation of the optimal observer-estimator pair, most of the analysis of Section 3.2 carries over.

In order to see that the structure of the solution is preserved, first observe that when $s_k = 0$, $\hat{\mu}_k^* = 0$, and for $s_k \geq 1$, the optimal estimator has the form:

$$\hat{\mu}_k^*(I_k^d) = E\{b_k|(s_k, t_k); x_k\}$$

Substituting this into the error expression, and following along the lines of Proposition 1, one can see that the optimal observer policy has the form

$$\mu_k(I_k^e) = \bar{\mu}((s_k, t_k); z_k)$$

In other words $\{(s_k, t_k); z_k\}$ is a sufficient statistics for the optimal policy $\mu_k^*(I_k^e)$.

Since μ_k depends explicitly only on the current measurement z_k, for $s_k \geq 1$, the search for an optimal encoder policy can be narrowed down to the class of policies of the form

$$\mu_k(I_k^e) = \bar{\mu}((s_k, t_k); z_k) = \begin{cases} z_k & \text{if } z_k \in \mathcal{T}_{(s_k, t_k)} \\ \text{NT} & \text{if } z_k \in \mathcal{T}_{(s_k, t_k)}^c \end{cases}$$

where $\mathcal{T}_{(s_k, t_k)}$ is a measurable set on $\mathcal{B}$, and is a function of (s_k, t_k). Since $x_k = \mu_k(I_k^e)$, for $s_k \geq 1$, we can write the optimal estimator as

$$\hat{\mu}_k((s_k, t_k); x_k) = \begin{cases} \frac{\sigma_b^2}{\sigma_b^2 + \sigma_v^2} z_k & \text{if } z_k \in \mathcal{T}_{(s_k, t_k)} \\ E\left\{ b_k | z_k \in \mathcal{T}_{(s_k, t_k)}^c \right\} & \text{if } z_k \in \mathcal{T}_{(s_k, t_k)}^c \end{cases}$$

We proceed as in Section 3.2, and write the dynamic programming recursion governing the evolution of the optimal estimation error as follows:

$$e_{(s,t)}^* = \min_{\mathcal{T}_{(s,t)}^c} \left\{ e_{(s-1,t-1)}^* P[z \in \mathcal{T}_{(s,t)}] + \sigma_b^2 + \left(\frac{\sigma_b^2}{\sigma_b^2 + \sigma_v^2}\right)^2 \int_{z \in \mathcal{T}_{(s,t)}} z^2 f_Z(z) dz \right.$$

$$\left. + e_{(s,t-1)}^* P[z \in \mathcal{T}_{(s,t)}^c] - 2\frac{\sigma_b^2}{\sigma_b^2 + \sigma_v^2} \int_{z \in \mathcal{T}_{(s,t)}} z E[b|z] f_Z(z) dz \right\}$$

where $f_Z(z) \sim N(0, \sigma_b^2 + \sigma_v^2)$, and $f_{B|Z}(b|z) \sim N(\frac{\sigma_b^2}{\sigma_b^2 + \sigma_v^2} z, \frac{\sigma_b^2 \sigma_v^2}{\sigma_b^2 + \sigma_v^2})$. The recursion can be simplified as

$$e_{(s,t)}^* = \min_{\mathcal{T}_{(s,t)}^c} \left\{ e_{(s-1,t-1)}^* + \sigma_b^2 - (e_{(s-1,t-1)}^* - e_{(s,t-1)}^*) \int_{z \in \mathcal{T}_{(s,t)}^c} f_Z(z) dz \right.$$

$$\left. - \left(\frac{\sigma_b^2}{\sigma_b^2 + \sigma_v^2}\right)^2 (\sigma_b^2 + \sigma_v^2) + \left(\frac{\sigma_b^2}{\sigma_b^2 + \sigma_v^2}\right)^2 \int_{z \in \mathcal{T}_{(s,t)}^c} z^2 f_Z(z) dz \right\}$$

---

[18] We also assume that the processes $\{b_k\}$ and $\{v_k\}$ are independent.

Following along the lines of Section 3.2, we restrict our search for an optimum set to simple intervals, i.e., $\mathcal{T}^c_{(s,t)} = [\alpha_{(s,t)}, \beta_{(s,t)}]$. The same analysis gives the optimum choice for $\beta_{(s,t)}$

$$\beta^*_{(s,t)} = \frac{\sigma_b^2 + \sigma_v^2}{\sigma_b^2} \sqrt{e^*_{(s-1,t-1)} - e^*_{(s,t-1)}}$$

and $\alpha^*_{(s,t)} = -\beta^*_{(s,t)}$. Substituting these values into the error recursion, we obtain the two-dimensional recursion for the estimation error:

$$e^*_{(s,t)} = e^*_{(s-1,t-1)} - \left[e^*_{(s-1,t-1)} - e_{(s,t-1)} - \frac{(\sigma_b^2)^2}{\sigma_b^2 + \sigma_v^2} \right]$$

$$\times \left[2\Phi \left(\frac{\sqrt{\sigma_b^2 + \sigma_v^2}\sqrt{e^*_{(s-1,t-1)} - e^*_{(s,t-1)}}}{\sigma_b^2} \right) - 1 \right]$$

$$- \frac{2\sigma_b^2}{\sqrt{2\pi(\sigma_b^2 + \sigma_v^2)}} \sqrt{e^*_{(s-1,t-1)} - e^*_{(s,t-1)}} e^{-\frac{\left(\frac{\sigma_b^2 + \sigma_v^2}{\sigma_b^2} \right)^2 (e^*_{(s-1,t-1)} - e^*_{(s,t-1)})}{2(\sigma_b^2 + \sigma_v^2)}}$$

$$+ \sigma_b^2 - \frac{(\sigma_b^2)^2}{\sigma_b^2 + \sigma_v^2}$$

Note that, for $\sigma_v^2 = 0$ this recursion simplifies to (23), which is the recursion for the perfect state measurements.

We can normalize the optimal estimation error by letting

$$\epsilon_{(s,t)} = \frac{\sigma_b^2 + \sigma_v^2}{(\sigma_b^2)^2} e^*_{(s,t)} \tag{26}$$

and rewrite the above recursion in a simpler form:

$$\epsilon_{(s,t)} = \epsilon_{(s-1,t-1)} - \left[\epsilon_{(s-1,t-1)} - \epsilon_{(s,t-1)} - 1 \right] \left[2\Phi \left(\sqrt{\epsilon_{(s-1,t-1)} - \epsilon_{(s,t-1)}} \right) - 1 \right]$$

$$- \frac{2}{\sqrt{2\pi}} \sqrt{\epsilon_{(s-1,t-1)} - \epsilon_{(s,t-1)}} e^{-\frac{\epsilon_{(s-1,t-1)} - \epsilon_{(s,t-1)}}{2}} + \frac{\sigma_v^2}{\sigma_b^2} \tag{27}$$

with the initial conditions

$$\epsilon(t,t) = \frac{\sigma_v^2}{\sigma_b^2} t, \quad \epsilon(0,t) = \left(1 + \frac{\sigma_v^2}{\sigma_b^2} \right) t, \quad \forall t \geq 0$$

and $\epsilon(s,t)$ is undefined for $s > t$.

We note that the recursion (27) reduces to the recursion (25), as the noise variance $\sigma_v^2 \to 0$.

3.3   Estimating a Gauss-Markov Process with Limited Measurements

In this section, we discuss the case when the source process is Markov

$$b_{k+1} = Ab_k + w_k$$

driven by an i.i.d. Gaussian process $\{w_k\}$ with zero-mean. The solution to this case is similar to the Gaussian i.i.d. case when the observer has access to the source output b_k without noise. The only difference is that, now the observer-estimator pair has to keep track of three variables (r_k, s_k, t_k), where r_k keeps track of the number of time units passed since the last use of the channel for transmission. A similar DP recursion, now in three dimensions, can be obtained.

Let r denote the number of time units passed since the last transmission of a source output. Reasoning as in Section 3.2, we can deduce that for $s \geq 1$, the optimal estimator has the form

$$\hat{\mu}((r,s,t); b_{N-t}) = \begin{cases} b_{N-t} & b_{N-t} \in \mathcal{T}_{(r,s,t)} \\ E\left\{b_{N-t} | b_{N-t} \in \mathcal{T}^c_{(r,s,t)}\right\} & b_{N-t} \in \mathcal{T}^c_{(r,s,t)} \end{cases}$$

With the estimator structure in place, the error recursion can be derived following along the lines of previous sections:

$$e^*_{(r,s,t)} = \min_{\mathcal{T}_{(r,s,t)}} \left\{ e^*_{(1,s-1,t-1)} P[b_{N-t} \in \mathcal{T}_{(r,s,t)}] + e^*_{(r+1,s,t-1)} P[b_{N-t} \in \mathcal{T}^C_{(r,s,t)}] \right.$$

$$\left. + \int_{b_{N-t} \in \mathcal{T}^c_{(r,s,t)}} [b_{N-t} - A^r b_{N-t-r}]^2 f_{b_{N-t}}(b_{N-t}) db_{N-t} \right\}$$

where $b_{N-t} \sim N\left(A^r b_{N-t-r}, \left(\sum_{k=1}^r A^{2(k-1)}\right) \sigma_b^2\right)$.

Now if we let $\mathcal{T}^c_{(r,s,t)} = [\alpha_{(r,s,t)} \ \beta_{(r,s,t)}]$, the optimal choices for the parameters $\alpha_{(r,s,t)}$ and $\beta_{(r,s,t)}$ are

$$\alpha^*_{(r,s,t)} = A^r b_{N-t-r} + \sqrt{e^*_{(1,s-1,t-1)} - e^*_{(r+1,s,t-1)}}$$

$$\beta^*_{(r,s,t)} = A^r b_{N-t-r} - \sqrt{e^*_{(1,s-1,t-1)} - e^*_{(r+1,s,t-1)}}$$

Substituting these choices back into the error recursion and simplifying yields

$$e^*_{(r,s,t)} = e^*(1,s-1,t-1) - \left[e^*_{(1,s-1,t-1)} - e^*_{(r+1,s,t-1)} - \left(\sum_{k=1}^r A^{2(k-1)}\right) \sigma_b^2 \right]$$

$$\times \left[2\Phi\left(\sqrt{\frac{e^*_{(1,s-1,t-1)} - e^*_{(r+1,s,t-1)}}{\sum_{k=1}^r A^{2(k-1)} \sigma_b^2}} \right) - 1 \right] \tag{28}$$

$$\frac{2\sqrt{\sum_{k=1}^r A^{2(k-1)} \sigma_b^2}}{2\pi} \sqrt{e^*_{(1,s-1,t-1)} - e^*_{(r+1,s,t-1)}} \, e^{-\frac{e^*_{(1,s-1,t-1)} - e^*_{(r+1,s,t-1)}}{2\sum_{k=1}^r A^{2(k-1)} \sigma_b^2}}$$

where we have made use of the fact that for a Gaussian random variable x with mean m and variance k^2, and for $a \geq 0$, we have the expression

$$\int_{m-\sqrt{a}}^{m+\sqrt{a}} (x-m)^2 f_x(x) dx = -\frac{2\sqrt{a}k}{\sqrt{2\pi}} e^{-\frac{a}{2k^2}} + k^2 \left(2\Phi\left(\frac{\sqrt{a}}{k}\right) - 1 \right)$$

The recursion (28) is defined for $r \geq 1$, and $0 \leq s \leq t$ with the boundary conditions given by

$$e^*_{(r,t,t)} = 0, \quad e^*_{(r,0,t)} = \left(\sum_{l=r}^{r+t-1} \sum_{k=1}^{l} A^{2(k-1)} \right) \sigma_b^2$$

Note that, as in Section 3.2, one can define the normalized estimation error by

$$\epsilon_{(r,s,t)} = \frac{1}{\left(\sum_{k=1}^{r} A^{2(k-1)} \right) \sigma_b^2} e^*_{(r,s,t)} \tag{29}$$

and simplify the recursion (28) further as follows:

$$\epsilon_{(r,s,t)} = \epsilon_{(1,s-1,t-1)} - \left[\epsilon_{(1,s-1,t-1)} - \epsilon_{(r+1,s,t-1)} - 1 \right]$$
$$\times \left[2\Phi\left(\sqrt{\epsilon_{(1,s-1,t-1)} - \epsilon_{(r+1,s,t-1)}} \right) - 1 \right]$$
$$- \frac{2}{\sqrt{2\pi}} \sqrt{\epsilon_{(1,s-1,t-1)} - \epsilon_{(r+1,s,t-1)}} e^{-\frac{\epsilon_{(1,s-1,t-1)} - \epsilon_{(r+1,s,t-1)}}{2}}$$

Note that when $r = 0$, this is the exact same recursion as in the case of estimating an i.i.d. Gaussian process with no measurement noise. The only difference between this case and the i.i.d. case is in scaling back into the original estimation error via (29). However, unlike the i.i.d. case, this recursion must be solved offline for all feasible (r, s, t) triplets, and a three-dimensional table has to be formed.

3.4   Illustrative Examples

Example 1

As an example for the case when the source is binary, i.e., $b_k \in \{0,1\}$, consider the problem of sequentially estimating a Bernoulli process of length N with M opportunities to transmit over a noiseless binary channel. This problem is a special case of the general problem we solved in Section 3.2. The probability distribution of the source is given, and say, without loss of any generality, that 1 is a more likely outcome than 0. In this case, the best observation policy is to start at time $k = 0$ not transmit the likely outcome 1, and to use the channel to transmit only the unlikely outcome 0. And the best estimation scheme is to employ the MAP estimator which estimates NT as 1, and 0 as 0, as long as $s_k \geq 1$. If $s_k = 0$, on the other hand, then the best estimator should estimate 1 regardless of the channel output.

Example 2

The second example is just solving the problem of Section 3.2 for $(s,t) = (1,2)$. So, the observer can use the channel for transmission only once, at time $k = 0$ or 1, and the observer and the estimator are jointly trying to minimize the average distortion (or estimation error):

$$e = E\left\{ (b_0 - \hat{b}_0)^2 + (b_1 - \hat{b}_1)^2 \right\}$$

where b_0, b_1 are i.i.d. Gaussian with zero mean, and variance σ_b^2. If we arbitrarily choose to transmit the first source output, or the second one, the estimation error would be

$$e^*_{no-observer} = \sigma_b^2$$

which is the best error that can be achieved without a decision maker that observes the source output. Now, suppose the observer is aware of the fact that the estimator knows the *a priori* distribution of b_0. So, it makes sense for the observer not to transmit the realized value of b_0 if this value happens to be close to the *a priori* estimate of it, which in this case is the mean value of b_0, i.e., zero.

Motivated by this intuition, the observer decides to adopt a policy in which it will not use the channel to transmit b_0 if it lies in an interval $[\alpha, \beta]$ around zero. Note that the decision for the second stage would already have been made once α and β are determined, because, if $b_0 \in [\alpha, \beta]$, then the observer cannot use the channel to transmit at time 1, and if $b_0 \notin [\alpha, \beta]$, there is no reason why it should not transmit at time 1.

Now, the optimization problem faced by the observer is to choose α and β such that the following error is minimized:

$$e_{(\alpha,\beta)} = \int_{\alpha}^{\beta} \left(b - E\left\{b | b \in [\alpha, \beta]\right\}\right)^2 f(b)db + \sigma_b^2 P\left\{b_0 \notin [\alpha, \beta]\right\}$$

where $f(b)$ is the standard Gaussian density. The solution can be easily obtained by checking the first and second order optimality conditions, and is given by

$$(\alpha^*, \beta^*) = (-\sigma_b, \sigma_b)$$

Thus, the observer should not use the channel to transmit the source output b_0 if it falls within one standard deviation of its mean. For these values of α and β, the optimal value of the estimation error can be calculated as

$$e_{(\alpha^*,\beta^*)} = \sigma_b^2 \left[1 - \sqrt{\frac{2}{\pi e}}\right]$$

Comparing this error to the no-observer policy, $e^*_{no-observer} = \sigma_b^2$, we see that there is an approximately $\sqrt{\frac{2}{\pi e}} \approx 48\%$ improvement in the estimation error.

Example 3

The third and final example we will discuss considers the following design problem. We are given a time-horizon of a fixed length N, say 100 For this $N = 100$ time units, we would like to sequentially estimate the state of a zero-mean, i.i.d. Gaussian process with unit variance. We have a design criterion which says that the *aggregate* estimation error should not exceed 20. The solution to this problem without an observer agent is to reveal 80 arbitrary observations to the estimator and achieve an aggregate estimation error of 20. Suppose, now we use the

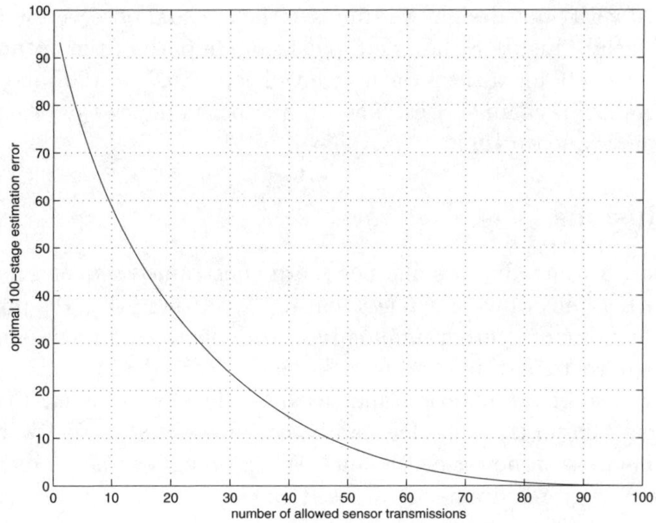

Fig. 7. Optimal 100-stage estimation error vs. the number of allowed channel uses

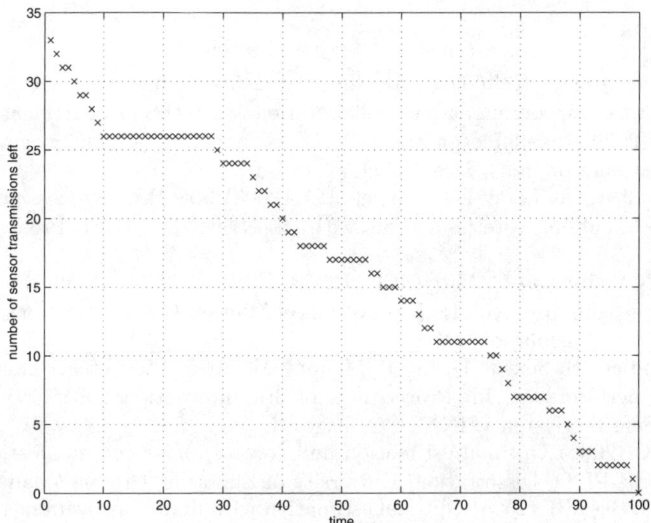

Fig. 8. A typical sample path of the number of channel uses left under the optimal observer-estimator policies. $(N, M) = (100, 34)$

optimal observer-estimator pair. In Figure 7, we plot the optimal value of the 100-stage estimation error for different values of M.

It is striking that a cumulative estimation error of 20 can be achieved with only 34 transmissions. This is approximately a $\frac{80-34}{80} \times 100 \approx 58\%$ improvement over the no-observer policy.

In order to verify our design, we simulate the optimal observer and estimator policies in Matlab. Figure 8 shows a typical sample path of the optimal number of channel uses left for a decision horizon of length $N = 100$, and a limited, $M = 34$, number of channel uses. The sample paths depend on the realization of the random sequence $\{b_k\}_0^{N-1}$.

4   Conclusions

In this paper, we introduced some new hard-constrained sequential estimation problems with applications in wireless sensing. We showed that the problems can be solved using dynamic-programming type arguments, and their solutions have a threshold characterization. The process models considered in this paper were idealized for ease of presentation and mathematical tractability. However, the basic thinking behind these models can be easily adopted to real-world wireless sensing problems with power constraints. When doing so, one needs to consider several other design requirements imposed on the system, such as network-level connectivity and time synchronization. Current research effort is directed towards developing algorithms that take some of these cross-layer design issues into account.

References

1. Industrial wireless technology for the 21st century, based on views of the industrial wireless community, in collaboration with U.S. Department of Energy (DOE), (2002) (available on the DOE's website at http://energy.gov/industry/ sensorsautomation/pdfs/wirelesstechnology.pdf)
2. Gutierrez J A, Calllaway E H, Barrett R L (2003) Low-Rate wireless personal area networks: Enabling wireless sensors with IEEE 802.15.4, IEEE Press, New York, NY
3. Stralen N V, Imer O C, Mitchell R, Evans S, Iyer S (2006) A multiband random access messaging protocol, In: Proceedings of Military Communications Conference (MILCOM), Washington, DC
4. Azimi-Sadjadi B, Sexton D, Liu P, Mahony M (2006) Interference effect on IEEE 802.15.4 performance, In: Proceedings of 3rd International Conference on Networked Sensing Systems (INNS), Chicago, IL
5. Imer O C (2005) Optimal estimation and control under communication network constraints, Ph.D. Dissertation, University of Illinois at Urbana-Champaign
6. Imer O C, Başar T (2005) Optimal estimation with limited measurements, In: Proceedings of 44th IEEE Conference on Decision and Control (CDC) and European Control Conference (ECC), Seville, Spain
7. Imer O C, Başar T (2006) Optimal control with limited controls, In: Proceedings of American Control Conference (ACC), Minneapolis, MN
8. Anderson B D, Moore J B (1979) Optimal filtering, Prentice-Hall, Inc., Englewood Cliffs, NJ
9. Cover T M, Thomas J A (1991) Elements of information theory, John Wiley & Sons, Inc., New York, NY
10. Bertsekas D P (1995) Dynamic programming and optimal control, Athena Scientific, Belmont, MA

The Important State Coordinates
of a Nonlinear System

Arthur J. Krener

University of California, Davis, CA
and Naval Postgraduate School, Monterey, CA
ajkrener@ucdavis.edu

Summary. We offer an alternative way of evaluating the relative importance of the state coordinates of a nonlinear control system. Our approach is based on making changes of state coordinates to bring the controllability and observability functions into input normal form. These changes of coordinates are done degree by degree and the resulting normal form is unique through terms of degree seven.

Keywords: Nonlinear Control Systems, Model Reduction.

1 The Problem

The theory of model reduction for linear control systems was initiated by B. C. Moore [6]. His method is applicable to controllable, observable and exponentially stable linear systems. The reduction is accomplished by making a linear change of state coordinates to simultaneously diagonalize the controllability and observability gramians and make them equal. The diagonal entries of the gramians are the singular values of the Hankel map from past inputs to future outputs. The reduction is accomplished by Galerkin projection onto the states associated to large singular values. The method is intrinsic, the reduced order model depends only on the dimension of the reduced state space.

Jonckheere and Silverman [4] extended Moore's methodology to controllable, observable but not necessarily stable linear system. Their method is based on simultaneously diagonalizing the positive definite solutions of the control and filtering Riccati equations and making them equal. The diagonal entries are called the characteristic values of the system and reduction is achieved by Galerkin projection onto the states associated to large characteristic values. The method is sometimes called LQG balancing and reduction. Two nice features of their approach is that it is applicable to unstable systems and LQG controller of the reduced order model is the Galerkin projection of the LQG controller of the high order model. This method is intrinsic. Mustafa and Glover [7] extended Jonckheere and Silverman using H^∞ rather than LQG methods. This method is intrinsic once the attenuation level, γ has been specified.

Moore's method was extended to asymptotically stable nonlinear systems by Scherpen [8]. Scherpen and Van der Schaft [10] extended Jonckheere and Silverman to nonlinear systems and Scherpen [9] extended Mustafa and Glover.

C. Bonivento et al. (Eds.): Adv. in Control Theory and Applications, LNCIS 353, pp. 161–170, 2007.
springerlink.com   © Springer-Verlag Berlin Heidelberg 2007

Unfortunately none of the nonlinear extensions are intrinsic, the reduced order model depends on choices made during the reduction process.

We build on the fundamental method of Scherpen and offer an alternative way of computing the reduced order model. Because of space limitations we shall restrict our attention to Moore's method and Scherpen's nonlinear generalization.

2 Linear Balancing and Reduction

2.1 Minimal Realizations

Consider an autonomous finite dimensional linear system

$$\begin{aligned} \dot{x} &= Fx + Gu \\ y &= Hx \end{aligned} \tag{1}$$

where $x \in I\!\!R^n$, $u \in I\!\!R^m$, $y \in I\!\!R^p$. The linear system (1) initialized at $x(-\infty) = 0$ defines a mapping from past inputs $\{u(s) : -\infty < s \leq 0\}$ to future outputs $\{y(t) : 0 \leq t < \infty\}$ called the Hankel map. The map factors through the state $x(0)$ at time $t = 0$. This map has infinite dimensional domain and infinite dimensional range but it factors through the finite dimensional state space $x(0) \in I\!\!R^n$ although n might be very large. The state space representation is a very succint way of describing an infinite dimensional mapping. One goal of model reduction is to reduce the state dimension as much as possible while keeping the essential features of the Hankel map.

The first step in linear model reduction is to check whether the system (1) is a minimal realization of the Hankel mapping and if it is not minimal then to reduce it to a minimal realization. This procedure is classical and goes back to Kalman and others circa 1960. We check whether the system is controllable, i.e., the system can be excited to any state $x(0)$ when started at $x(-\infty) = 0$ by using an appropriate control trajectory $\{u(s) : -\infty < s \leq 0\}$. This will be possible iff F, G is a controllable pair, i.e., the smallest F-invariant subspace $\mathcal{V}_c$ containing the columns of G is the whole state space. If the system cannot be excited to every state then we should restrict the state space to $\mathcal{V}_c$. The restricted system is controllable and has the same Hankel map.

Then we check whether this reduced system is observable in the sense that any changes in the initial condition $x(0)$ can be detected by changes in the resulting output tajectory. The system is observable iff H, F is an observable pair, i.e., the largest F-invariant subspace $\mathcal{V}_u$ contained in the kernel of H is zero. If the system is not observable then $x(0)$ can be perturbed in the directions of $\mathcal{V}_u$ without changing the output trajectory. To make the system observable we must project $\mathcal{V}_u$ to zero. The projecteded system is observable and has the same Hankel map.

In summary, a linear system (1) is a minimal realization (of smallest state dimension) of the Hankel map iff it is controllable and observable. Any realization can be made minimal by restricting to its controllable directions and projecting out its unobservable directions.

2.2   Linear Input-Output Balancing

So far we have discussed linear systems that exactly realize the Hankel map. B. C. Moore [6] considered reduced order systems that approximately realize the Hankel map. His basic intuition was that we should ignore directions that are difficult to reach and that don't affect the output much.

To quantify these ideas, he introduced the controllablity and observability functions of the system. The controllability function is

$$\pi_c(x^0) = \inf \frac{1}{2} \int_{-\infty}^{0} |u(s)|^2 \, ds \tag{2}$$

subject to the system dynamics (1) and

$$x(-\infty) = 0, \quad x(0) = x^0.$$

If $\pi_c(x^0)$ is large then it takes a lot of input energy to excite the system in the direction x^0 and so this direction might be ignored in a reduced order model.

The observability function is

$$\pi_o(x^0) = \frac{1}{2} \int_{0}^{\infty} |y(t)|^2 \, dt \tag{3}$$

subject to the system dynamics (1) and

$$x(0) = x^0, \quad u(t) = 0.$$

If $\pi_o(x^0)$ is small then changes in this direction lead to small changes in the output energy and so this direction might be ignored in a reduced order model.

If F is Hurwitz, F, G is a controllable pair and H, F is an observable pair then it is not hard to see that

$$\pi_c(x) = \frac{1}{2} x' P_c^{-1} x, \quad \pi_o(x) = \frac{1}{2} x' P_o x$$

for some positive definite matrices P_c, P_o that are the unique solutions of the linear Lyapunov equations,

$$0 = FP_c + P_cF' + GG'$$
$$0 = F'P_o + P_oF + H'H.$$

P_c, P_o are called the controllablity and observability gramians of the system.

Moore realized that large and small are relative terms and one needs scales to measure such things. This can be accomplished by using one gramian to scale the other and vice versa. Trivally there is a linear change of state coordinates so that, in the new coordinates also denoted by x,

$$P_c = P_o = \begin{bmatrix} \sigma_1 & & 0 \\ & \ddots & \\ 0 & & \sigma_n \end{bmatrix}$$

where $\sigma_1 \geq \sigma_2 \geq \ldots \geq \sigma_n > 0$. These are called the Hankel singular values and they are the nonzero singular values of the Hankel map.

A reduced model can be obtained by only keeping the states corresponding to large σ_i. More pecisely suppose $\sigma_k >> \sigma_{k+1}$, let x_1 denote the first k coordinates of x and x_2 denote the remaining $n - k$ coordinates. We partition the system matrices accordingly

$$\begin{bmatrix} \dot{x}_1 \\ \dot{x}_2 \end{bmatrix} = \begin{bmatrix} F_{11} & F_{12} \\ F_{21} & F_{22} \end{bmatrix} \begin{bmatrix} x_1 \\ x_2 \end{bmatrix} + \begin{bmatrix} G_1 \\ G_2 \end{bmatrix} u$$
$$y = \begin{bmatrix} H_1 & H_2 \end{bmatrix} \begin{bmatrix} x_1 \\ x_2 \end{bmatrix} \tag{4}$$

The reduced model is then obtained by Galerkin projection onto the x_1 subspace,

$$\dot{x}_1 = F_{11} x_1 + G_1 u$$
$$y = H_1 x_1. \tag{5}$$

Notice several things. Viewed abstractly model reduction of a linear system involves injection and a surjection that is similar to minimal realiztion theory. The major difference is that in the former we need a sense of scale on $I\!R^n$ that is supplied to one gramian by the other. In minimal realization theory we did not need a scale because a direction is either controllable or not, a direction is either unobservable or not.

The eigenvalues of F play an indirect role in the reduction process. By assumption they are all in the open left half plane. It is very hard to excite the system in a direction corresponding to a very stable eigenvalue and so π_c tends to be very large in such a direction. Moreover, a state direction corresponding to a very stable eigenvalue tends to damp out quickly and so it has very little output energy as measured by π_o. Hence the very stable directions of F tend to correspond to small Hankel singular values and they tend to drop out of the reduced model.

3   Nonlinear Balancing and Reduction

Scherpen [8] generalized Moore to affine nonlinear systems of form

$$\dot{x} = f(x) + g(x)u$$
$$y = h(x). \tag{6}$$

where the unforced dynamics $u = 0$ is asymptotically stable. She defined the controllability and observability functions (2, 3) as did Moore subject to the nonlinear system (6).

She noted that if π_c is smooth then it satisfies the Hamilton-Jacobi-Bellman equation

$$0 = \frac{\partial \pi_c}{\partial x}(x)f(x) + \tfrac{1}{2}\left(\frac{\partial \pi_c}{\partial x}(x)g(x)\right)\left(\frac{\partial \pi_c}{\partial x}(x)g(x)\right)'$$

and if it π_o smooth then it satisfies the Lyapunov equation

$$0 = \frac{\partial \pi_o}{\partial x}(x)f(x) + \tfrac{1}{2}|h(x)|^2. \tag{7}$$

Suppose that the system has a Taylor series expansion

$$\dot{x} = f(x) + g(x)u = Fx + Gu + O(x,u)^2$$
$$y = h(x) = Hx + O(x)^2. \tag{8}$$

If F is Hurwitz, F, G is a controllable pair, and H, F is a observable pair then it is not hard to prove that there exists locally smooth, positive definite solutions to the above PDEs and

$$\pi_c(x) = \frac{1}{2}x'P_c^{-1}x + O(x)^3$$

$$\pi_o(x) = \frac{1}{2}x'P_o x + O(x)^3$$

where P_c, P_o are the controllability and observability gramians defined above.

So far nonlinear balancing looks very much like linear balancing but, in general, there is not a nonlinear change of state coordinates that simultaneously "diagonalizes" both $\pi_c(x)$ and $\pi_o(x)$.

So Scherpen invoked the Morse lemma to show that after a nonlinear change of state coordinates

$$\pi_c(x) = \frac{1}{2}|x|^2,$$
$$\pi_o(x) = \frac{1}{2}x'Q(x)x, \qquad Q(0) = P_o.$$

Then after a further nonlinear change of coordinates $\pi_c(x)$ is unchanged and

$$\pi_o(x) = \frac{1}{2}x' \begin{bmatrix} \tau_1(x) & & 0 \\ & \ddots & \\ 0 & & \tau_n(x) \end{bmatrix} x$$

where $\tau_i(x)$ are called the singular value functions. It is not hard to see that $\tau_i(0) = \sigma_i^2$ where σ_i are the Hankel singular values of the linear part of the system.

The Hankel singular values σ_i of the linear part of the system are intrinsic and hence so are their squares, $\tau_i(0)$. But the singular value functions $\tau_i(x)$ are not [3]. For example, choose any two distinct indices $i \neq j$ and any $c \in \mathbb{R}$. Define $\bar{\tau}_i(x) = \tau_i(x) + cx_j^2$, $\bar{\tau}_j(x) = \tau_j(x) - cx_i^2$ and $\bar{\tau}_k(x) = \tau_k(x)$ otherwise. Then $\pi_c(x)$ is unchanged and

$$\pi_o(x) = \frac{1}{2}x' \begin{bmatrix} \bar{\tau}_1(x) & & 0 \\ & \ddots & \\ 0 & & \bar{\tau}_n(x) \end{bmatrix} x.$$

Scherpen's next step was to make an additional change of coordinates so that if x is in a coordinate direction $x = (0, \ldots, x_i, \ldots, 0)$ then

$$\pi_c(x) = \frac{1}{2}\bar{\sigma}_i(x_i)^{-1}x_i^2$$

$$\pi_o(x) = \frac{1}{2}\bar{\sigma}_i(x_i)x_i^2$$

where $\bar{\sigma}_i(0) = \sigma_i$ and $\bar{\sigma}_i(x_i)^2 \approx \tau_i(x)$.

Scherpen obtain a reduced order model by neglecting states with small $\bar{\sigma}_i(x)$. Suppose for all $x \in \mathcal{X}$, a neighborhood of $0 \in I\!\!R^n$,

$$\bar{\sigma}_1(x) \geq \ldots \geq \bar{\sigma}_k(x) >> \bar{\sigma}_{k+1}(x) \geq \ldots \geq \bar{\sigma}_n(x) > 0$$

then as before we partition $x = (x_1, x_2)$ and Galerkin project onto the states coorresponding to large $\bar{\sigma}_i(0)$.

Unfortuantely this approach to obtaining a reduced order model is not intrinsic. The resulting reduced order system is not independent of the particular coordinate changes that led to it. Also it depends on the choice of singular value functions $\tau_i(x)$.

One nice feature of this approach is that the controllability function of the reduced order model is the restriction of the controllability function of the full order model. However this is not true for the observability functions but they do agree to $O(x)^3$.

4   The New Approach

Following Moore and Scherpen we consider the optimal control problem of steering from $x = 0$ at $t = -\infty$ to an arbitrary x at $t = 0$ while minimizing the energy of the input

$$\pi_c(x) = \inf \frac{1}{2} \int_{-\infty}^{0} |u|^2 dt$$

for the system

$$\begin{aligned} \dot{x} &= f(x, u) = Fx + Gu + f^{[2]}(x, u) + \ldots \\ y &= h(x) = Hx + h^{[2]}(x) + \ldots. \end{aligned} \tag{9}$$

where $f^{[d]}(x, u), h^{[2]}(x)$ denotes homogeneous polynomials of degree d. Scherpen only considered systems affine in u but it is an easy generalization to the above.

If F is Hurwitz and F, G is a controllable pair then there is an unique, locally smooth and positive definite optimal cost $\pi_c(x)$ and an unique, locally smooth optimal control $u = \kappa(x)$ which solve the HJB equations

$$0 = \frac{\partial \pi_c}{\partial x}(x)f(x, \kappa(x)) - \frac{1}{2}|\kappa(x)|^2 \tag{10}$$

$$\kappa(x) = \left(\frac{\partial \pi_c}{\partial x}(x)\frac{\partial f}{\partial u}(x, \kappa(x))\right)' \tag{11}$$

Moreover, following Al'brecht [1], the Taylor series of $\pi_c(x), \kappa(x)$ can be computed term by term from the Taylor series of $f(x, u)$,

$$\pi_c(x) = \frac{1}{2}x'P_c^{-1}x + \pi_c^{[3]}(x) + \ldots + \pi_c^{[r]}(x) + O(x)^{r+1}$$

$$\kappa(x) = Kx + \kappa^{[2]}(x) + \ldots + \kappa^{[r-1]}(x) + O(x)^r$$

where $P_c > 0$ and $K = G'P_c^{-1}$ are the controllability gramian and the optimal feedback of the linear part of the system.

As before we also consider the output energy released by the system when it starts at an arbitrary x at $t = 0$ and decays to 0 as $t \to \infty$,

$$\pi_o(x) = \frac{1}{2}\int_0^\infty |y|^2 dt.$$

If F is Hurwitz and H, F is an observable pair then there is a unique locally smooth and positive definite solution $\pi_o(x)$ to the corresponding Lyapunov equation

$$0 = \frac{\partial \pi_o}{\partial x}(x)f(x) + \frac{1}{2}h'(x)h(x) \tag{12}$$

Again the Taylor series of $\pi_o(x)$ can be computed term by term from the Taylor series of f and h,

$$\pi_o(x) = \frac{1}{2}x'P_o x + \pi_o^{[3]}(x) + \ldots + \pi_o^{[r]}(x) + O(x)^{r+1}$$

where $P_o > 0$ is the observability gramianof the linear part of the system.

From [6], [8] we know that we can choose a linear change of coordinates so that in the new coordinates also denoted by x

$$\pi_c(x) = \frac{1}{2}|x|^2 + \pi_c^{[3]}(x) + O(x)^4$$

$$\pi_o(x) = \frac{1}{2}x'\begin{bmatrix} \tau_1 & & 0 \\ & \ddots & \\ 0 & & \tau_n \end{bmatrix} x + \pi_o^{[3]}(x) + O(x)^4$$

where the so called singular values $\tau_1 \geq \tau_2 \geq \ldots \geq \tau_n > 0$ are the ordered eigenvalues of $P_o P_c$. If this holds then we say that the system is in *input normal form of degree one*. Because of space limitations we shall restrict our attention to the generic case where the singular values $\tau_1 > \tau_2 > \ldots > \tau_n > 0$ are distinct.

A system with distinct singular values is in *input normal form of degree d* if

$$\pi_c(x) = \frac{1}{2}\sum_{i=1}^n x_i^2 + O(x)^{d+2}$$

$$\pi_o(x) = \frac{1}{2}\sum_{i=1}^n \eta_i^{[0:d-1]}(x_i)x_i^2 + O(x)^{d+2} \tag{13}$$

where $\eta_i^{[0:d-1]}(x_i) = \tau_i + \ldots$ is a polynomial in x_i with terms of degrees 0 through $d - 1$. They are called the squared singular value polynomials of degree $d - 1$.

There is also an *output normal form of degree d* where the forms of $\pi_c(x)$ and $\pi_o(x)$ are reversed.

The proof of the following is omitted because of page limitations. The full details can be found in [5].

Theorem. Suppose the system (9) is C^r, $r \geq 2$ with controllable, observable and exponentially stable linear part. If the τ_i are distinct and if $d < r - 1$ then there is a change of state coordinates that takes the system into input normal form of degree d (13). The change of coordinates that achieves the input normal form of degree d is not necessarily unique but the input normal form of degree $d \leq 6$ is unique. If f, h are odd functions the input normal form of degree $d \leq 12$ is unique.

The differences between input normal form of degree d and Scherpen's normal form are threefold. First the former is only approximate through terms of degree $d + 1$ while the latter is exact. The second difference is that in the former the parameters $\eta_i^{[0:d-1]}(x_i)$ only depend on x_i while in the latter the parameters $\tau_i(x)$ can depend on all the components of of x. Thirdly the parameters $\eta_i^{[0:d-1]}(x_i)$ of the former are unique if $d \leq 6$ while the parameters $\tau_i(x)$ of the latter are not unique except at $x = 0$ [3] .

Recently Fujimoto and Scherpen [2] have shown the existence of a normal form where π_c is one half the sum of squares of the state coordinates and

$$\frac{\partial \pi_o}{\partial x_i}(x) = 0 \ \text{ iff } \ x_i = 0. \tag{14}$$

It is closer to our input normal form of degree d (13) which has similar properties. The controllability function π_c is one half the sum of squares of the state coordinates through terms of degree $d + 1$ and

$$\frac{\partial \pi_o}{\partial x_i}(x) = O(x)^{d+1}$$

if $x_i = 0$. But the normal form of Fujimoto and Scherpen is not unique while the input normal form of degree $d \leq 6$ is unique.

Notice that if a system with distinct singular values $\tau_i = \tau_i(0)$ is in input normal form of degree d then its controllability and observability functions are "diagonalized" through terms of degree $d+1$. They contain no cross terms where one coordinate multiplies a different coordinate. This is reminiscent of the balancing of linear systems by B. C. Moore [6].

For linear systems the singular value τ_i is a measure of the importance of the coordinate x_i. The "input energy" needed to reach the state x is $\pi_c(x)$ and the "output energy" released by system from the state x is $\pi_o(x)$. The states that are most important are those with the most "output energy" for fixed "input energy". Therefore in constructing the reduced order model, Moore kept the states with largest τ_i for they have the most "output energy" per unit "input energy".

In Scherpen's generalization [8] of Moore, the singular value functions $\tau_i(x)$ measure the importance of the state x_i. To obtain a reduced order model, she assumed $\tau_i(x) > \tau_j(x)$ whenever $1 \leq i \leq k < j \leq n$ and x is in a neighborhood of the origin. Then she kept the states $x_1, \ldots, x_k$ in the reduced order model. But the $\tau_i(x)$ are not unique.

For nonlinear systems in input normal form of degree d, the polynomial $\eta_i^{[0:d-1]}(x_i)$ is a measure of the importance of the coordinate x_i for moderate sized x. If the τ_i are distinct and $d \leq 6$ then $\eta_i^{[0:d-1]}(x_i)$ is unique. The leading coefficient of this polynomial is the singular value τ_i so in constructing a reduced order model we will want to keep the states with the largest τ_i. But τ_i can be small yet $\eta_i^{[0:d-1]}(x_i)$ can be large for moderate sized x_i. If we are interested in capturing the behavior of the system for moderate sized inputs, we may also want to keep such states in the reduced order model.

To obtain a reduced order model we proceed as follows. We start by making a linear change of coordinates to take the system into input normal form of degree 1. If the singular values are distinct this change of coordinates is uniquely determined up to the signs of the coordinates. In other words replacing x_i by $-x_i$ does not change the input normal form of degree 1. Next one computes the Taylor series expansions to degree $d+1$ of the controllability and observability functions, $\pi_c(x)$, $\pi_o(x)$. Then degree by degree one makes changes of state coordinates to bring the system into input normal form of degree d. The input normal form of degree d are intrinsic but the changes of state coordinates that achieve are not. We defer for a later paper [5] the discussion of which changes should be used . Suppose that the input energies that we shall use are all less than $\frac{c^2}{2}$ for some constant $c > 0$. Then we expect the system to operate in $|x| < c$ where x are the input normal coordinates of degree d. We compare the sizes of $\eta_i^{[0:d-1]}(x_i)$ for $|x_i| < c$ and split them into two categories, large and small. The reduced order model is obtained by Galerkin projection onto the coordinates corresponding to the large $\eta_i^{[0:d-1]}(x_i)$.

5 Conclusion

We have developed a way of finding state coordinates that lend themselves to measuring there relative importance. The measure of importance is unique up to degree 6 (degree 12 for odd systems). Unfortunately the coordinates are not unique beyond degree one. Since a reduced order model is obtained by Galerkin projection in these coordinates, it is not unique. Further research is needed to clarify these issues.

Research supported in part by NSF DMS-0505677.

References

1. Al'brecht E G (1961) On the optimal stabilization of nonlinear systems, PMM-Journal of Applied Mathematics and Mechanics, 25:1254–1266
2. Fujimoto K, Scherpen J M A (2005) Nonlinear Input-Normal Realizations Based on the Differential Eigenstructure of Hankel Operators, IEEE Transaction on Automatic Control, 50:2–18

3. Grey W S, Scherpen J M A (2001) On the Nonuniqueness of Singular Value Functions and Balanced Nonlinear Realizations, Systems and Control Letters, 44:219–232

4. Jonckheere E A, Silverman L M (1983) A New Set of Invariants for Linear Systems-Application to Reduced Order Compensator Design, IEEE Transaction on Automatic Control, 28:953–964

5. Krener A J (2006) Normal Forms for Reduced Order Modeling of Nonlinear Control Systems, In preparation

6. Moore B C (1981) Principle Component Analysis in Linear Systems: Controllability, Observability and Model Reduction, IEEE Transaction on Automatic Control, 26:17–32

7. Mustafa D, Glover K (1991) Controller Reduction by H_∞ Balanced Truncation, IEEE Transaction on Automatic Control,36:668–682

8. Scherpen J M A (1993) Balancing for Nonlinear Systems, Systems and Control Letters, 21:143–153

9. Scherpen J M A (1996) H_∞ Balancing for Nonlinear Systems, International Journal of Robust and Nonlinear Control, 6:645–668

10. Scherpen J M A, van der Schaft A J (1994) Normalized Coprime Factorizations and Balancing for Unstable Nonlinear Systems, International Journal of Control, 60:1193-1222

On Decentralized and Distributed Control of Partially-Observed Discrete Event Systems

Stéphane Lafortune

Department of Electrical Engineering and Computer Science
University of Michigan, Ann Arbor, Michigan, USA
stephane@eecs.umich.edu

Summary. This paper surveys recent work of the author with several collaborators, principally Feng Lin, Weilin Wang, and Tae-Sic Yoo; they are kindly acknowledged. Decentralized control of discrete event systems, where local controllers cannot explicitly communicate in real-time, is considered in the first part of the paper. Then the problem of real-time communication among a set of local discrete-event controllers (or diagnosers) is discussed. The writing is descriptive and is meant to inform the reader about important conceptual issues and some recently-completed or on-going research efforts.

Keywords: Discrete Event Systems, Decentralized control, Communicating Controllers.

1 Introduction

Distributed dynamic systems are pervasive in today's technological society. They are often referred to as "networked systems" when the different system modules (i.e., components) are able to communicate with one another. Distributed networked systems occur in many different application areas nowadays, ranging from communication and transportation to manufacturing, building automation, computing, software, automotive, and aerospace, to mention but a few key areas. The development of appropriate control architectures and associated controller design algorithms for such complex systems are crucial and challenging tasks, due in particular to the distribution of the sensors and actuators and the potentially large size of the entire system. An important element in the design of integrated control strategies for complex engineering and computing systems is the consideration of logical specifications that must be enforced by high-level supervisory control modules regarding safety, liveness, diagnosability, modularity, reconfigurability, and fault tolerance. This is the realm of discrete-event system and control theory, where the system behavior is abstracted in terms of event-driven transitions that cause changes to the discrete states of the modules in the system.

This paper surveys some recent work of the author with several collaborators (principally Feng Lin, Weilin Wang, and Tae-Sic Yoo) on decentralized and

C. Bonivento et al. (Eds.): Adv. in Control Theory and Applications, LNCIS 353, pp. 171–184, 2007.
springerlink.com

distributed control of discrete event systems (DES). It is based on the invited lecture of the author at the "2006 CASY Workshop on Advances in Control Theory and Applications" held in Bertinoro, Italy, in May 2006. The writing is descriptive and is meant to inform the reader about important conceptual issues and some recently-completed or on-going research efforts. Precise technical formulations and associated results can be found in the references cited. For the benefit of the reader with little familiarity with the area of DES, Section 2 discusses some salient features of the theory of supevisory control theory of DES. The survey that follows consists of two parts. Decentralized control, where local controllers cannot explicitly communicate in real-time, is considered in Section 3. Then, the problem of real-time communication among a set of local controllers (or diagnosers) is discussed in Section 4. Some comments on the decentralization of information in networked systems conclude this introductory section.

1.1 Decentralized Information Structures

Distributed networked systems occur commonly in engineering systems [19]. The distributed networked systems considered in this paper possess a *decentralized information structure* in the sense that sensors do not (or cannot in general) report their observations to all sites. That is, each site may only monitor directly a subset of the entire set of sensor readings. Sites may be able to communicate in order to share "raw" data (sensor readings) or "processed" data (e.g., "observer" or "diagnoser" states); hence, the terminology "networked systems" is often used in this case. In the setting considered in this paper, all sites participate in the same control task; the respective controllers must work in concert. Hence, this is a instance of "cooperative control."

Communication networks and sensor networks are by nature networked systems with decentralized information. The above-described system structures also arise in many types of "centralized" engineering systems where distributed control and monitoring is deemed preferable for practical reasons, such as reconfigurability, reliability, safety, scalability, and security. In addition, distributed implementations of monitoring and control functions often permit to perform maintenance and reconfiguration tasks at the subsystem level. Many problems in automotive control systems, building automation systems, and automated manufacturing systems for instance fall in that category. Automobiles nowadays have a large number of microprocessors for performing the necessary control tasks for the different subsystems, such as engine control, transmission control, antilock braking, steering, climate control, entertainment, and so forth. The same occurs in automated manufacturing, where machines in a line or a cell each have their own controller modules. Today's "smart building" technologies incorporate several networked control systems within a building, and often across multiple buildings, in order to provide enhanced security, reduce energy consumption, and increase the comfort of the occupants. In these application areas, monolithic control implementations are impractical for the reasons described above. The MoBIES program [36] provides several examples of system/software technologies that are being developed for integration, analysis, and control of distributed/modular embedded systems. All of the above considerations motivate the research discussed in Sections 3 and 4.

2 Supervisory Control Theory

The modeling formalism considered in this paper is based on regular languages and their associated finite-state automata representations. This is arguably the most common and widely used formalism in DES nowadays. The control objective is to design supervisory controllers that are provably correct with respect to a formal specification, despite the presence of uncontrollable and unobservable events. This control paradigm is generally known as *supervisory control theory* in the area of Systems and Control and it was initiated by Ramadge and Wonham [41]; see [67] and Chapter 3 in [10] for textbook treatments of supervisory control theory and [42, 57] for excellent surveys of this theory. The notions of uncontrollable events and unobservable events are used to model the limitations of the actuators and sensors, respectively. The control problems that can be addressed using the concepts and algorithmic techniques of supervisory control theory include *safety* and so-called *nonblocking* specifications. Safety specifications pertain to the avoidance of illegal states or of illegal subsequences of events. The nonblocking property is a form of liveness that is captured using the notion of marked states in the system and the concept of nonblocking supervisors; see [10]. Roughly speaking, nonblockingness implies the absence of "deadlocks" and "livelocks" and guarantees that the system performs the tasks at hand to completion.

The principal advantage of the framework of supervisory control theory is that it "separates" the "uncontrolled system" from the "controller". In this context, one is able to analyze the system-theoretic properties of the system under consideration together with safety, nonblockingness, and diagnosability requirements imposed on it. The most important properties studied to-date are: *controllability* properties (in view of the presence of uncontrollable events), *observability* properties (in view of the presence of unobservable events), *non-conflict* properties (pertaining to the issue of nonblockingness), *coobservability* properties (in the context of decentralized-information systems), and *diagnosability* properties (in the context of detection and isolation of significant unobservable events). (Chapter 3 of [10] may be consulted for precise definitions of these notions; the original references are, respectively, [41, 30, 68, 12, 51, 53].) These properties arise in the necessary and sufficient conditions for the existence of controllers that achieve the given safety, nonblockingness, and diagnosability specifications. After analysis of these properties with respect to the set of specifications, algorithms exist to "automatically" *synthesize* controllers that are *guaranteed* to be correct with respect to these specifications and the system model. Thus, it is not necessary to perform an additional phase of verification for the controlled system.

With the exception of the large body of literature on the control of Petri nets (see, e.g., [18, 37]), the author is not aware of another approach for designing supervisory control laws for dynamic systems with uncontrollable and unobservable events that is comparable to supervisory control theory in terms of the above-mentioned analytical and computational results. Other state-based approaches to modeling and analyzing controlled discrete event systems (e.g., more general types of state-transition systems, such as [2, 16, 26]), or the elegant work of Hoare, Milner, and others on process-algebraic models (e.g., [4, 17, 21, 32, 33, 34, 35]), yield more expressive and compact models of the controlled system behavior that may be preferable for verification purposes (i.e., verification of the properties of a *controlled system*). However, these formalisms do not generally enjoy the same analytical power as supervisory control theory does for the study of system-theoretic properties or for controller synthesis. In fact, supervisory control theory has been the subject of recent attention in theoretical computer science research [3, 39] and some of its results have been recast in more general logic-based frameworks; at present however, these generalizations do not support general-purpose controller synthesis algorithms.

The key results of supervisory control theory have been used in a variety of areas for understanding better the salient features of supervisory control software in engineering and computing dynamic systems (the references mentioned below are representative but not meant to be exhaustive): database concurrency control [27]; feature interactions in telecommunications networks [11, 58]; protocol verification and synthesis in communication networks [49, 50]; protocol conversion and gateway synthesis in computer networks [20, 23]; logic control in automated manufacturing systems [5, 9, 31, 14]. In particular, diagnosability theory of DES has been successfully applied in heating, ventilation, and air conditioning systems [55], document processing systems [52], and intelligent transportation systems [54], among other areas.

Several software tools have been developed for creating, manipulating, and analyzing DES modeled by automata and for implementing the main results of supervisory control theory. The proceedings of the *8th International Workshop on Discrete Event Systems - WODES'06* (available through IEEE) contain papers describing six such tools, among them the tool DESUMA developed by Laurie Ricker and the author [43, 13].

3   Decentralized Control

Considerable progress has been made in the last decade in the field of DES regarding the development of a comprehensive theory for designing *decentralized* controllers. The problem formulation is as follows. Several controllers, termed *supervisors* and denoted by S_i, $i = 1, \ldots, n$, act together to control a given system, denoted by G, in order to enforce a given logical specification (incorporating safety and nonblockingness). Each supervisor knows the entire system model G and the specification. However, the supervisors see and control different (complementary) aspects of the behavior of the system. Let the event set of G be denoted by E and let E_o be the set of events of E that are observable (by one of more supervisors); similarly, let E_c be the set of events of E

that are controllable (by one of more supervisors) Associated with each supervisor S_i are $E_{i,o} \subset E_o$ and $E_{i,c} \subset E_c$, its observable and controllable event set, respectively. The goal is to design the set of S_is so that their joint control actions result in a controlled system that is guaranteed to satisfy the specification. Many works have addressed this general problem formulation, among them [29, 12, 66, 51, 8, 40, 24, 62, 56, 44, 28, 38, 69].

One key challenge in decentralized control is an observational one: the supervisors see a different aspect of the system behavior since they observe different sets of events through their sensors. Another challenge is the decentralization of the control authority. Clearly, these two issues are coupled. The conditions that capture informational and control constraints at each supervisor are known under the name of *coobservability* conditions. Several notions of coobservability have been characterized according to the assumptions made about the specifics of the decentralized control architecture employed. These types of coobservability include C&P-coobservability, D&A-coobservability, and mixed versions of these where C&P and D&A notions are combined; see, e.g., [51, 40, 69]. Coobservability is one of the necessary and sufficient conditions for the existence of S_i, $i = 1, \ldots, n$, such that the controlled system satisfies the given specification *exactly*, in the sense of language equality. (The other necessary and sufficient conditions are the same ones that appear in the centralized supervisory control problem under full observation: controllability and relative-closure of the specification.)

The different notions of coobservability capture the effect of (i) the fusion rule employed at the actuator when its associated controllable event is controllable by more than one supervisors and (ii) the default control action issued by a supervisor when its knowledge is insufficient to unambiguously determine the next control action. In a nutshell, in C&P-coobservability, (i) the fusion rule is conjunction of enabled events and (ii) the default action is to enable (permissive). In D&A-coobservability, (i) the fusion rule is disjunction of enabled events and (ii) the default action is to disable (anti-permissive). It is important to emphasize that the above notions of coobservability all assume architectures where there is *no* explicit real-time communication among supervisors. The computational complexity of verifying C&P-coobservability is addressed in [48, 46]; similar results hold for D&A-coobservability [69].

The recent work on *conditional coobservability* in [70, 71] is now discussed. In the control architecture considered in [70], each supervisor makes inferences about the knowledge and/or actions of other supervisors when computing its own control actions. In addition, the architecture allows the use of more complicated fusion rules at each actuator for the events that are controlled by two or more supevisors. More precisely, the supervisors are allowed to make *four decisions*: "enable", "disable", "enable if nobody disables", and "disable if nobody enables". In addition, it is decided *a priori* that some controllable events should be disabled by default and the remaining controllable events should be enabled by default if no local decision (of any of the four above types) is made over those events. We refer to this architecture as the *architecture with conditional decisions* or simply the *conditional architecture.*

There are many motivations for considering the conditional architecture. Arguably the most important one is the following result proved independently in [28] and [61]. In the basic decentralized architecture (with fusion by conjunction of enabled events and permissive default rule), if the given specification does not possess the property of C&P-coobservability, then the decision problem "Does there exist a solution to the decentralized control problem that is both safe and nonblocking?" is *undecidable.* This motivates the goal of relaxing C&P-coobservability and identifying weaker notions of coobservability that are satisfied by larger classes of specification languages and are decidable. Allowing supervisors to issue conditional decisions is one way of achieving that goal.

In [70], three notions of *conditional coobservability* are introduced to characterize the necessary and sufficient conditions for the existence of a set of supervisors that jointly achieve a given desired specification language in the context of the conditional architecture. They are conditional C&P-coobservability, conditional D&A-coobservability, and a joint version of these two simply called conditional coobservability. It is shown that the notions of conditional coobservability generalize the "unconditional" ones and are decidable. The synthesis of supervisors that implement the conditional architecture for specification languages that are coobservable is the topic of [71]. The work in [45] also considers inferencing and conditional decisions, but in the framework of knowledge theory [15].

While the work reported in [70, 71] considers one level of conditional decisions, it is possible to extend the approach to multiple levels of inferecing by enlarging the class of conditional decisions and suitably refining the fusion rule. Recent work in this regard is reported in [25]. It is known however that even if one allows an "infinite" number of levels of inferencing, the class of specifications that can be achieved is still a proper subset of the class of architectures that can be achieved if a single centralized supervisor, that has access to the entire sets E_o and E_c, is used.

One could argue that one limitation of decentralized architectures is the fact that they do not allow real-time communication among supervisors. Allowing real-time communication could greatly enhance the classes of specifications that can be achieved under control; in fact, by communicating every observation to all supervisors, centralized architectures could be replicated. However, in many applications, such distributed architectures could be costly in terms of communications required. The next section discusses issues that arise in distributed control with communication.

4   Distributed Control

In order to address objectives of distributed implementation, reconfigurability, and fault tolerance in networked systems, as well as in order to tackle the problem of undecidability in decentralized control, it becomes necessary to consider explicitly real-time communication among sites of the distributed system, namely among the various supervisors, when solving control problems for systems with decentralized information. This means moving significantly beyond the scope of current decentralized discrete-event control theory and the existing notions of coobservability discussed in the preceding section. In principle, sites could be required to exchange all of their (raw or processed) data, thereby transforming a decentralized-information problem into a centralized-information one. However, this is not practical since it is often imperative for reliability and reconfigurability reasons to avoid a centralized implementation of control. Moreover, there are many instances of networked systems in wireless communications, transportation, telephony, and so forth, where communication among nodes (or agents) in the system is costly due to one or more of the following considerations: power, bandwidth, security, or network topology. For these reasons, there is a long-standing interest in minimizing communication among sites in distributed dynamic systems. In this regard, it becomes imperative to first determine which communications among local controllers are "essential" for achieving a given global specification. Then, in the second step, the goal is to introduce some amount of redundancy in the communication strategy in order to fulfill the objectives of reconfiguration and fault tolerance.

The joint design of control and communication strategies in decentralized-information systems is well known to be very difficult for most classes of dynamic systems, due to the "dual role" of control and the mutual dependency of control and communication (see, e.g., [22]). Of course, the same difficulties arise in DES (see, e.g., [6, 7]). There has been a large amount of interest lately in solving joint *control and communication* problems for cooperative networked systems, principally in the context of continuous-state models of dynamic systems; see, e.g., [1]. This problem can take many forms depending upon the modeling formalism chosen to describe the system, the structural assumptions about the distribution of the sensors, the decentralization of information in the system, and the cost function to be minimized regarding communication. Interest in this problem in the context of distributed DES is relatively recent and the emerging approaches are discussed in the survey paper [63].

The works reviewed in [63] concern the minimization of communication in the context of *diagnosis* or *control* of DES where the information is decentralized. The recent work in [59, 60] is also of interest. It solves the problem of turning on/off sensors dynamically for achieving diagnosability for DES in the context of a general formulation based on information structures.

Due to the mutual coupling in partially-observed systems between the state estimation policy, the communication policy, and, in the case of control problems, the control policy, most of the work in the area of distributed systems with communicating agents (i.e., controllers or diagnosers) attacks the problem by forcing a *separation* of estimation, communication, and control. For instance, a specific minimum-communication problem where the diagnosis and control policies are assumed fixed and given *a priori* was formulated and solved in [47] for DES modeled by finite-state automata. An algorithm that finds a *minimal* set of communication policies among two agents that exchange event occurrences was proposed. The notion of minimality is a logical one: a communication policy is minimal if removing one or more communications of event occurrences in the dynamic evolution of the system renders the otherwise feasible solution *infeasible* (see [47] for a precise definitions). The solution in [47] suffers from computational difficulties and to a certain amount of asymmetry in the solution procedure.

In recent work reported in [64, 65], it is shown that it is possible under certain conditions to synthesize communication strategies for a set of communicating agents in *polynomial-time in the number of states in the system model and in the number of events*. This is an important result and to the best of our knowledge the first of its kind for communication problems in DES. The problem considered in [64, 65] borrows some of the features of the problem treated in [47], including a similar notion of minimality of communication. However, the approach is more general than that in [47] in many respects: number of agents, communication structure, state disambiguation, and algorithmic procedure. This more general formulation is one of the contributions of this work. In order to obtain a more computationally-efficient solution procedure than that of [47], additional assumptions are made on the structure of the system as compared with those made in [47], specifically regarding the absence of cycles other than self-loops in G.

Consider a distributed networked DES modeled by automaton G. There are n agents observing the behavior of G using their own sets of sensors. The agents may be supervisors or diagnosers. The agents are able to communicate among each other with negligible delay as compared with the dynamics of the system. Agents communicate event occurrences to each other. The communication structure considered is general and allows agent j to immediately relay to agent k information it just received from agent i about an event occurrence. This formulation is adopted for the sake of generality and the solution procedure proposed is easily adaptable to more restricted contexts. The agents are working as a team to accomplish some given task: monitoring, diagnosis, or control. For this purpose, they need to be able to distinguish unambiguously among certain *pairs of states* in the state space of G; the pairs that need to be distinguished are assumed specified at the outset. This requirement is called the *state disambiguation condition* and it necessitates the exchange of information about event occurrences in real time among agents. Moreover, for the purpose of realizing the communication policy to be designed for each agent, the agent will need to distinguish further pairs of states if these have different communication decisions associated with them; this latter requirement is called *feasibility* of the communication policy. We discuss these two requirements in the following paragraphs.

Feasibility is a key concept that arises in situations involving communicating agents and it can be paraphrased as follows: "I have to know enough to tell you what you expect me to tell you and only when you expect me to tell it to you." Here, what an agent "knows" is not only a function of its own observations, but also of the communications it receives from other agents. As was pointed out in prior works (see, e.g., [6, 47]), the interdependence of the communication policies of the agents for the purpose of feasibility makes any type of minimization-of-communication problem quite intricate, no matter what modeling formalism it is cast into. Indeed, an interesting example presented in [64] shows a "lack of monotonicity" of feasible solutions when communications of event occurrences are removed one-by-one. Namely, it is shown that in some cases the communications of event occurrences A and B cannot be removed individually but can be jointly removed. This is a counter-intuitive result as it says that "seeing less is better"! To the best of our knowledge, this problem had not been previously identified in the DES literature.

In order to illustrate the phenomenon of lack of monotonicity in a simple context, consider the DES G whose language model is

$$\mathcal{L}(G) = \overline{\{a_1 a_2, e_u a_1 a_2\}}$$

where the event e_u is unobservable. In this case, the two traces $s_1 = a_1 a_2$ and $s_2 = e_u a_1 a_2$ cannot be distinguished. In particular, the occurrence of unobservable event e_u cannot be diagnosed since the observed trace $a_1 a_2$ is inconclusive. However, let us suppose that the occurrence of a_1 after e_u can no longer be observed, while the other occurrence of a_1 remains observed. (For instance, one might have decided to remove the communication to the observer site of the occurrence of a_1 after e_u). In this case, s_1 and s_2 *can* now be distinguished, since the new projected version of s_1 is $a_1 a_2$ and the new projected version of s_2 is a_2. The two traces can be distinguished since their projections are different. In other words, if the occurrence of a_1 after e_u can no longer be observed, then the occurrence or not of unobservable event e_u can be diagnosed: we know that e_u did not occur if we see $a_1 a_2$ and we know that e_u did occur if we see a_2. Thus, in this example, seeing less leads to knowing more!

The difficulties resulting from the above lack of monotonicity of solutions was avoided in [47] because the algorithm in there essentially proceeds by exhaustive enumeration of candidate solutions in a certain range (hence its high computational cost). In [64, 65] it is shown that it is possible to eliminate the undesirable lack of monotonicity by making suitable structural assumptions on G and examining the states one-by-one in a certain order. One such assumption is the absence of cycles other than self-loops in the graphical representation of G. A very important benefit of overcoming the lack of monotonicity is that it is now possible to synthesize communication policies in *polynomial-time complexity in the size of the state space of G* for a large class of minimum-communication problems. This is in contrast to all prior works on minimum-communication problems [63]. At this point, it remains to understand more clearly the sources of the lack of monotonicity and to identify other structural assumptions on G that permit its elimination. Then one can seek to exploit monotonicity properties for the development of efficient algorithms for the synthesis of communication policies.

The results in [64, 65] lead us to propose the following overall approach for solving the distributed controller synthesis problem under a global specification.

Algorithm for distributed control of DES

1. Build a model of the entire distributed system on the basis of the global specification. Call the resulting model G_{gs}.

   Since we are concerned with a global specification (typically, high-level), it is expected that a lot of the internal behavior of each system component can be abstracted away. In such situations, building the monolithic model G_{gs} will not be computationally prohibitive.

2. Solve the control problem for the global specification as a "full-observation partial-controllability" problem, assuming that each controller fully observes the behavior of G_{gs}, but taking into account the distribution of the actuators in the system.

   The state disambiguation condition mentioned in the above discussion of the minimum-communication problem will arise as a consequence of the solution of the control problem.

3. Synthesize communication policies among the resulting set of controllers subject to the state disambiguation and feasibility requirements using the results in [64].

   This will guarantee that the control policies obtained in the second step are "implementable" and "correct." An implementable control policy is one where the information available to each controller given its own sensors and the communications it receives satisfies the feasibility condition of the minimum communication problem. A set of control policies is correct if it satisfies the global specification. The set of control policies that ensure that the global specification is satisfied specifies the state disambiguation condition used to solve the minimum-communication problem.

The above steps represent a new paradigm for solving control problems for decentralized-information systems. It is a departure from prior works that are based on the various concepts of coobservability. We believe that this paradigm is better-suited to deal with many classes of networked systems.

5   Conclusion

The purpose of this paper was to discuss, in a descriptive non-technical manner, some important issues that arise in decentralized and distributed control of DES. Research results of the author, in joint work with Feng Lin, Weilin Wang, and Tae-Sic Yoo, were emphasized. Many challenges remain in future research on these problems. Among them we mention: (i) a better understanding and characterization of the boundary between decidable and undecidable controller synthesis problems in decentralized control architectures; (ii) the identification of other classes of systems where structural assumptions lead to computationally-efficient algorithms for the synthesis of communication policies in distributed control; and (iii) the ability to move away from monolithic models and judiciously exploit the modular structure of the system when solving controller synthesis problems for distributed systems.

Acknowledgement

It is a pleasure to thank Feng Lin, Weilin Wang, and Tae-Sic Yoo. The organizers of the "2006 CASY Workshop on Advances in Control Theory and Applications" are gratefully acknowledged. The research reviewed in this paper was supported in part by NSF grants CCR-0082784 and CCR-0325571 and by ONR grant N0001-14-03-1-0232.

References

1. Antsaklis P, Baillieul (Eds) J (2004) Special issue on Networked Control Systems, IEEE Transaction on Automatic Control, 49(9)
2. Arnold A (1994) Finite transition systems, International Series in Computer Science, Prentice-Hall
3. Arnold A, Vincent A, Walukiewicz I (2003) Games for synthesis of controllers with partial observation, Theoretical Computer Science, 303(1):7–34
4. Baeten J C M, Weijland W P (1990) Process Algebra, Volume 18 of Cambridge Tracts in Theoretical Computer Science, Cambridge University Press
5. Balemi S, Hoffmann G J, Gyugyi P, Wong-Toi H, Franklin G F (1993) Supervisory control of a rapid thermal multiprocessor, IEEE Transaction on Automatic Control, 38(7):1040–1059
6. Barrett G, Lafortune S (2000) Decentralized supervisory control with communicating controllers, IEEE Transaction on Automatic Control, 45(9):1620–1638
7. Barrett G, Lafortune S (2000) On the separation of estimation and control in discrete-event systems, In: Proceeding of 39th IEEE Conference on Decision and Control, pages 2258–2259
8. Bergeron A (1995) Sharing out control in distributed processes, Theoretical Computer Science, 139:163–186
9. Brandin B A (1996) The real-time supervisory control of an experimental manufacturing cell, IEEE Transactions on Robotics and Automation, 12(1):1–14
10. Cassandras C G, Lafortune S (1999) Introduction to Discrete Event Systems, Kluwer Academic Publishers
11. Chen Y L, Lafortune S, Lin F (1997) Resolving feature interactions using modular supervisory control with priorities, In: Feature Interactions in Telecommunications IV, pages 108–122, IOS Press
12. Cieslak R, Desclaux C, Fawaz A, Varaiya P (1988) Supervisory control of discrete-event processes with partial observations, IEEE Transaction on Automatic Control, 33(3):249–260
13. DESUMA A software tool integrating GIDDES and UMDES, `http://www.eecs.umich.edu/umdes/toolboxes.html`
14. Endsley E, Almeida E, Tilbury D (2006) Modular finite state machines: Development and application to reconfigurable manufacturing cell controller generation, Control Engineering Practice To appear
15. Fagin R, Halpern J Y, Moses Y, Vardi M Y (1995) Reasoning about Knowledge, MIT Press
16. Harel D, Politi M, editors (1998) Modeling Reactive Systems with Statecharts: The Statemate Approach, Wiley
17. Hoare C A R (1985) Communicating Sequential Processes, International Series in Computer Science, Prentice-Hall, Englewood Cliffs, NJ

18. Holloway L, Krogh B, Giua A (1997) A survey of Petri net methods for controlled discrete event systems, Discrete Event Dynamic Systems: Theory and Applications, 7(2):151–190

19. Hristu-Varsakelis D, Levine W (2005) Handbook of Networked and Embedded Control Systems, Birkhäuser

20. Inan K (1994) Supervisory control: Theory and application to the gateway synthesis problem, Technical report, Electrical and Electronics Department, Middle East Technical University, Turkey

21. Inan K M, Varaiya P P (1989) Algebras of discrete event models, Proceedings of the IEEE, 77(1):24–38

22. Kumar P R, Varaiya P (1986) Stochastic Systems. Estimation, Identification, and Adaptive Control, Prentice-Hall

23. Kumar R, Nelvagal S, Marcus S I (1997) A discrete event systems approach for protocol conversion, Discrete Event Dynamical Systems: Theory and Applications, 7(3):295–315

24. Kumar R, Shayman M A (1997) Centralized and decentralized supervisory control of nondeterministic systems under partial observation, SIAM Journal of Control and Optimization, 35(2):363–383

25. Kumar R, Takai S (2005) Inference-based ambiguity management in decentralized decision-making: Decentralized control of discrete event systems, In: Proceedings of the 44th IEEE Conference on Decision and Control, pages 3480–3485

26. Kurshan R P (1994) Computer-Aided Verification of Coordinating Processes: The Automata-Theoretic Approach, Princeton University Press

27. Lafortune S (1988) Modeling and analysis of transaction execution in database systems, IEEE Transaction on Automatic Control, 33(5):439–447

28. Lamouchi H, Thistle J G (2000) Effective control systhesis for DES under partial observations, In: Proceedings of the 39th IEEE Conference on Decision and Control, pages 22–28

29. Lin F, Wonham W M (1988) Decentralized supervisory control of discrete-event systems, Information Sciences, 44:199–224

30. Lin F, Wonham W M (1988) On observability of discrete-event systems, Information Sciences, 44:173–198

31. Lucas M R, Endsley E W, Tilbury D M (1999) Coordinated logic control for reconfigurable machining systems, In: Proceedings of 1999 American Control Conference

32. Milner R (1980) A Calculus of Communicating Systems, Springer-Verlag

33. Milner R (1989) Communication and Concurrency, International Series in Computer Science, Prentice-Hall

34. Milner R (1993) The polyadic pi-calculus: A tutorial, In: Logic and Algebra of Specification (Marktoberdorf, 1991), pages 203–246, Springer

35. Milner R, Parrow J, Walker D (1992) A calculus of mobile processes, I, Information and Computation, 100(1):1–40

36. MoBIES Model-Based Integration of Embedded Software program, http://www.rl.af.mil/tech/programs/MoBIES/

37. Moody J O, Antsaklis P J (1998) Supervisory Control of Discrete Event Systems Using Petri nets, Kluwer Academic Publishers

38. Overkamp A, van Schuppen J (2000) Maximal solutions in decentralized supervisory control, SIAM Journal of Control and Optimization, 39(2):492–511

39. Pinchinat S, Riedweg S (2005) A decidable class of problems for control under partial observation, Information Processing Letters, 95(4):454–460

40. Prosser J H, Kam M, Kwatny H G (1997) Decision fusion and supervisor synthesis in decentralized discrete-event systems, In: Proceedings of 1997 American Control Conference, pages 2251–2255

41. Ramadge P J, Wonham W M (1987) Supervisory control of a class of discrete event processes, SIAM Journal of Control and Optimization, 25(1):206–230

42. Ramadge P J, Wonham W M (1989) The control of discrete event systems, Proceedings of the IEEE, 77(1):81–98

43. Ricker L, Lafortune S, Genc S (2006) DESUMA: A tool integrating GIDDES and UMDES, In: Proceedings of the 8th International Workshop on Discrete Event Systems - WODES'06, pages 392–393

44. Ricker S L, Rudie K (2000) Know means no: Incorporating knowledge into discrete-event control systems, IEEE Transaction on Automatic Control, 45(9):1656–1668

45. Ricker S L, Rudie K (2003) Knowledge is a terrible thing to waste: using inference in discrete-event control problems, In: Proceedings of 2003 American Control Conference, pages 2246–2251

46. Rohloff K, Yoo T S, Lafortune S (2003) Deciding coobservability is PSPACE-complete, IEEE Transaction on Automatic Control, 48(11):1995–19998

47. Rudie K, Lafortune S, Lin F (2003) Minimal communication in a distributed discrete-event system, IEEE Transaction on Automatic Control, 48(6):957–975

48. Rudie K, Willems J C (1995) The computational complexity of decentralized discrete-event control problems, IEEE Transaction on Automatic Control, 40(7):1313–1318

49. Rudie K, Wonham W M (1990) Supervisory control of communicating processes, L. Logrippo, R. L. Probert, and H. Ural, editors, Protocol Specification, Testing and Verification X, pages 243–257, North-Holland

50. Rudie K, Wonham W M (1992) Protocol verification using discrete-event systems, In: Proceedings of 31st IEEE Conference on Decision and Control

51. Rudie K, Wonham W M (1992) Think globally, act locally: Decentralized supervisory control, IEEE Transaction on Automatic Control, 37(11):1692–1708

52. Sampath M (2001) A hybrid approach to failure diagnosis of industrial systems, In: Proceedings of 2001 American Control Conference

53. Sampath M, Sengupta R, Lafortune S, Sinnamohideen K, Teneketzis D (1995) Diagnosability of discrete event systems, IEEE Transaction on Automatic Control, 40(9):1555–1575

54. Sengupta R (2001) Discrete-event diagnostics of automated vehicles and highways, In: Proceedings of 2001 American Control Conference

55. Sinnamohideen K (2001) Discrete-event diagnostics of heating, ventilation, and air-conditioning systems, In: Proceedings of 2001 American Control Conference

56. Takai S (1998) On the language generated under fully decentralized supervision, IEEE Transaction on Automatic Control, 43(9):1253–1256

57. Thistle J G (1996) Supervisory control of discrete event systems, Mathematical and Computer Modelling, 23(11/12):25–53

58. Thistle J G, Malhamé R P, Hoang H H, Lafortune S (1997) Feature interaction modeling, detection and resolution: A supervisory control approach, In: Feature Interactions in Telecommunications IV, pages 93–107, IOS Press

59. Thorsley D, Teneketzis D (2004) Active acquisition of information for diagnosis of discrete event systems, In: Proceedings of the Allerton Conference on Control, Communication, and Computing

60. Thorsley D, Teneketzis D (2006) Diagnosis of cyclic discrete-event systems using active acquisition of information, In: Proceedings of 8th International Workshop on Discrete Event Systems (WODES'06), pages 248–255

61. Tripakis S (2001) Undecidable problems of decentralized observation and control, In: Proceedings of 40th IEEE Conference on Decision and Control, pages 4104–4109

62. van Schuppen J (1998) Decentralised supervisory control with information structures, In: Proceedings of the 1998 International Workshop on Discrete Event Systems (WODES'98), pages 36–41

63. van Schuppen J H (2004) Decentralized control with communication between controllers, Blondel V D and Megretski A, editors, Unsolved Problems in Mathematical Systems and Control Theory, pages 144–150, Princeton University Press, Princeton

64. Wang W (2006) Optimization of Communication and Coverage in Classes of Distributed Systems, PhD thesis, Department of Electrical Engineering and Computer Science, University of Michigan

65. Wang W, Lafortune S, Lin F (2006) A polynomial algorithm for minimizing communication in a distributed discrete event system with a central station, In: Proceedings of 45th IEEE Conference on Decision and Control

66. Willner Y, Heyman M (1991) Supervisory control of concurrent discrete event systems, International Journal of Control, 54(5):1143–1169

67. Wonham W M Supervisory Control of Discrete-Event Systems, University of Toronto, Revised 2005.07.01, Available at http://www.control.toronto.edu/people/profs/wonham/wonham.html

68. Wonham W M, Ramadge P J (1998) Modular supervisory control of discrete-event systems, Mathematics of Control, Signals and Systems, 1(1):13–30

69. Yoo T S, Lafortune S (2002) A general architecture for decentralized supervisory control of discrete-event systems, Discrete Event Dynamic Systems: Theory and Applications, 12(3):335–377

70. Yoo T S, Lafortune S (2004) Decentralized supervisory control with conditional decisions: Supervisor existence, IEEE Transaction on Automatic Control, 49(11):1886–1904

71. Yoo T S, Lafortune S (2005) Decentralized supervisory control with conditional decisions: Supervisor synthesis, IEEE Transaction on Automatic Control, 50(8):1205–1211

A Unifying Approach to the Design of Nonlinear Output Regulators

Lorenzo Marconi[1] and Alberto Isidori[1,2]

[1] CASY-DEIS, Università of Bologna, Italy
`lmarconi@deis.unibo.it`
[2] DIS, Sapienza - Università di Roma, Italy
and ESE, Washington University, St. Louis, USA
`isidori@ese.wustl.edu`

Summary. The goal of this paper is to propose a unique vision able to frame a number of results recently proposed in literature to tackle problems of output regulation for nonlinear systems. This is achieved by introducing the so-called *asymptotic internal model property* as the crucial property which, if fulfilled, leads to the design of the regulator for a fairly general class of nonlinear systems satisfying a proper minimum-phase condition. It is shown that recent frameworks based upon the use of nonlinear high-gain and adaptive observer techniques for the regulator design can be cast in this setting. A recently proposed technique for output regulation without immersion is also framed in these terms.

1 Introduction

This paper focuses on the problem of output regulation for nonlinear systems, namely the problem of designing an output feedback controller able to offset the effect of exogenous signals, which could be references to be tracked and/or disturbances to be rejected, generated by an autonomous system usually referred to as the "exosystem".

Besides specific meaningful control applications (see [3], [17]) which motivate the formulation and the interest of the problem at hand, the historical importance of the output regulation theory relies in fact that it studies control design paradigms which directly employ the a-priori knowledge of the environment in which the plant operates (provided, in the classical framework, by the structure of the exosystem) to obtain regulators with guaranteed asymptotic performances. This has led to the concept, of paramount importance in the linear (see [13]) as well as nonlinear ([18]) control theory, of *internal model* and to the identification of design procedures for *internal-model based regulators*.

It is a well-known fact that the ability of solving the problem at issue passes through the fulfillment of two key properties which should be achieved by a candidate controller. The first is the so-called "internal model property", required to any regulator solving the problem at hand (see [4]), which is related to the ability of generating, by means of the regulator's output, all the possible 'feed-forward inputs" which force an identically zero regulation error and, in turn, to guarantee the existence of a zero-error manifold which is invariant for the closed-loop

C. Bonivento et al. (Eds.): Adv. in Control Theory and Applications, LNCIS 353, pp. 185–200, 2007.
springerlink.com   © Springer-Verlag Berlin Heidelberg 2007

dynamics. Additionally, the candidate controller is required to exhibit a further crucial property asking that the "zero error manifold" is asymptotically stable for the closed-loop dynamics with a domain of attraction which, according to the specific output regulation problem dealt with, can be local or global.

In the case of linear systems, it is a well-known fact (see [12]-[13]) that a crucial role in the ability of fulfilling simultaneously the previous two properties is played by the so-called "non-resonance condition" asking, in plain words, that zeros of the controlled plant are disjoint from modes of the exosystem. This, under the obvious additional assumption that the plant is controllable, allows one to obtain constructive design procedures of internal model-based regulators.

For nonlinear systems the problem, as expected, is much more involved and challenging. This has motivated a number of works published on the related literature in the last fifteen years or so, all relying on the formulation of the so-called "immersion assumption" which consists in the requirement that the dynamical system defining all possible "feed-forward inputs" which force an identically zero regulation error be "immersed" into a system exhibiting certain structural properties. In this respect the past literature witnessed a steady development of less stringent assumptions: immersion into a linear *known* observable system (see [16], [20], [6], [23]), immersion into a linear *un-known* (but linearly parameterized) system ([24]), immersion into a linear system having a nonlinear output map ([7]), immersion into a nonlinear system linearizable by output injection ([10]), immersion into a system in canonical observability form ([5]), immersion into a system in a nonlinear adaptive observability form ([8], [9]), are only a few examples testifying the richness and liveliness of the past literature on this topic. This escalation of even more general and less restrictive conditions is then culminated in the result [21], in which, by taking advantage of the observer theory pioneered in [19] and developed in [1], it has been shown how the immersion assumption can be completely dropped. It must be noted, though, that the results in [21] lead to a non-constructive regulation theory and that immersion assumptions are still needed if one is willing to practically implement the regulator.

The goal of this paper is to put a bit in order this rich and apparently untidy scenario of contributions, by proposing a unique vision able to frame a number of previously mentioned results. More specifically we introduce the so-called *asymptotic internal model property* as the crucial property which, if fulfilled, allows one to design the regulator for a fairly general class of minimum-phase systems and we show how a number of (apparently un-correlated) immersion assumptions proposed so far, can be thought as conditions under which the asymptotic internal model property can be achieved. In particular we show how in the frameworks proposed in [5] and [8] (and all the frameworks encompassed by these works) the asymptotic internal model property can be constructively fulfilled and thus the regulator constructively designed. We also show how the recent framework of output regulation without immersion proposed in [21] can be cast in these terms.

The work is organized as follows. In the next section we present the framework of the problem. The definition of "asymptotic internal model property" and the result claiming that this property is sufficient for the regulator design are deferred in Section 3. The practical fulfilment of the asymptotic internal model property under the frameworks of [5], [8] and [21] is discussed in Section 4 while Section 5 concludes with final remarks.

2   The Framework

In this paper we consider nonlinear systems modeled by equations of the form

$$\begin{aligned}
\dot{z} &= f_0(w, z) + f_1(w, z, e_1)e_1 \\
\dot{e}_1 &= e_2 \\
&\;\;\vdots \\
\dot{e}_{r-1} &= e_r \\
\dot{e}_r &= q(w, z, e_1, \ldots, e_r) + u \\
e &= e_1 \\
y &= \mathrm{col}(e_1, \ldots, e_r),
\end{aligned} \tag{1}$$

with state $(z, e_1, \ldots, e_r) \in R^n \times R^r$, control input $u \in R$, regulated output $e \in R$, measured output $y \in R^r$, in which the exogenous (disturbance) input $w \in R^s$ is generated by an exosystem

$$\dot{w} = s(w). \tag{2}$$

The functions $f_0(\cdot), f_1(\cdot), q(\cdot), s(\cdot)$ in (1) and (2) are assumed to be at least continuously differentiable. The initial conditions of (1) range on a set $Z \times E$, in which Z is a fixed *compact* subset of R^n and $E = \{(e_1, \ldots, e_r) \in R^r : |e_i| \leq c\}$, with c a fixed number. The initial conditions of the exosystem (2) range on a *compact* subset W of R^s. In this framework we address the so-called problem of output regulation which consists in the design of an output feedback regulator of the form

$$\begin{aligned}
\dot{\zeta} &= \varphi(\zeta, y) \\
u &= \gamma(\zeta, y)
\end{aligned} \tag{3}$$

such that, in the corresponding closed loop system (1)-(3), *for all initial conditions $w(0) \in W$ and $(z(0), e_1(0), \ldots, e_r(0)) \in Z \times E$ trajectories are bounded in forward time* and $\lim_{t \to \infty} e(t) = 0$.

Augmenting (1) with (2) yields a system which, viewing u as input and e as output, has relative degree r. The associated "augmented" zero dynamics, which is forced by the control

$$c(w, z) = -q(w, z, 0, \ldots, 0), \tag{4}$$

is given by

$$\begin{aligned}
\dot{w} &= s(w) \\
\dot{z} &= f_0(w, z).
\end{aligned} \tag{5}$$

In what follows, we assume that system (5) satisfies the following three assumptions.

Assumption (i): the set W is (forward and backward) invariant for (2).   ∎

Note that, since W is invariant for $\dot{w} = s(w)$, the closed cylinder

$$\mathcal{C} := W \times R^n$$

is locally invariant for (5). Hence, it is natural regard (5) as a system defined on $\mathcal{C}$ and endow the latter with the subset topology.

Assumption (ii): there exists a compact subset $\mathcal{Z}$ of $\mathcal{C}$ which contains the positive orbit of the set $W \times Z$ under the flow of (5) and the resulting omega-limit set $\omega(W \times Z)$ satisfies

$$(w, z) \in \mathcal{C}, \quad |(w, z)|_{\omega(W \times Z)} \leq d_0 \quad \Rightarrow \quad z \in Z \tag{6}$$

where d_0 is a positive number.   ∎

As a remark on the above hypotheses, note that, since the positive orbit of the set $W \times Z$ under the flow of (5) is bounded, the set $\omega(W \times Z)$, namely the ω-limit set of $W \times Z$ under the flow of (5), is a nonempty, compact and invariant subset of $\mathcal{C}$ which uniformly attracts all trajectories of (5) with initial conditions in $W \times Z$. It can also be shown (as in [4]) that for every $w \in W$ there is $z \in R^n$ such that $(w, z) \in \omega(W \times Z)$. In what follows, for convenience, we introduce the notation

$$\mathcal{A} := \omega(W \times Z).$$

Condition (6) in assumption (ii) implies that $\mathcal{A}$, besides uniformly attracting trajectories of (5) originating from $W \times Z$, is also stable in the sense of Lyapunov (see [15]). In the next assumption we strengthen this property by also requiring the set $\mathcal{A}$ is locally exponentially stable.

Assumption (iii): there exist $M \geq 1$, $\lambda > 0$ such that

$$(w_0, z_0) \in \mathcal{C}, \quad |(w_0, z_0)|_{\mathcal{A}} \leq d_0 \quad \Rightarrow \quad |(w(t), z(t))|_{\mathcal{A}} \leq M e^{-\lambda t} |(w_0, z_0)|_{\mathcal{A}}$$

in which $(w(t), z(t))$ denotes the solution of (5) passing through (w_0, z_0) at time $t = 0$.   ∎

For sake of simplicity, in the next part of the paper we address the problem at hand under assumptions (i) - (ii) - (iii) in the simplified case of plants with relative degree $r = 1$, i.e. in the special case in which system (1) is a system of the form

$$\begin{aligned}
\dot{z} &= f_0(w, z) + f_1(w, z, e)e \\
\dot{e} &= q(w, z, e) + u \\
y &= e.
\end{aligned} \tag{7}$$

As shown in [8], this can be done without loss of generality, since a wise use of the tools proposed in [25] allows one to reduce the higher relative case to an equivalent problem of output regulation for a system of form (7). The interested reader is referred to these references for details which are omitted here for reasons of space.

3   The Asymptotic Internal Model Property

We begin by rewriting the zero dynamics of the augmented system (2), (7), given by

$$\dot{w} = s(w)$$
$$\dot{z} = f_0(w, z)\,, \qquad (8)$$

in the more compact form

$$\dot{\mathbf{z}} = \mathbf{f}_0(\mathbf{z}) \qquad (9)$$

where $\mathbf{z} := \mathrm{col}(w, z)$. Moreover, consistently with this notation, we rewrite the term $q(w, z, e)$ in (7) as

$$q(w, z, e) = \mathbf{q}_0(\mathbf{z}) + q_1(\mathbf{z}, e)e$$

in which $\mathbf{q}_0(\mathbf{z}) = q(w, z, 0)$, and we denote by $\mathbf{Z} := W \times Z$ the compact set where the initial condition $\mathbf{z}(0)$ is supposed to range. In view of this the overall system (2), (7) is rewritten as

$$\dot{\mathbf{z}} = \mathbf{f}_0(\mathbf{z}) + \mathbf{f}_1(\mathbf{z}, e)e$$
$$\dot{e} = \mathbf{q}_0(\mathbf{z}) + q_1(\mathbf{z}, e)e + u \qquad (10)$$

where $\mathbf{f}_1(\mathbf{z}, e) = \mathrm{col}(0, f_1(w, z, e))$ and the initial conditions $(\mathbf{z}(0), e(0))$ range in the set $\mathbf{Z} \times E$.

System (10) being affine in the control input u, it seems natural to look for a controller having a similar structure, namely a controller of the form

$$\dot{\xi} = \varphi(\xi) + \psi(\xi)v$$
$$u = \gamma(\xi) + v \qquad (11)$$

with state $\xi \in R^d$, in which v is a residual control input, to be eventually chosen as a function of the measured output e. For consistency with the earlier assumptions, the initial condition $\xi(0)$ of (11) is allowed to range on a fixed compact set $\varXi$ of R^d. Here $\varphi(\cdot)$, $\psi(\cdot)$ and $\gamma(\cdot)$ are at least continuous functions to be determined.

The main result of the section is to show that, if the triplet $\{\varphi(\xi), \psi(\xi), \gamma(\xi)\}$ possesses what we now define as *asymptotic internal model* property, the choice of the residual control v in (11) as

$$v = -ke$$

solves the problem of output regulation, provided that the gain coefficient k is large enough.

Definition 1. *The triplet* $\{\varphi(\xi), \psi(\xi), \gamma(\xi)\}$ *has the* asymptotic internal model *property if there exists a* C^1 *map* $\tau : \mathcal{Z} \to R^d$ *such that:*

(i) the vector fields $\mathbf{f}_0|_{\mathcal{A}}$ *and* φ *are* τ-*related, namely*

$$\frac{\partial \tau(\mathbf{z})}{\partial \mathbf{z}} \mathbf{f}_0(\mathbf{z}) = \varphi(\tau(\mathbf{z})) \qquad \forall\, \mathbf{z} \in \mathcal{A}\,, \qquad (12)$$

and

$$\mathbf{q}_0(\mathbf{z}) + \gamma \circ \tau(\mathbf{z}) = 0 \qquad \forall \, \mathbf{z} \in \mathcal{A} \, ; \tag{13}$$

(ii) in the composite system

$$\dot{\mathbf{z}} = \mathbf{f}_0(\mathbf{z})$$
$$\dot{\xi} = \varphi(\xi) - \psi(\xi)[\gamma(\xi) + \mathbf{q}_0(\mathbf{z})] \tag{14}$$

the set

$$\mathrm{graph}(\tau|_{\mathcal{A}}) = \{(\mathbf{z}, \xi) : \mathbf{z} \in \mathcal{A}, \, \xi = \tau(\mathbf{z})\}$$

uniformly and locally exponentially attracts $\mathbf{Z} \times \Xi$.

Note that conditions (12) and (13), in the terminology of [11], simply express the property that the restriction to $\mathcal{A}$ of the autonomous system with output

$$\dot{\mathbf{z}} = \mathbf{f}_0(\mathbf{z}), \qquad y = \mathbf{q}_0(\mathbf{z}) \tag{15}$$

is *immersed* into the system

$$\dot{\xi} = \varphi(\xi), \qquad y = \gamma(\xi) \, . \tag{16}$$

As a remark to this definition, note that the conditions indicated in (i) imply the invariance of the compact set $\mathrm{graph}(\tau|_{\mathcal{A}})$ under the flow of (14). If condition (ii) also holds, the set in question can be identified with $\omega(\mathbf{Z} \times \Xi)$, the limit set of $\mathbf{Z} \times \Xi$ under the flow of (14), as shown below in the proof of Lemma 1. The use of the adjective "asymptotic" in previous definition is meant to highlight the fact that it is only the set $\mathcal{A}$ – which, in turn, characterizes the *asymptotic* behavior of the augmented zero dynamics (9) – that matters in the conditions (i) and (ii).

We postpone to the next section the presentation of relevant cases in which a controller which possesses the asymptotic internal model property can be constructively designed. In this section we are mostly interested to the conceptual result that properties (i), (ii) and (iii) involving the augmented zero dynamics and the asymptotic internal model property of the triplet $\{\varphi(\xi), \psi(\xi), \gamma(\xi)\}$ are indeed sufficient for solving the problem in question with a regulator of the form (11). This is formally stated and proved in the next lemma.

Lemma 1. *Pick compact sets* $\mathbf{Z}$, E *and* Ξ *for the initial conditions of the closed-loop system (2), (7), (11). Assume that (i)-(ii)-(iii) hold and that the triplet* $\{\varphi, \psi, \gamma\}$ *has the asymptotic internal model property. Assume, in addition, that the vector field* $\psi(\xi)$ *is complete. Then there exists* $k^{\star} > 0$ *such that for all* $k \geq k^{\star}$ *the controller (11) with* $v = -ke$ *solves the problem of output regulation.*

Proof. Consider the closed-loop system

$$\dot{\mathbf{z}} = \mathbf{f}_0(\mathbf{z}) + \mathbf{f}_1(\mathbf{z}, e)e$$
$$\dot{e} = \mathbf{q}_0(\mathbf{z}) + q_1(\mathbf{z}, e)e + \gamma(\xi) + v$$
$$\dot{\xi} = \varphi(\xi) + \psi(\xi)v$$

which, regarded as a system with input v and output e, has relative degree 1 and zero dynamics given by

$$\dot{\mathbf{z}} = \mathbf{f}_0(\mathbf{z})$$
$$\dot{\xi} = \varphi(\xi) - \psi(\xi)(\gamma(\xi) + \mathbf{q}_0(\mathbf{z})) . \tag{17}$$

The first crucial step to prove the lemma is to show that the trajectories of (17) originating from $\mathbf{Z} \times \Xi$ are bounded and the consequent ω-limit set $\omega(\mathbf{Z} \times \Xi)$ is precisely $\mathrm{graph}(\tau|_{\mathcal{A}})$. To this end note that boundedness of the trajectories is a consequence of requirement (ii) in the definition of the asymptotic internal model property. To show that $\omega(\mathbf{Z} \times \Xi) = \mathrm{graph}(\tau|_{\mathcal{A}})$ note that, by the triangular structure of (17), it turns out that

$$\omega(\mathbf{Z} \times \Xi) \subset \mathcal{C} .$$

Furthermore, by requirement (i) in definition 1, it follows that $\mathrm{graph}(\tau|_{\mathcal{A}})$ is an invariant set for (17) and thus $\mathrm{graph}(\tau|_{\mathcal{A}}) \subset \omega(\mathbf{Z} \times \Xi)$. To prove that $\mathrm{graph}(\tau|_{\mathcal{A}}) \equiv \omega(\mathbf{Z} \times \Xi)$ we proceed by contradiction. For, suppose that there exists $(\mathbf{z}_0', \xi_0') \in \omega(\mathbf{Z} \times \Xi)$ such that

$$|(\mathbf{z}_0', \xi_0')|_{\mathrm{graph}(\tau|_{\mathcal{A}})} = c > 0 \tag{18}$$

and denote by $(\mathbf{z}'(t), \xi'(t))$ the solution of (17) at time t passing through $(\mathbf{z}_0', \xi_0')$ at time $t = 0$. As $\omega(\mathbf{Z} \times \Xi)$ is (backward) invariant and compact there exists a number $K_1 > 0$ such that

$$|(\mathbf{z}'(t), \xi'(t))|_{\mathrm{graph}(\tau|_{\mathcal{A}})} \leq K_1 \qquad \text{for all } t \leq 0 . \tag{19}$$

Now note that by uniform attractiveness in requirement (ii) of definition 1, it turns out that for all positive $K_2 \leq K_1$ there exists $T > 0$ such that for all $(\mathbf{z}_0, \xi_0) \in \mathbf{Z} \times \Xi$ satisfying

$$|(\mathbf{z}_0, \xi_0)|_{\mathrm{graph}(\tau|_{\mathcal{A}})} \leq K_1 \tag{20}$$

the trajectory $(\mathbf{z}(t), \xi(t))$ of (14) passing through $(\mathbf{z}_0, \xi_0)$ at time $t = 0$ is such that

$$|(\mathbf{z}(T), \xi(T))|_{\mathrm{graph}(\tau|_{\mathcal{A}})} \leq K_2 . \tag{21}$$

Moreover local exponential stability in the second requirement of the previous definition implies the existence of positive d, M, λ such that for all $(\mathbf{z}_0, \xi_0)$ satisfying

$$|(\mathbf{z}_0, \xi_0)|_{\mathrm{graph}(\tau|_{\mathcal{A}})} \leq d$$

the trajectory is such that

$$|(\mathbf{z}(t), \xi(t))|_{\mathrm{graph}(\tau|_{\mathcal{A}})} \leq M e^{-\lambda t} |(\mathbf{z}_0, \xi_0)|_{\mathrm{graph}(\tau|_{\mathcal{A}})} .$$

Combining the previous two properties with K_2 chosen so that $K_2 \leq d$ and T consequently, it is possible to check that

$$|(\mathbf{z}(0), \xi(0))|_{\mathrm{graph}(\tau|_{\mathcal{A}})} \leq K_1 \qquad \Rightarrow$$

$$|(\mathbf{z}(t), \xi(t))|_{\mathrm{graph}(\tau|_{\mathcal{A}})} \leq \bar{M} e^{-\lambda t} |(\mathbf{z}(0), \xi(0))|_{\mathrm{graph}(\tau|_{\mathcal{A}})}$$

where $\bar{M} := \max\{M, K_1/e^{-\lambda T}\}$. From this, choosing T' such that $\bar{M} e^{-\lambda T'} K_1 \leq 0.5c$ and using (19), it turns out that

$$\begin{aligned} |(\mathbf{z}_0', \xi_0')|_{\mathrm{graph}(\tau|_{\mathcal{A}})} &\leq \bar{M} e^{-\lambda T'} |(\mathbf{z}'(-T'), \xi'(-T'))|_{\mathrm{graph}(\tau|_{\mathcal{A}})} \\ &\leq \bar{M} e^{-\lambda T'} K_1 \leq 0.5c \end{aligned}$$

which contradicts (18). This proves that $\mathrm{graph}(\tau|_{\mathcal{A}}) \equiv \omega(\mathbf{Z} \times \Xi)$.

Let now $\Phi_\psi^t(\xi)$ denote the flow of the complete vector field $\psi(\xi)$ and consider the (partial) change of coordinates

$$\eta := \Phi_\psi^{-e}(\xi) \,.$$

Since, by definition,

$$\frac{\partial \Phi_\psi^{-e}(\xi)}{\partial e} + \frac{\partial \Phi_\psi^{-e}(\xi)}{\partial \xi} \psi(\xi) = \frac{\partial \Phi_\psi^{-e}(\Phi_\psi^e(\eta))}{\partial e} = 0$$

and $\Phi_\psi^0(\eta) = \eta$, it is easy to see that the system in the new coordinates reads as

$$\begin{aligned} \dot{p} &= f(p) + \ell(p, e) \\ \dot{e} &= q(p) + r(p, e) + v \end{aligned} \tag{22}$$

in which $p := \mathrm{col}(\mathbf{z}, \eta)$,

$$f(p) = \begin{pmatrix} \mathbf{f}_0(\mathbf{z}) \\ \varphi(\eta) - \psi(\eta)(\mathbf{q}_0(\mathbf{z}) + \gamma(\eta)) \end{pmatrix} \qquad q(p) = \mathbf{q}_0(\mathbf{z}) + \gamma(\eta)$$

and $\ell(p, e)$, $r(p, e)$ are suitably defined smooth functions of their arguments such that $\ell(p, 0) = r(p, 0) = 0$ for all p. In particular, by the first part of the proof, it turns out that the zero dynamics $\dot{p} = f(p)$ of (22) posses an uniformly attractive (locally exponentially) compact set on which $q(p)$ is identically zero. From this and the choice $v = -ke$ the claim of the Lemma follows by high-gain results such as the ones proposed in [21] (see Theorems 2 and 3 in the quoted reference). ∎

4   Achieving the Asymptotic Internal Model Property

Goal of this section is to present relevant cases, taken from existing literature, in which a controller satisfying the asymptotic internal model property can be identified. By bearing in mind the definition, the property in question is easily seen to be related to the capability of reproducing, by means of the output $\gamma(\xi)$ of the system $\dot{\xi} = \varphi(\xi) - \psi(\xi)(\gamma(\xi) + \mathbf{q}_0(\mathbf{z}))$, the *asymptotic* behavior of the output $\mathbf{q}_0(\mathbf{z})$ of the system $\dot{\mathbf{z}} = \mathbf{f}_0(\mathbf{z})$. This, in particular, shows up through the two requirements detailed in the definition: the first which asks for the existence of the invariant compact set $\mathrm{graph}(\tau|_{\mathcal{A}})$ for the two systems on which the two outputs coincide, and the second in which this set is required to be (locally exponentially) attractive for the composite system (14).

It is apparent that the problem in question is intimately related to the problem of designing nonlinear observers for the system $\dot{\mathbf{z}} = \mathbf{f}_0(\mathbf{z})$ with output $y = \mathbf{q}_0(\mathbf{z})$. As a matter of fact, with an eye to the composite system (14), one would be tempted to design the triplet (φ, ψ, γ) looking at the ξ-subsystem as an observer of the $\mathbf{z}$-subsystem, with $\varphi(\xi)$ playing the role of observer dynamics, $\psi(\xi)$ the role of output injection "gain" and $\gamma(\xi) + \mathbf{q}_0(\mathbf{z})$ the role of innovation term. In the following part of the section we follow this intuition by showing how, indeed, the theory of nonlinear observers is helpful to identify a triplet (φ, ψ, γ) satisfying the asymptotic internal model property. More specifically, in the next two subsections, we show how the theory of *nonlinear high gain observers* (see [14]) and, respectively, *nonlinear adaptive observers* (see [2], [22]) can be successfully employed to obtain *constructive design procedures* for the triplet in question. In doing this we follow design procedures which have been proposed respectively in [5] and [8]. These results rely upon the extra observability assumption that the "observed" system $(\mathbf{f}_0, \mathbf{q}_0)$ is immersed into a system in nonlinear uniform observability form and, respectively, nonlinear adaptive observability form.

Furthermore, in subsection 3.3, following [21], the observer theory of [19] (see also [1]) is taken as theoretical tool to present a result claiming the existence of the triplet in question without any specific observability assumption. As opposite to the procedures presented in subsections 3.1 and 3.2, this result is *not constructive* but it applies to a fairly general class of nonlinear systems only fulfilling hypotheses (i)-(iii) in section 2 without any extra immersion assumption.

4.1   Nonlinear Immersion (See [5])

Assume the existence of an integer $d > 0$, of a locally Lipschitz function $f : R^d \to R$ such that, for any $\mathbf{z} \in \mathcal{A}$, the solution $\mathbf{z}(t)$ of passing through $\mathbf{z}$ at time $t = 0$ is such that the function $\rho(t) := \mathbf{q}_0(\mathbf{z}(t))$ satisfies

$$\rho^{(d)}(t) = f(\rho(t), \rho^{(1)}(t), \dots, \rho^{(d-1)}(t))$$

for all $t \in R$.

Let $\tau' : \mathcal{Z} \to R^d$ be the map defined as

$$\tau'(\mathbf{z}) := \mathrm{col}(\mathbf{q}_0(\mathbf{z}), L_{\mathbf{f}_0}\mathbf{q}_0(\mathbf{z}), \dots, L_{\mathbf{f}_0}^{d-1}\mathbf{q}_0(\mathbf{z})) \tag{23}$$

and let $f_c : R^d \to R$ be a function with compact support which agrees with $f(\cdot)$ on $\tau'(\mathcal{A})$, namely

$$f_c|_{\tau'(\mathcal{A})} = f|_{\tau'(\mathcal{A})} \qquad \text{and} \qquad |f_c(s)| \le K < \infty \qquad \text{for all } s \in R^d .$$

Then, it easy to check that the properties indicated in item (i) of the definition are fulfilled by choosing

$$\varphi(\xi) = \begin{pmatrix} \xi_2 \\ \vdots \\ \xi_d \\ f_c(\xi_1, \xi_2, \ldots, \xi_d) \end{pmatrix}, \qquad \gamma(\xi) = \xi_1, \qquad (24)$$

with $\tau(\mathbf{z}) = \tau'(\mathbf{z})$. Comparing this construction with the remark after Definition 1 we observe, in particular, that system (15) is *immersed into a system which is uniformly observable*, in the sense of [14] (even though system (15) might not have had such a property). It is precisely this that makes it possible to choose $\psi(\xi)$ in such a way that also the property indicated in item (ii) of the definition can be achieved.

As a matter of fact, the property in question is achieved by choosing

$$\psi(\xi) = D_k \begin{pmatrix} c_0 \\ \vdots \\ c_{d-1} \end{pmatrix}$$

where $D_k = \mathrm{diag}(k, k^2, \cdots, k^d)$, k is a design parameter, and the c_i's are such that the polynomial $\lambda^d + c_0 \lambda^{d-1} + \cdots + c_{d-1} = 0$ is Hurwitz, as formally proved in Lemmas 1 and 2 of [5] to which the interested reader is referred for details.

It is worth noting that the assumption in question clearly covers the interesting (and widely addressed in the recent past literature, see [16]) case in which the function $f(\cdot)$ is linear, namely the case in which (15) is immersed into a linear observable system. In this case, although the choice indicated above is clearly still valid, a more direct way of designing the regulator is to use $f(\cdot)$ instead of $f_c(\cdot)$ in the definition of $\varphi(\xi)$, to set $\psi(\xi) = G$ and simply choose G in such a way that $\dot{\xi} = \varphi(\xi) - G\gamma(\xi)$ is a stable linear system.

4.2   Adaptive Immersion (See [8])

Implicit in the setup of the problem of output regulation is the possibility that the vector w of exogenous inputs includes a set of uncertain constant parameters. The latter can be uncertain parameters in the model of the controlled plant (1) but also uncertain parameters affecting the dynamics of some other exogenous inputs. In this case, in fact, one can still consider a set (w_1, w_2) of exogenous inputs obeying

$$\dot{w}_1 = s_1(w_1, w_2)$$
$$\dot{w}_2 = 0$$

in which $s_1(w_1, w_2)$ explicitly depends on w_2. If this is the case, it is unlikely that an assumption such as the one introduced at the beginning of the earlier section is going to be fulfilled, and different scenarios have to be considered. A an obvious option would be to assume the existence of a function $f : R^d \times R^q \to R$ and of a map $\theta : \mathcal{A} \to R^q$ such that, for any $\mathbf{z} \in \mathcal{A}$, the solution $\mathbf{z}(t)$ of passing through $\mathbf{z}$ at time $t = 0$ is such that the functions $\rho(t) := \mathbf{q}_0(\mathbf{z}(t))$ and $\theta(t) := \theta(\mathbf{z}(t))$ satisfy

$$\rho^{(d)}(t) = f(\rho(t), \rho^{(1)}(t), \ldots, \rho^{(d-1)}(t), \theta(t)) \quad \text{and} \quad \theta^{(1)}(t) = 0$$

for all $t \in R$. In this case, though, while the immersion property (i) is easily fulfilled (exactly as in the previous case), it becomes quite hard to have property (ii) fulfilled. In order to make this possible, some extra (stringent) assumptions, on the function f, must be imposed.

Note that, if the hypothesis indicated above holds, system (15) is immersed into the $(d + q)$ – dimensional system

$$\dot{\eta} = \begin{pmatrix} \eta_2 \\ \vdots \\ \eta_d \\ f(\eta_1, \eta_2, \ldots, \eta_d, \theta) \end{pmatrix}, \qquad y = \eta_1 \tag{25}$$

$$\dot{\theta} = 0 \tag{26}$$

via the pair of maps

$$\eta = \tau'(\mathbf{z}), \qquad \theta = \theta(\mathbf{z}),$$

in which $\tau'(\mathbf{z})$ is the map defined in (23). The assumption that we make now is that there is a globally defined diffeomorphism $\tilde{\eta} = \Phi(\eta)$ that changes system (25) into a system in adaptive observability form, in the sense of [22], namely a system of the form

$$\dot{\tilde{\eta}} = A\tilde{\eta} + \phi(C\tilde{\eta}) + \Omega(C\tilde{\eta})\theta, \qquad y = C\tilde{\eta} \tag{27}$$

in which A, C is an observable pair, and $\phi : R \to R^d$ and $\Omega : R \to R^{d \times q}$ are smooth functions. Conditions under which this is possible are well-known and can be found, for instance, in [22]. Note that, if this assumption holds, the map $\tilde{\tau}(\mathbf{z}) := \Phi(\tau'(\mathbf{z}))$ satisfies

$$\frac{\partial \tilde{\tau}}{\partial \mathbf{z}} \mathbf{f}_0(\mathbf{z}) = A\tilde{\tau}(\mathbf{z}) + \phi(C\tilde{\tau}(\mathbf{z})) + \Omega(C\tilde{\tau}(\mathbf{z}))\theta(\mathbf{z}), \qquad \mathbf{q}_0(\mathbf{z}) = C\tilde{\tau}(\mathbf{z}). \tag{28}$$

This being said, we define now the triplet $\{\varphi(\xi), \psi(\xi), \gamma(\xi)\}$ as follows (see [8])

$$\xi = \mathrm{col}(\xi_1, \xi_2, \xi_3) \qquad \text{with } \xi_1 \in R^d, \ \xi_2 \in R^q, \ \xi_3 \in R^{d-1} \times R^q,$$

$$\varphi(\xi) = \begin{pmatrix} A\xi_1 + \phi_c(C\xi_1) + \Omega_c(C\xi_1)\xi_2 - M(\xi_3)\mathrm{dzv}_\ell(\xi_2) \\ -\mathrm{dzv}_\ell(\xi_2) \\ F\xi_3 + G\Omega_c(C\xi_1) \end{pmatrix},$$

$$\tag{29}$$

$$\psi(\xi) = \begin{pmatrix} H(\xi_3, \xi_1) \\ \beta(\xi_3, \xi_1) \\ 0 \end{pmatrix}, \qquad \gamma(\xi) = C\xi_1$$

in which $\phi_c(\cdot)$ and $\Omega_c(\cdot)$ denote functions with compact support which agree with $\phi(\cdot)$ and $\Omega(\cdot)$ on $C\tilde{\tau}(\mathcal{A})$, $F \in R^{d-1 \times d-1}$ and $G \in R^{d-1 \times d}$ are chosen as

$$F = \begin{pmatrix} -b_2 & 1 & \cdots & 0 & 0 \\ \cdot & \cdot & \cdots & \cdot & \cdot \\ -b_{d-1} & 0 & \cdots & 0 & 1 \\ -b_d & 0 & \cdots & 0 & 0 \end{pmatrix}, \qquad G = \begin{pmatrix} -b_2 & 1 & \cdots & 0 & 0 & 0 \\ \cdot & \cdot & \cdots & \cdot & \cdot & \cdot \\ -b_{d-1} & 0 & \cdots & 0 & 1 & 0 \\ -b_d & 0 & \cdots & 0 & 0 & 1 \end{pmatrix} \qquad (30)$$

and $M(\cdot)$, $\beta(\cdot,\cdot)$, $H(\cdot,\cdot)$ as

$$M(\xi_3) = \begin{pmatrix} 0 \\ \xi_3 \end{pmatrix}, \qquad \begin{aligned} \beta^{\mathrm{T}}(\xi_3,\xi_1) &= CAM(\xi_3) + C\Omega_c(C\xi_1), \\ H(\xi_3,\xi_1) &= M(\xi_3)\beta(\xi_3,\xi_1) + K, \end{aligned}$$

where the b_i's, $i = 2, \cdots, d$, and K are design parameters. Finally, $\mathrm{dzv}_\ell(\cdot)$ is the vector-valued *dead-zone* function defined as

$$\mathrm{dzv}_\ell(\mathrm{col}(s_1,\ldots,s_d)) = \mathrm{col}(\mathrm{dz}_\ell(s_1),\ldots,\mathrm{dz}_\ell(s_d)) \qquad (31)$$

in which $\mathrm{dz}_\ell(\cdot)$ is any continuously differentiable function satisfying

$$\mathrm{dz}_\ell(x) = \begin{cases} 0 & \text{if } |x| \leq \ell \\ x & \text{if } |x| \geq \ell + 1. \end{cases}$$

Lengthy, but not difficult, computations can be used to check that if the coefficients b_i's are chosen so that the matrix F is Hurwitz and ℓ so that $\ell \geq \max_{\mathbf{z} \in \mathcal{A}} |\theta(\mathbf{z})|$, then the map

$$\tau(\mathbf{z}) = \mathrm{col}(\tilde{\tau}(\mathbf{z}), \theta(\mathbf{z}), \sigma(\mathbf{z})) \quad \text{where} \quad \sigma(\mathbf{z}) = \int_{-\infty}^{0} e^{-Fs} G\Omega(C\tilde{\tau}(\mathbf{z}(s,\mathbf{z}))) ds$$

$$(32)$$

is such that $\mathrm{graph}(\tau|_{\mathcal{A}})$ is invariant for $\dot{\xi} = \varphi(\xi)$ and $\mathbf{q}_0|_{\mathcal{A}} = \gamma \circ \tau|_{\mathcal{A}}$ and thus the first requirement in the Definition 1 is fulfilled. Furthermore it can be proved that, if K is appropriately chosen, $\mathrm{graph}(\tau|_{\mathcal{A}})$ also uniformly (and locally exponentially) attracts $\mathbf{Z} \times \Xi$ under the flow of (14), namely that the triplet (29) also fulfills the second requirement of Definition 1. The result in question is presented in the next proposition, whose proof – which relies upon a persistence of excitation condition – can be found in [8] .

Proposition 1. *Fix* $(\varphi(\xi),\gamma(\xi),\psi(\xi))$ *as in (29) and* $\tau(\mathbf{z})$ *as in (32). Set* $b = \mathrm{col}(1,b_2,\ldots,b_{q_1})$ *and choose*

$$K = Ab + \lambda b$$

with λ *a design parameter. If for all* $\mathbf{z}_0 \in \mathcal{A}$ *the following implication is true (persistence of excitation condition)*

$$\varsigma^{\mathrm{T}}\beta\left(\sigma\left(\mathbf{z}(t,\mathbf{z}_0)\right),\, \tilde{\tau}\left(\mathbf{z}(t,\mathbf{z}_0)\right)\right) = 0 \quad \forall t \geq 0 \qquad \Rightarrow \qquad \varsigma \equiv 0,$$

then there exists $\lambda^\star > 0$ *such that for all* $\lambda \geq \lambda^\star$ *the set* $\mathrm{graph}(\tau|_{\mathcal{A}})$ *uniformly (locally exponentially) attracts* $\mathbf{Z} \times \Xi$ *under the flow of (14).*

It is interesting to note that the analysis discussed above covers also the particular case in which the exosystem state $\mathbf{z}$ includes a vector ϱ of *constant* uncertain parameters ranging in a compact set $P \subset R^p$ and there exists a differentiable map $\tau' : \mathcal{Z} \to R^d$ such that[1]

$$\frac{\partial \tau'(\mathbf{z})}{\partial \mathbf{z}} \mathbf{f}_0(\mathbf{z}) = S(\varrho)\tau'(\mathbf{z})$$
$$\mathbf{q}_0(\mathbf{z}) = \Gamma(\varrho)\tau'(\mathbf{z}) \tag{33}$$

in which $(S(\varrho), \Gamma(\varrho)) \in R^{d\times d} \times R^{1\times d}$ is an observable pair for all $\varrho \in \mathcal{P}$. In fact, note that, since the pair $(S(\varrho), \Gamma(\varrho))$ is observable for all ϱ, standard arguments can be used to show that there exist a nonsingular matrix $M(\varrho) \in R^{d\times d}$, a column vector $L(\varrho) \in R^{d\times 1}$, and an observable (parameter independent) pair $(A, C) \in R^{d\times d} \times R^{1\times d}$ such that

$$M(\varrho)S(\varrho)M(\varrho)^{-1} = A + L(\varrho)C$$
$$\Gamma(\rho)M(\varrho)^{-1} = C \,.$$

From this, it turns out that relation (28) holds with $q = d$, $\phi(s) = 0$, $\Omega(s) = sI_d$, $\tau(\mathbf{z}) = M(\varrho)\tau'(\mathbf{z})$ and $\theta(\mathbf{z}) = L(\varrho)$. This, in particular, shows that general design procedure leading to chioce of the triplet $(\varphi(\xi), \gamma(\xi), \psi(\xi))$ in (29) can be successfully adopted. It is interesting to note, however, that in this particular case the general procedure detailed above can be simplified to obtain the triplet fulfilling the asymptotic internal model property in a more direct and effective way. How this is possible is explained in the following (see [9] for details).

Let $(F, G) \subset R^{d\times d} \times R^{1\times d}$ be an arbitrary controllable pair with F Hurwitz and let $T(\varrho)$ denote the unique nonsingular solution of the Sylvester equation

$$FT(\varrho) - T(\varrho)S(\varrho) = -G\Gamma(\varrho)$$

and $\Psi(\varrho)$ the row vector $\Psi(\varrho) = \Gamma(\varrho)T^{-1}(\varrho)$. By bearing in mind the definition (31), set $\xi = \mathrm{col}(\xi_1, \xi_2)$ with $\xi_1 \in R^d$ and $\xi_2 \in R^d$, and choose the triplet as

$$\varphi(\xi) = \begin{pmatrix} (F + G\xi_2^{\mathrm{T}})\xi_1 \\ -\mathrm{dzv}_\ell(\xi_2) \end{pmatrix}, \qquad \gamma(\xi) = \xi_2^{\mathrm{T}}\xi_1, \qquad \psi(\xi) = \begin{pmatrix} G \\ \xi_1 \end{pmatrix}. \tag{34}$$

Simple, though lengthy, algebra can be used to show that if ℓ is chosen so that

$$\ell \geq \max_{\varrho \in P} |\Psi^{\mathrm{T}}(\varrho)|$$

then the first requirement of Definition 1 is satisfied by the triplet (34) through the map

$$\tau(\mathbf{z}) = \begin{pmatrix} T(\varrho)\,\tau'(\mathbf{z}) \\ \Psi^{\mathrm{T}}(\varrho) \end{pmatrix}, \tag{35}$$

in which, as also stressed above, the constant parameters ϱ can be though as trivial components of $\mathbf{z}$. Moreover also the second requirement of Definition 1 can be shown to be satisfied provided that a persistence of excitation condition,

---

[1] This scenario is representative of the important case in which $\mathbf{q}_0(\mathbf{z}(t))$ is the sum of a finite number of periodic signals of uncertain amplitude, phase and frequency (see [24]).

detailed in the next proposition, is fulfilled. For the proof of this proposition the interested reader is referred to [9].

Proposition 2. *Fix $(\varphi(\xi), \gamma(\xi), \psi(\xi))$ as in (34) and $\tau(\mathbf{z})$ as in (35). If there exist positive T and K such that*

$$\int_t^{t+T} \tau'(\mathbf{z}(s, \mathbf{z}_0))\, \tau'^{\mathrm{T}}(\mathbf{z}(s, \mathbf{z}_0))\, ds \geq KI$$

for all $t \geq 0$ and for all $\mathbf{z}_0 \in \mathcal{A}$, then the set $\mathrm{graph}(\tau|_{\mathcal{A}})$ uniformly (locally exponentially) attracts $\mathbf{Z} \times \Xi$ under the flow of (14).

4.3 Asymptotic Internal Model Property Without Immersion (See [21])

In this section we follow the theory presented in [21] to show that no immersion assumptions are needed at all in order to fulfill the asymptotic internal model property. As opposite to the frameworks discussed in the previous two subsections, this kind theory leads to non constructive results for the design of the regulator which, furthermore, is not guaranteed to be smooth.

Let $(F, G) \in R^{d \times d} \times R^{d \times 1}$ be a controllable pair and set

$$\varphi(\xi) = F\xi + G\gamma(\xi), \qquad \psi(\xi) = G$$

with $\gamma : R^d \to R$ a continuous function to be chosen in such a way that the proposed triplet has the required properties.

With this choice it turns out that the composite system (14) assumes the form

$$\begin{aligned}
\dot{\mathbf{z}} &= \mathbf{f}_0(\mathbf{z}) \\
\dot{\xi} &= F\xi - G\mathbf{q}_0(\mathbf{z})\,.
\end{aligned} \tag{36}$$

The first step instrumental to prove that the triplet in question can be made to satisfy the asymptotic internal model property is presented in the following proposition whose only requirement is that the matrix F is Hurwitz. For the proof the reader is referred to [21].

Proposition 3. *Consider system (36) under the assumption (i)-(ii)-(iii) in Section 2. There exists an $\ell > 0$ such that if the eigenvalues of F have real parts lower than $-\ell$, then the map*

$$\tau(\mathbf{z}) = \int_{-\infty}^0 e^{-Fs} G\mathbf{q}_0(\mathbf{z}(s, \mathbf{z}))ds \tag{37}$$

is differentiable, satisfies

$$\frac{\partial \tau}{\partial \mathbf{z}} \mathbf{f}_0(\mathbf{z}) = F\tau(\mathbf{z}) - G\mathbf{q}_0(\mathbf{z}) \qquad \forall\, \mathbf{z} \in \mathcal{A}, \tag{38}$$

and it is such that the set $\mathrm{graph}(\tau|_{\mathcal{A}})$ is locally exponentially stable for (36) with a domain of attraction containing $\mathbf{Z} \times \Xi$.

It is worth stressing that the requirement of choosing F, besides being Hurwitz, with a certain stability margin fixed by the integer ℓ, represents only a technical assumption needed to guarantee differentiability of the function τ (see the proof of Proposition 2 in [21]). In this sense the assumption in question must be not confused with an "high gain" requirement on the choice of F. Note, moreover, that the function γ and the dimension d of the pair (F, G) do not play any role in establishing this result.

By its own the previous result only guarantees the fulfillment of the requirement (ii) in definition 1, namely the existence of the exponentially stable set $\mathrm{graph}(\tau|_\mathcal{A})$ for system (14) but it says nothing regarding requirement (i). In this respect it is easy to realize that also requirement (i) is fulfilled in the case the function γ can be designed in such a way that (13) is satisfied. As a matter of fact, by bearing in mind that $\varphi(\xi) = F\xi + G\gamma(\xi)$, (12) reads as

$$\frac{\partial \tau}{\partial \mathbf{z}} \mathbf{f}_0(\mathbf{z}) = F\mathbf{z} - G\gamma \circ \tau(\mathbf{z}) \qquad \forall \mathbf{z} \in \mathcal{A}$$

which, if (13) holds, reduces to (38). This, along with Proposition 3, yields that the proposed triplet satisfies the asymptotic internal model property in the case the function $\gamma(\cdot)$ can be chosen so that (13) holds. Here is where the dimension d of the pair (F, G) plays a role, as formalized in the next proposition whose proof can be found in [21].

Proposition 4. *Let*

$$d \geq 2(n + s) + 2$$

and (F, G) be a controllable pair, with F chosen as indicated in Proposition 3. Then there exist a continuous function $\gamma : R^d \to R$ fulfilling (13), with $\tau(\cdot)$ the map defined by (37).

5   Conclusions

The *Asymptotic Internal Model Property* has been introduced as natural property to be achieved in order to solve a problem of nonlinear output regulation in the case high-gain error feedback techniques are used as stabilization tool. It has been shown how a number of results and frameworks recently proposed in the related literature can be re-formulated in these terms by thus presenting a common vision able to frame apparently different design techniques. Specifically, results relying on the use of nonlinear high-gain, nonlinear adaptive, and nonlinear Luenberger observers design techniques have been re-interpreted in the proposed framework. It is expected that the proposed property can be useful to identify other relevant cases in which the regulator can be successfully and constructively designed.

References

1. Andrieu V, Praly L (2006) On the existence of a Kazantis-Kravaris/Luenberger observer, SIAM Journal on Control and Optimization, 45: 432–456
2. Bastin G, Gevers M R (1988) Stable adaptive observers for nonlinear time varying systems, IEEE Transactions on Automatic Control, 33: 650–657

3. Bonivento C, Isidori A, Marconi L, Paoli A (2004) Implicit fault tolerant control: application to induction motors, Automatica, 40: 355-371.
4. Byrnes C I, Isidori A (2003) Limit sets, zero dynamics and internal models in the problem of nonlinear output regulation, IEEE Transactions on Automatic Control, 48: 1712–1723
5. Byrnes C I, Isidori A (2004) Nonlinear internal models for output regulation, IEEE Transactions on Automatic Control, 49: 2244–2247
6. Byrnes C I, Delli Priscoli F, Isidori A, Kang W (1997) Structurally stable output regulation of nonlinear systems, Automatica, 33: 369–385
7. Chen Z, Huang J (2004) Global robust servomechanism problem of lower triangular systems in the general case, Systems and Control Letters, 52: 209–220
8. Delli Priscoli F, Marconi L, Isidori A (2006) A new approach to adaptive nonlinear regulation, SIAM Journal on Control and Optimization, 45: 829–855
9. Delli Priscoli F, Marconi L, Isidori A (2006) Nonlinear observers as nonlinear internal models, Systems and Control Letters, 55: 640–649
10. Delli Priscoli F (2004) Output regulation with nonlinear internal models, Systems and Control Letters, 53: 177–185
11. Fliess M, Kupka I (1983) A finitness criterion for nonlinear input-output differential systems, SIAM Journal on Control and Optimization, 21: 721–728
12. Francis B A (1977) The linear multivariable regulator problem, SIAM Journal on Control and Optimization, 14: 486–505
13. Francis BA, Wonham WM (1976) The internal model principle of control theory, Automatica, 12: 457–465
14. Gauthier J P, Kupka I (2001), Deterministic Observation Theory and Applications. Cambridge University Press
15. Hale J K, Magalhães L T, Oliva W M (2002) Dynamics in Infinite Dimensions. Springer Verlag, New York
16. Huang J, Lin C F (1994) On a robust nonlinear multivariable servomechanism problem, IEEE Transactions on Automatic Control, 39: 1510–1513
17. Isidori A, Marconi L, Serrani A (2003) Robust Autonomous Guidance: An Internal Model-based Approach. Springer Verlag London, Limited series Advances in Industrial Control
18. Isidori A, Byrnes C I (1990) Output Regulation of Nonlinear Systems, IEEE Transactions on Automatic Control, 25: 131–140
19. Kazantzis K, Kravaris C (1998) Nonlinear Observer Design Using Lyapunov's Auxiliary Theorem, Systems and Control Letters, 34: 241-247
20. Khalil H (1994)) Robust Servomechanism Output Feedback Controllers for Feedback Linearizable Systems, Automatica, 30: 587–1599
21. Marconi L, Praly L, Isidori A (2006) Output Stabilization via Nonlinear Luenberger Observers, SIAM Journal on Control and Optimization, To appear
22. Marino R, Tomei P (1992) Global Adaptive Observers for Nonlinear Systems via Filtered Transformations, IEEE Transactions on Automatic Control, 37: 1239–1245
23. Serrani A, Isidori A, Marconi L (2000) Semiglobal Output Regulation for Minimum-Phase Systems, International Journal on Robust and Nonlinear Control, 10, pp. 379–396
24. Serrani A, Isidori A, Marconi L (2001) Semiglobal Nonlinear Output Regulation with Adaptive Internal Model, IEEE Transactions on Automatic Control , 46: 1178-1194
25. Teel A R, Praly L (1995) Tools for Semiglobal Stabilization by Partial State and Output Feedback, SIAM Journal on Control and Optimization, 33: 1443–1485

Controller Design Through Random Sampling: An Example[*]

Maria Prandini[1], Marco C. Campi[2], and Simone Garatti[1]

[1] Dipartimento di Elettronica e Informazione - Politecnico di Milano, P.zza Leonardo da Vinci 32, 20133 Milano, Italia
{prandini,sgaratti}@elet.polimi.it
[2] Dipartimento di Elettronica per l'Automazione - Università di Brescia, via Branze 38, 25123 Brescia, Italia
campi@ing.unibs.it

Summary. In this chapter, we present the *scenario approach*, an innovative technology for solving convex optimization problems with an infinite number of constraints. This technology relies on random sampling of constraints, and provides a powerful means for solving a variety of design problems in systems and control. Specifically, the virtues of this approach are here illustrated by focusing on optimal control design in presence of input saturation constraints.

Keywords: Constrained Control, Noise Rejection, Convex Optimization, Scenario Optimization, Randomized Methods.

1 Introduction

Many problems in systems and control can be formulated as optimization problems, often times of *convex* type, [1, 2]. Convexity is appealing since 'convex' - as opposed to 'non-convex' - means 'solvable' in many cases.

In practical problems, an often-encountered feature is that the environment is uncertain, i.e. some elements and/or variables are not known with precision. This leads naturally to *robust* convex optimization. Similarly, design against uncertain signals and/or disturbances gives rise to optimization of the robust type.

A robust convex optimization problem is expressed in mathematical terms as

$$\text{RCP} : \min_{x \in \mathbb{R}^n} g(x) \text{ subject to:} \tag{1}$$

$$f_\delta(x) \leq 0, \ \forall \delta \in \Delta,$$

where δ is the uncertain parameter, and $g(x)$ and $f_\delta(x)$ are convex functions in the n-dimensional optimization variable x for every δ within the uncertainty set Δ. An example of formalization of a control problem as RCP is provided in the next section.

---

[*] This work is supported by MIUR (Ministero dell'Istruzione, dell'Universita' e della Ricerca) under the project *New methods for Identification and Adaptive Control for Industrial Systems*.

C. Bonivento et al. (Eds.): Adv. in Control Theory and Applications, LNCIS 353, pp. 201–212, 2007.
springerlink.com   © Springer-Verlag Berlin Heidelberg 2007

Often times, Δ is a set containing an infinite number of instances. If e.g. δ represents the uncertain gain of a plant and this gain is known to take on value in some interval, Δ is such an interval. In the example discussed in this chapter, Δ is the infinite set of possible disturbances entering a given system.

Problems with a finite number of optimization variables and an infinite number of constraints are called *semi-infinite* optimization problems in the mathematical programming literature. It is well known that these problems are difficult to solve and they have been proven NP-hard in many cases, [3, 4, 5, 6].

In [7, 8], an innovative technology called 'scenario approach' has been introduced to deal with semi-infinite convex programming at a very general level. The main thrust of this technology is that solvability can be obtained through random sampling of constraints provided that a probabilistic relaxation of the worst-case robust paradigm of (1) is accepted. When dealing with problems in systems and control, the scenario approach opens up new avenues for working out solutions in many different contexts.

The scenario approach is presented in this chapter in an easy-to-follow manner by way of an example in optimal control with input saturation constraints.

2   An Optimal Control Problem with Constraints

Consider the following control problem: given a linear system affected by a disturbance belonging to some class, design a feedback controller that attenuates the effect of the disturbance on the system output, while avoiding saturation of the control action due to actuator limitations.

Although quite standard in practice, this design problem is generally difficult to solve because of the presence of saturation constraints, and trial-and-error solutions are often adopted.

In this section, we illustrate a new approach to address this control problem in a systematic and optimal way. As we shall see, the proposed design methodology relies on the re-formulation of the problem as a robust convex optimization program by adopting an appropriate parametrization of the controller. Solvability of this robust convex optimization program is then attained through the scenario optimization technology.

2.1   Problem Formulation

We consider a discrete time linear system with scalar input and scalar output, $u(t)$ and $y(t)$, governed by the following equation:

$$y(t) = G(z)u(t) + d(t), \tag{2}$$

where $G(z)$ is a stable transfer function and $d(t)$ is an additive disturbance.

Our objective is to determine a feedback control law

$$u(t) = C(z)y(t) \tag{3}$$

such that the disturbance $d(t)$ is optimally attenuated for every realization of $d(t)$ in some set of possible realizations $\mathcal{D}$, and such that the control input keeps within certain saturation limits. For example, $\mathcal{D}$ can be the set of step functions with specified maximum amplitude or the set of sinusoids with frequency in a certain range. A precise formalization of the optimization problem is next given.

Consider the finite-horizon 2-norm $\sum_{t=1}^{M} y(t)^2$ of the closed-loop system output. This norm quantifies the effect of the disturbance $d(t)$. For simplicity, we here consider (2) and (3) initially at rest, namely $G(z)u(t)$ represents an infinite backwards expansion $\sum_{j=1}^{\infty} g_j u(t-j)$ where $u(t-j) = 0$ for $t - j \leq 0$, and similarly for $C(z)y(t)$.

The goal is to minimize the worst-case disturbance effect

$$\max_{d(t)\in\mathcal{D}} \sum_{t=1}^{M} y(t)^2, \tag{4}$$

while maintaining the control input $u(t)$ within a saturation limit u_{bound}:

$$\max_{1\leq t\leq M} |u(t)| \leq u_{\mathrm{bound}}, \ \forall d(t) \in \mathcal{D}. \tag{5}$$

Controller $C(z)$ is expressed in terms of an Internal Model Control (IMC) parametrization, [9]:

$$C(z) = \frac{Q(z)}{1 + G(z)Q(z)}, \tag{6}$$

where $G(z)$ is the system transfer function and $Q(z)$ is a free-to-choose transfer function (see Figure 1).

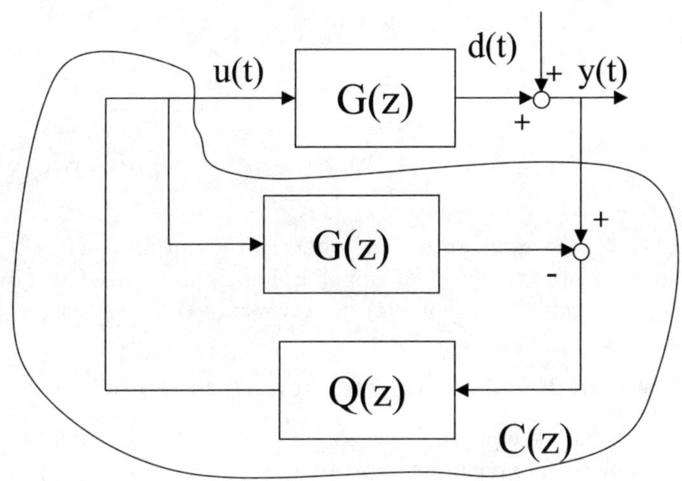

Fig. 1. The IMC parameterization

Expression of $C(z)$ in (6) is totally generic, in that, given a $C(z)$, a $Q(z)$ can be always found generating that $C(z)$ through expression (6). The advantage of (6) is that the set of all controllers that closed-loop stabilize $G(z)$ is simply obtained from (6) by letting $Q(z)$ vary over the set of all stable transfer functions (see [9] for more details).

With (6) in place, the control input $u(t)$ and the controlled output $y(t)$ are given by:

$$u(t) = Q(z)d(t) \tag{7}$$
$$y(t) = [G(z)Q(z) + 1]d(t). \tag{8}$$

The distinctive feature of these expressions is that $u(t)$ and $y(t)$ are affine in $Q(z)$. Consequently, (4) is a convex cost in $Q(z)$ and (5) are convex constraints.

In the sequel, we refer to the case where $Q(z)$ is selected from a family of stable transfer functions linearly parameterized in $\gamma := [\gamma_0\ \gamma_1\ \ldots, \gamma_k]^T \in \mathbb{R}^{k+1}$, i.e.

$$Q(z) = \gamma_0\beta_0(z) + \gamma_1\beta_1(z) + \gamma_2\beta_2(z) + \cdots + \gamma_k\beta_k(z), \tag{9}$$

where $\beta_i(z)$'s are pre-specified stable transfer functions. Note that linearity in γ is important because, due to convexity of (4) and (5) in $Q(z)$, it translates into convexity of the problem in γ.

A common choice for the $\beta_i(z)$'s functions is to set them equal to pure 'delays': $\beta_i(z) = z^{-i}$, leading to

$$Q(z) = \gamma_0 + \gamma_1 z^{-1} + \gamma_2 z^{-2} + \cdots + \gamma_k z^{-k}.$$

Another possibility is to let $\beta_i(z)$'s be Laguerre polynomials, [10, 11].

The control design problem can now be precisely formulated as follows:

$$\min_{\gamma, h \in \mathbb{R}^{k+2}} h \quad \text{subject to:} \tag{10}$$

$$\sum_{t=1}^{M} y(t)^2 \leq h, \ \forall d(t) \in \mathcal{D}, \tag{11}$$

$$\max_{1 \leq t \leq M} |u(t)| \leq u_{\text{bound}}, \ \forall d(t) \in \mathcal{D}. \tag{12}$$

Due to (11), h represents an upper bound to the output 2-norm $\sum_{t=1}^{M} y(t)^2$ for any realization of $d(t)$. Such an upper bound is minimized in (10) under the additional constraint (12) that $u(t)$ does not exceed the saturation limits.

2.2   Rewriting Problem (10)–(12) in a More Explicit Form

By (7) and (8) and the parametrization of $Q(z)$ in (9), the input and the output of the controlled system can be expressed as

$$u(t) = \big(\gamma_o\beta_0(z) + \ldots + \gamma_k\beta_k(z)\big)d(t) \tag{13}$$
$$y(t) = G(z)\big(\gamma_o\beta_0(z) + \ldots + \gamma_k\beta_k(z)\big)d(t) + d(t). \tag{14}$$

Let us define the following vectors containing filtered versions of the disturbance $d(t)$:

$$\phi(t) := \begin{bmatrix} \beta_0(z)d(t) \\ \beta_1(z)d(t) \\ \vdots \\ \beta_k(z)d(t) \end{bmatrix} \quad \text{and} \quad \psi(t) := \begin{bmatrix} G(z)\beta_0(z)d(t) \\ G(z)\beta_1(z)d(t) \\ \vdots \\ G(z)\beta_k(z)d(t) \end{bmatrix}. \tag{15}$$

Then, (13) and (14) can be re-written as

$$u(t) = \phi(t)^T \gamma$$
$$y(t) = \psi(t)^T \gamma + d(t),$$

and $\sum_{t=1}^{M} y(t)^2 = \gamma^T A \gamma + B\gamma + C$, where

$$A = \sum_{t=1}^{M} \psi(t)\psi(t)^T \quad B = 2\sum_{t=1}^{M} d(t)\psi(t)^T \quad C = \sum_{t=1}^{M} d(t)^2 \tag{16}$$

are matrices that depend on $d(t)$ only.

With all these positions, (10)–(12) rewrites as

$$\min_{\gamma,h\in\mathbb{R}^{k+2}} h \text{ subject to:} \tag{17}$$

$$\gamma^T A \gamma + B\gamma + C \leq h, \ \forall d(t) \in \mathcal{D}$$
$$- u_{\text{bound}} \leq \phi(t)^T \gamma \leq u_{\text{bound}}, \ \forall t \in \{1, 2, \ldots, M\}, \ \forall d(t) \in \mathcal{D}.$$

Compared with the general form (1), the optimization variable x is here (γ, h) and has size $n = k + 2$, and the uncertain parameter δ is the disturbance realization $d(t)$ taking value in the set $\Delta = \mathcal{D}$. Note that, given $d(t)$, quantities A, B, C, and $\phi(t)$ are fixed so that the first constraint in (17) is quadratic, while the others are linear.

Typically, the set $\mathcal{D}$ of disturbance realizations has infinite cardinality. Hence, problem (17) is a semi-infinite convex optimization problem.

2.3   Randomized Solution Through the Scenario Technology

As already pointed out in the introduction, semi-infinite convex optimization problems like (17) are difficult to solve. The idea of the scenario approach is that solvability can be recovered if some relaxation in the concept of solution is accepted. In the context of our control design problem, this means requiring that the constraints in (17) are satisfied for all disturbance realizations but a small fraction of them (*chance-constrained* approach).

The scenario approach goes as follows. Since we are unable to deal with the wealth of constraints in (17), we concentrate attention on just a few of them and extract at random N disturbance realizations $d(t)$ according to some probability distribution P introduced over $\mathcal{D}$. This probability distribution should reflect the likelihood of the different disturbance realizations. If no hint is available on which realization is more likely to occur, then the uniform distribution can be adopted. Only these extracted instances ('scenarios') are considered in the scenario optimization:

SCENARIO OPTIMIZATION

extract N independent identically distributed realizations $d(t)_1$, $d(t)_2$, ..., $d(t)_N$ from $\mathcal{D}$ according to P. Then, solve the scenario convex program:

$$\text{SCP}_N : \quad \min_{\gamma, h \in \mathbb{R}^{k+2}} \ h \text{ subject to:} \qquad (18)$$

$$\gamma^T A_i \gamma + B_i \gamma + C_i \leq h, \ i = 1, \ldots, N,$$

$$-u_{\text{bound}} \leq \phi(t)_i^T \gamma \leq u_{\text{bound}}, \ \forall t \in \{1, 2, \ldots, M\},$$

$$i = 1, \ldots, N,$$

where A_i, B_i, C_i, and $\phi(t)_i$ are as in (16) and (15) for $d(t) = d(t)_i$.

Letting (γ_N^*, h_N^*) be the solution to SCP_N, γ_N^* returns the designed controller parameter.

The implementation of the scenario optimization requires that one picks N realizations of the disturbance and computes A_i, B_i, C_i, and $\phi(t)_i$ in correspondence of the extracted realizations. Since these quantities are artificially generated (that is they are not actual measurements coming from the system, but, instead, they are computer-generated), the proposed control design methodology can as well be seen as a *simulation-based approach*.

SCP_N is a standard convex optimization problem with a finite number of constraints, and therefore easily solvable. On the other hand, it is spontaneous to ask: what kind of solution is one provided by SCP_N? Specifically, what can we claim regarding the behavior of the designed control system for all other disturbance realizations, those we have not taken into consideration while solving the control design problem?

The above question is of the 'generalization' type in a learning-theoretic sense: we want to know whether and to what extent the solution generalizes in constraints satisfaction, from seen constraints to unseen ones. Certainly, any generalization result calls for some structure as no generalization is possible if no structure linking what has been seen to what has not been seen is present. The formidable fact in the context of convex optimization is that - by underlying hidden links - the solution of SCP_N always generalizes well, with no extra assumptions.

We have the following theorem (see Corollary 1 in [8]).

Theorem 1. *Select a 'violation parameter' $\epsilon \in (0,1)$ and a 'confidence parameter' $\beta \in (0,1)$. Let $n = k + 2$.*

If

$$N = \left\lceil \frac{2}{\epsilon} \ln \frac{1}{\beta} + 2n + \frac{2n}{\epsilon} \ln \frac{2}{\epsilon} \right\rceil \tag{19}$$

($\lceil \cdot \rceil$ denotes the smaller integer greater than or equal to the argument), then, with probability no smaller than $1 - \beta$, the solution (γ_N^*, h_N^*) to (18) satisfies all constraints of problem (17) with the exception of those corresponding to a set of disturbance realizations whose probability is at most ϵ. ∎

Let us read through the statement of this theorem in some detail. If we neglect the part associated with β, then, the result simply says that, by sampling a number of disturbance realizations as given by (19), the solution (γ_N^*, h_N^*) to (18) violates the constraints corresponding to other realizations with a probability that does not exceed a *user-chosen* level ϵ. This corresponds to say that – for other, unseen, $d(t)$'s – constraints (11) and (12) are violated with a probability at most ϵ. From (11) we therefore see that the found h_N^* provides an upper bound for the output 2-norm $\sum_{t=1}^{M} y(t)^2$ valid for any realizations of the disturbance with exclusion of at most an ϵ-probability set, while (12) guarantees that, with the same probability, the saturation limits are not exceeded.

As for the probability $1 - \beta$, one should note that (γ_N^*, h_N^*) is a random quantity because it depends on the randomly extracted disturbance realizations. It may happen that the extracted realizations are not representative enough (one can even stumble on an extraction as bad as selecting N times the same realization!). In this case no generalization is certainly expected, and the portion of unseen realizations violated by (γ_N^*, h_N^*) is larger than ϵ. Parameter β controls the probability of extracting 'bad' realizations, and the final result that (γ_N^*, h_N^*) violates at most an ϵ-fraction of realizations holds with probability $1 - \beta$.

In theory, β plays an important role and selecting $\beta = 0$ yields $N = \infty$. For any practical purpose, however, β has very marginal importance since it appears in (19) under the sign of logarithm: we can select β to be such a small number as 10^{-10} or even 10^{-20}, in practice zero, and still N does not grow significantly.

3 Numerical Example

A simple example illustrates the controller design procedure.

With reference to (2), let

$$G(z) = \frac{0.2}{z - 0.8},$$

and let the additive output disturbance be a piecewise constant signal that varies from time to time, at a low rate, of an amount bounded by some given constant. Specifically, let the set of admissible realizations $\mathcal{D}$ consists of piecewise constant signals changing at most once over any time interval of length 50, and taking value in $[-1, 1]$.

As for the IMC parametrization $Q(z)$ in (9), we choose $k = 1$ and $Q(z) = \gamma_0 + \gamma_1 z^{-1}$.

A control design problem (10)–(12) is considered with $M = 300$, and for two different values of the saturation limit u_{bound}: 10 and 1. Probability P is implicitly assigned by the recursive equation

$$d(t + 1) = \big(1 - \mu(t)\big)d(t) + \mu(t)v(t + 1),$$

initialized with $d(1) = v(1)$, where $\mu(t)$ is a $\{0, 1\}$-valued process ($\mu(t) = 1$ at times where a jump occurs), and $v(t)$ is a sequence of i.i.d. random variables uniformly distributed in $[-1, 1]$ ($v(t)$ is the new $d(t)$ value). $\mu(t)$ is generated according to

$$\mu(t) = \alpha(t) \prod_{k=1}^{50} \big(1 - \mu(t - k)\big),$$

initialized with $\mu(0) = \mu(-1) = \cdots = \mu(-49) = 0$, where $\alpha(t)$ is a sequence of i.i.d. $\{0, 1\}$-valued random variables taking value 1 with probability 0.01. An admissible realization of $d(t)$ in $\mathcal{D}$ is reported in Figure 2.

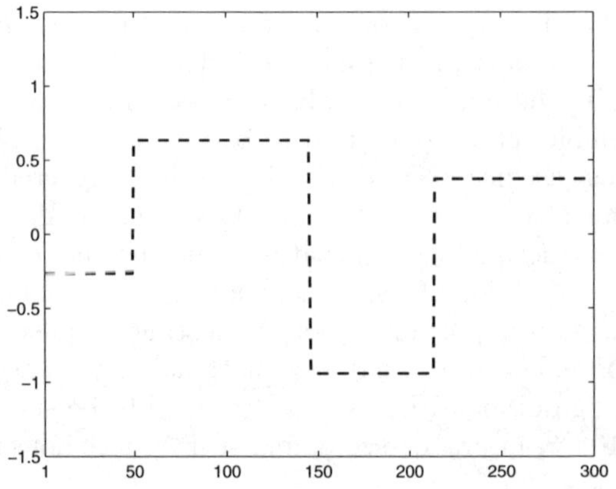

Fig. 2. A disturbance realization

In the scenario approach we let $\epsilon = 5 \cdot 10^{-2}$ and $\beta = 10^{-10}$. Correspondingly, N given by (19) is $N = 1370$.

From Theorem 1, with probability no smaller than $1 - 10^{-10}$, the obtained controller achieves the minimum of $\sum_{t=1}^{M} y(t)^2$ over all disturbance realizations, except a fraction of them of size smaller than or equal to 5%. At the same time, the control input $u(t)$ is guaranteed not to exceed the saturation limit u_{bound} except for the same fraction of disturbance realizations.

3.1   Simulation Results

For $u_{\text{bound}} = 10$, we obtained $Q(z) = -4.9931 + 4.0241z^{-1}$ and, correspondingly, the transfer function $F(z) = 1 + Q(z)G(z)$ between $d(t)$ and $y(t)$ (closed-loop sensitivity function) was

$$F(z) = 1 + (-4.993 + 4.024z^{-1})\frac{0.2}{z - 0.8} \simeq 1 - z^{-1}.$$

The pole-zero plot of $F(z)$ is in Figure 3.

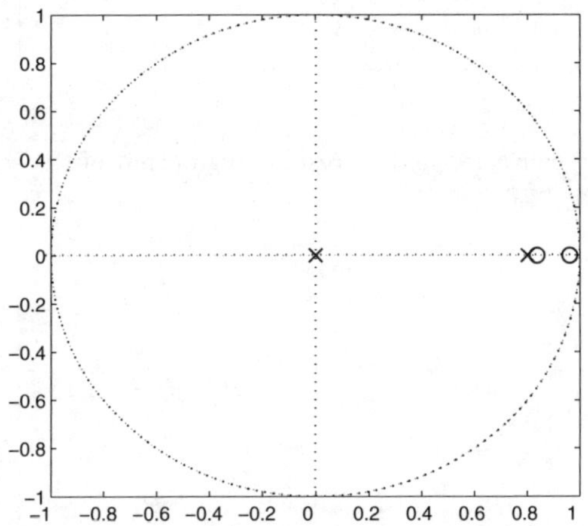

Fig. 3. Pole-zero plot of $F(z)$ when $u_{\text{bound}} = 10$. The poles are plotted as x's and the zeros are plotted as o's.

Since $y(t) = F(z)d(t) \simeq d(t) - d(t-1)$, then, when $d(t)$ has a step variation, $y(t)$ changes of the same amount and, when the disturbance gets constant, $y(t)$ is immediately brought back to zero and maintained equal to zero until the next step variation in $d(t)$ (see Figure 4). The obtained solution that $F(z)$ is approximately a FIR (Finite Impulse Response) of order 1 with zero DC-gain is not surprising considering that $d(t)$ varies at a low rate.

In the controller design just described, the limit $u_{\text{bound}} = 10$ played no role in that constraints $-u_{\text{bound}} \leq \phi(t)_i^T \gamma \leq u_{\text{bound}}$ in problem (18) were not active at the found solution. As u_{bound} is decreased, the saturation limits become more stringent and affect the solution.

For $u_{\text{bound}} = 1$, the following scenario solution was found $Q(z) = -0.991 + 0.011z^{-1}$, which corresponds to the sensitivity function:

$$F(z) = 1 + (-0.991 + 0.011z^{-1})\frac{0.2}{z - 0.8} \simeq \frac{z - 0.9960}{z - 0.8}.$$

The pole-zero plot of $F(z)$ is in Figure 5, while Figure 6 represents $y(t)$ obtained through this new controller for the same disturbance realization as in Figure 4.

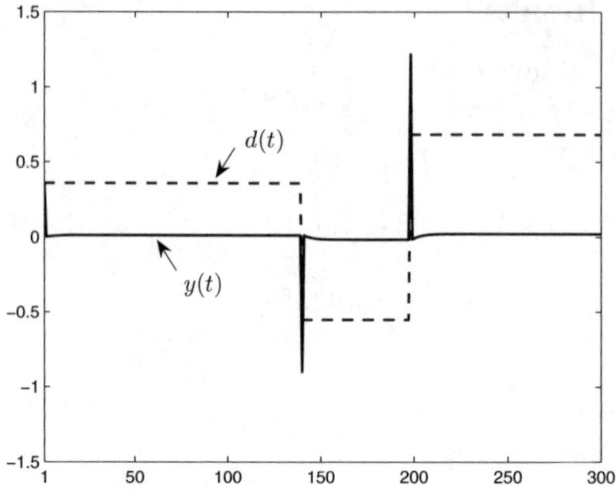

Fig. 4. Disturbance realization and corresponding output of the controlled system for $u_{\mathrm{bound}} = 10$

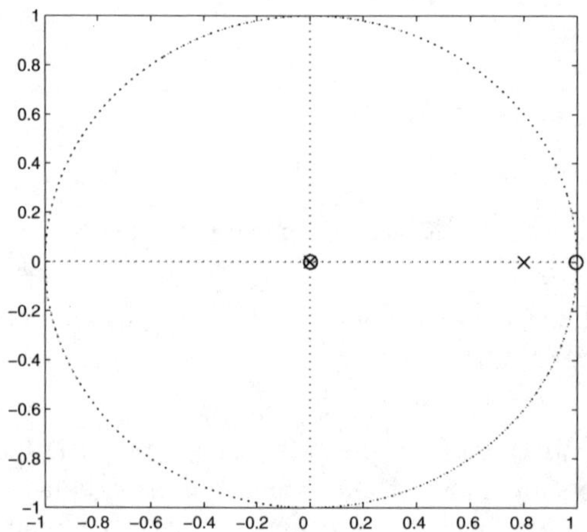

Fig. 5. Pole-zero plot of $F(z)$ when $u_{\mathrm{bound}} = 1$. The poles are plotted as x's and the zeros are plotted as o's.

Note that the time required to bring $y(t)$ back to zero after a disturbance jump is now longer than 1 time unit, owing to saturation constraints on $u(t)$.

The optimal control cost value h_N^* is $h_N^* = 9.4564$ for $u_{\mathrm{bound}} = 10$ and $h_N^* = 27.4912$ for $u_{\mathrm{bound}} = 1$. As expected, the control cost increases as u_{bound} becomes more stringent.

The numerical example of this section is just one instance of application of the scenario approach to controller selection. The introduced methodology is of general applicability to diverse situations with constraints of different type, presence of reference signals, etc.

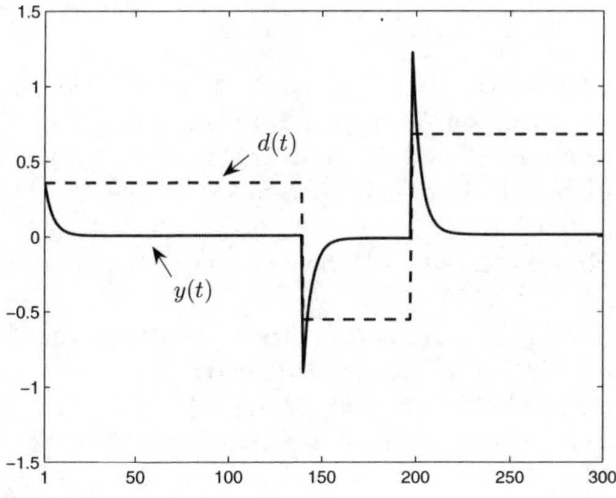

Fig. 6. Disturbance realization and corresponding output of the controlled system for $u_{\mathrm{bound}} = 1$

4   Conclusions: A Final Glance over the Scenario World

In this chapter, we considered an optimal disturbance rejection problem with limitations on the control action and showed how it can be effectively addressed by means of the so-called scenario technology. This approach basically consists of the following main steps:

- reformulation of the problem as a robust (usually with infinite constraints) *convex* optimization problem;
- randomization over constraints and resolution (by means of standard numerical methods) of the so obtained *finite* optimization problem;
- evaluation of the constraint satisfaction level of the obtained solution through Theorem 1.

The applicability of the scenario methodology is certainly not limited to optimal disturbance rejection problems and, indeed, this same methodology has been applied to a number of different endeavors in systems and control.

Robust control, for example, is a natural setting for the scenario approach, since robust control performance requirements can be often translated into optimization with an infinite number of constraints. The reader is referred to [12, 13, 8], where the scenario methodology has been applied to robust stabilization, LPV (Linear Parameter Varying) control, and robust pole assignment.

Another setting in which the scenario approach proved powerful is the identification of interval predictor models (i.e. models returning a prediction interval instead of a single prediction value), [14, 15]. Here, constraints are given by observed data and optimization is performed to shrink the interval model as tightly as possible around data.

Finally, the scenario approach is currently being applied to system identification through an innovative min-max perspective.

References

1. Goodwin G. C., Seron M. M., De Doná J. A. (2005) Constrained Control and Estimation: an Optimisation Approach, Springer-Verlag, New York.
2. Boyd S., El Ghaoui L., Feron E., Balakrishnan V. (1994) Linear Matrix Inequalities in System and Control Theory, SIAM Studies in Applied Mathematics, Philadelphia.
3. Ben-Tal A., Nemirovski A. (1998) Robust convex optimization, Mathematical Operational Research, 23(4):769–805.
4. Blondel V. D., Tsitsiklis J. N. (2000) A survey of computational complexity results in systems and control, Automatica, 36: 1249–1274.
5. Braatz R. P., Young P. M., Doyle J. C., Morari M. (1994) Computational complexity of μ calculation, IEEE Transactions on Automatic Control, 39(5):1000–1002.
6. Nemirovski A. (1993) Several NP-hard problems arising in robust stability analysis, SIAM Journal on Matrix Analysis and Application, 6:99–105.
7. Calafiore G., Campi M. C. (2005) Uncertain convex programs: randomized solutions and confidence levels, Mathematical Programming, Ser. A 102:25–46.
8. Calafiore G., Campi M. C. (2006) The scenario approach to robust control design, IEEE Transactions on Automatic Control 51(5):742–753.
9. Morari M., Zafiriou E. (1989) Robust process control, Prentice Hall, Englewood Cliffs, New Jersey.
10. Wahlberg B. (1991) System identification using Laguerre models, IEEE Transactions on Automatic Control, 36:551–562.
11. Wahlberg B., Hannan E. (1993) Parameteric signal modelling using Laguerre filters, The Annals of Applied Probabililty, 3:467–496.
12. Calafiore G., Campi M. C. (2003) Robust convex programs: randomized solutions and application in control, In: Proceedings of the 42nd IEEE Conference on Decision and Control, Maui, Hawaii.
13. Calafiore G., Campi M. C. (2004) A new bound on the generalization rate of sampled convex programs, In: Proceedings of the 43rd IEEE Conference on Decision and Control, Atlantis, Paradise Island, Bahamas.
14. Calafiore G., Campi M. C. (2003) A learning theory approach to the construction of predictor models, Discrete and Continuous Dynamical Systems, supplement volume:156–166.
15. Calafiore G., Campi M. C., Garatti S. (2005) Identification of reliable predictor models for unknown systems: a data-consistency approach based on learning theory, In: Proceedings of the 16th IFAC World Congress, Prague, Czech Republic.

Digital Control of High Performance Power Supplies for a Synchrotron Light Source*

Carlo Rossi, Andrea Tilli, and Manuel Toniato

Center of Research on Complex Automated Systems CASY, University of Bologna
{crossi,atilli,mtoniato}@deis.unibo.it

Summary. Design and control of Power Supplies (PSs) feeding the magnets of a Synchrotron Light Source have to match severe specifications; high accuracy in the range of ppm in output current tracking is required for the correct operation of the magnets, while a Power Factor (PF) close to the unit is demanded at the input section due to the high power involved.

In this paper an advanced control strategy is presented for a particular kind of Quadrupole Magnet Power Supply, where variable output current has to be imposed. The case of the "switch-mode" multilevel power converter for booster quadrupole magnets of the DIAMOND synchrotron radiation facility under construction at the Harwell Chilton Science Campus, Didcot, has been considered.

High accuracy in the tracking of the desired output current reference is reached by means of a digital internal model-based controller. A multivariable controller is adopted in order to ensure current balancing between the stages of the multilevel converter.

Front-end topology selection, proper dimensioning and control design are exploited to guarantee high power factor and low harmonic distortion of the input currents, and to avoid low-frequency components related to the quadrupole magnets' oscillating currents. For this purpose, confined oscillatory behavior imposed to the voltage of the DC-link capacitors plays a key role.

Simulations and experimental validations are reported that confirm the expected results.

Keywords: Internal Model Control, Multilevel AC-AC Converters, Digital Control of Power Converters.

1 Introduction

Recent advances in many fields as medicine, chemistry, electronics and nano-technologies have promoted the design and construction of many third generation synchrotron radiation facilities at the intermediate energies of 2.5-3.5 GeV [1, 2] worldwide. Synchrotron radiation is an extremely intense and coherent light beam emitted when charged particles traveling close to the speed of the light are bent by a magnetic field generated by multi-pole magnets as dipoles, quadrupoles and sextupoles. The design and control of Power Supplies (PSs) feeding the

---

* This work has been partially supported by OCEM S.p.A. and Diamond Light Source Ltd.

C. Bonivento et al. (Eds.): Adv. in Control Theory and Applications, LNCIS 353, pp. 213–237, 2007.
springerlink.com

magnets have to match two main specifications: an high accuracy in current tracking (due to the requirements on the magnetic fields) and a Power Factor (PF) close to the unit (due to high power involved).

Two classical solutions for variable currents PSs are the direct connection between the booster magnets and the local electricity distribution by means of a transformer and the "White Circuit", which adopts an inductive/capacitive resonant scheme. The first one was early considered, for example, for the DIAMOND Synchrotron [3], Didcot, Oxfordshire and for the booster of BESSY II, Berlin [4] but was soon discarded in both cases because of its large costs. The second solution is utilized to empower the booster of the aforementioned BESSY II, and the one of SSRL, Stanford [5].

In the last decade the availability of fast high-power switching devices has dramatically increased, permitting to consider different type of topologies for high power applications and revaluating the "Switch Mode technology". The "Switch-Mode technology" is a multilevel architecture made up of a series and/or parallel connection of many lower power modules. This solution is well established for ring magnets PSs with required constant current [6, 7], whereas it is the most innovative architecture for booster magnet PSs in which variable current are expected.

A breakthrough for the "Switch Mode" was the solution proposed by Jenni and his coauthors [8] for the Swiss Light Source (SLS). Its success made the switching solution the first choice for the synchrotron manufacturing companies as DIAMOND Ltd Company [2]. Another company that has already developed a similar solution for its booster dipole PS is CANDLE, Yerevan, Armenia [1].

The SLS control solutions of [8] were shown in latter works: [9, 10]. In particular, [9] describes the features of a digital PI regulator for current control where the aim is to ensure a good tracking for a biased sine-wave current reference. This PI solution represents the digital version of the widespread analog controllers already presented in literature [7]. The digital solution is becoming widespread because it allows the implementation of more complex and sophisticated control algorithms able to ensure good reference tracking, robustness to parameters variations from thermal effects and aging, and less sensitivity to noise. For instance, in [11] and [12] Pett and his coauthors adopt a modern RST approach and a digital PII plus feed-forward action to comply with a requirement of an accuracy of 1 ppm (part per million).

The other problem that control has to face with is to get a PF close to the unit. The requirement of a variable current running through the magnet involves an exchange of reactive energy between the magnet system and the PS. Without counter measures, this leads to a strong pulsation on the DC-link capacitor voltage. An high distorted current is drawn into the mains and an high PF can not be achieved. To cope with this problem in [8] 12-pulse bridges and buck converters, properly controlled by means of a pole placement, are inserted [10].

Although definitively interesting, the solutions proposed in [9] and in [10] leave open problems that have to be faced. The digital PI solution of [9] is a simple approach that can be substituted by more modern algorithms while in [10] the ultimate goal of a constant current flow from the mains is not achieved.

Aim of this paper is to present an advanced control strategy for a particular kind of quadrupole magnet PS. The case of the booster quadrupole magnet power converter of the DIAMOND synchrotron radiation facility under construction at the Harwell Chilton Science Campus, Didcot, has been considered [13]. The PS adopted in this case-study exploits a switch mode solution. Very high accuracy in the tracking of the desired current reference is reached by means of a digital internal model-based controller. The circuit and the control architecture of the front-end system is carefully considered. In particular, to achieve an high PF, the task of the input section is twofold: to guarantee low harmonic distortion of the current drawn from the line and to avoid low frequency components (usually referred to as "subharmonics"), related to the quadrupole magnet oscillating current. In order to comply with these requirements, 12-pulse bridges and booster circuits are adopted. In particular, dimensioning and control design of the booster controller effectively allows to fulfill the requirement of constant input power from the line, while stationary oscillations are imposed to the magnet. For this purpose, confined oscillatory behavior imposed to the DC-link voltage of the booster stage plays a key role.

The paper is organized as follows. In Section 2, the overall system is described: the control requirements, the structure of the adopted PS and the features of the input and output section. In Sections 4 and 3, motivations which lead to the adopted control design approaches are deeply discussed and the proposed control solutions are presented. Simulation results find address in Section 5 while experimental results are shown and discussed in Section 6.

2   Control Specification and System Analysis

2.1   Control Specification

Control specifications concern the following topics.

1. Current reference. The magnet has to track a sinusoidal biased current bounded within the range $2A - 200A$ expressed as:

$$i^*_{lm}(t) = I_0 + (I_{AC}\sin(2\pi f_r t) + I_{AC}) \tag{1}$$

   with $I_0 = 2A$, I_{AC} variable from $0A$ to $99A$ and $f_r = 5Hz$. An accuracy equal to $\pm 50ppm$ of the rated current, i.e. a current tracking error smaller than $10mA$, is required.
2. The PS topology has to adopt a switching solution. This requirement calls for a specification on the current ripple accuracy; a limit of $\pm 10ppm$ of the rated current, i.e. $2mA$, is demanded.
3. The connection between PS and mains has to be characterized by a PF close to the unit and low current distortion.

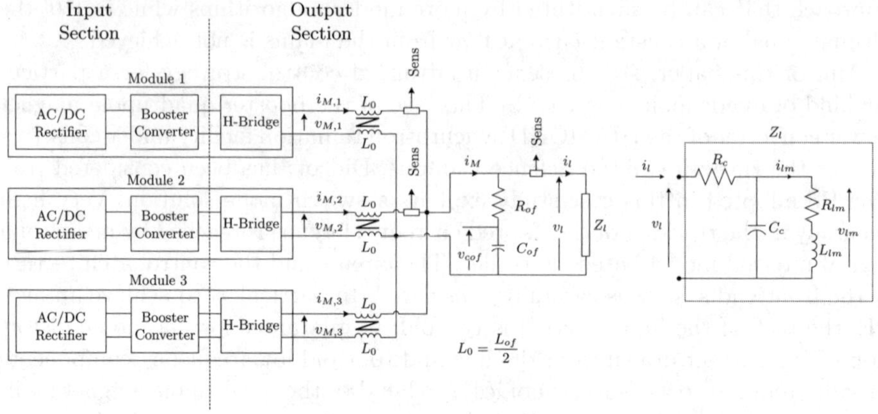

Fig. 1. Power supply scheme

2.2 System Analysis

The PS architecture, depicted in Fig. 1, consists of an input section and an output one connected with the magnet load, Z_l. A current sharing topology is implemented by three modules, each one exploiting an AC/DC rectifier, a booster converter and an H-Bridge. The sum of the three output currents is filtered by an output filter connected to the magnet load and composed by the inductors L_{of}, the capacitor C_{of} and the resistor R_{of}.

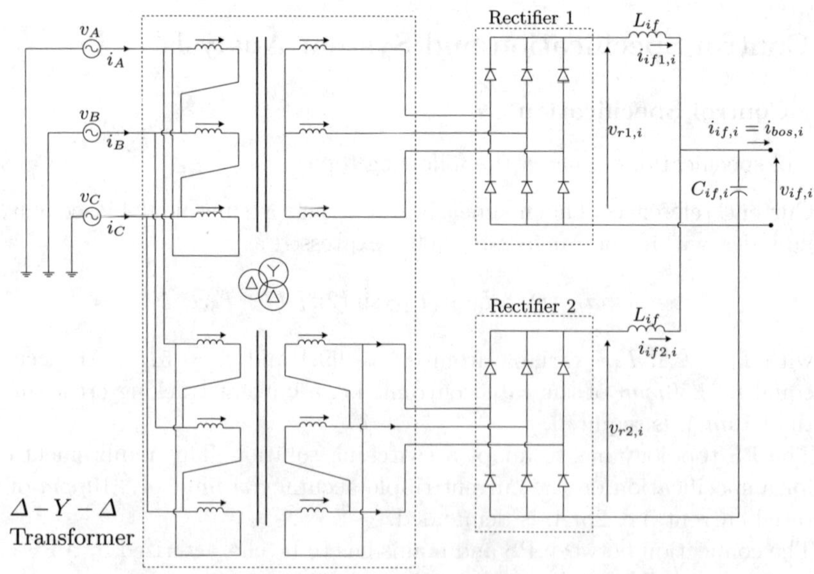

Fig. 2. i-th module: AC/DC rectifier electrical scheme

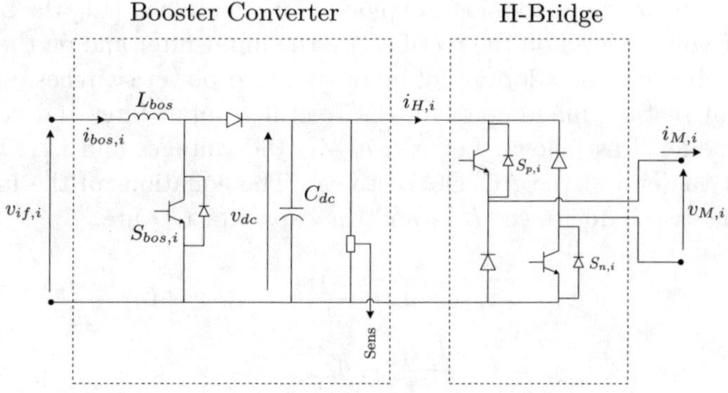

Fig. 3. i-th module: Booster Converter and H-bridge scheme

Input Section

To ensure a good PF, the distortion of mains current and voltage waveforms and the displacement between mains current vector and mains voltage vector have to be as low as possible. AC/DC rectifiers exploiting a 12-pulse bridge in their front-end can fulfill this need. Three devices are used instead of a unique one in order to avoid parasitic currents and to ensure galvanic isolation. The electrical scheme of Fig. 2 sketches the main features of the converters. In ideal conditions the voltage $v_{if,i}$ delivered by this device is constant and ripple free as well as the currents running through the inductances L_{if}. Such type of current ensures correct operation both for the rectifiers and the transformer and the distortion of the currents $i_{A,i}$, $i_{B,i}$ and $i_{C,i}$ is kept small. Conversely, when the ripple of the current $i_{if1,i}$ and $i_{if2,i}$ is appreciable, the distortion of the mains currents grows. In the worst case the ripple is such that the current flowing trough the diodes reaches negative values turning them off. Hence a worse PF has to be tolerated.

The capacitor C_{if} cannot be directly connected to the output section since the current reference to be tracked calls for an energy exchange between the magnet and the PS that, without counter measures, leads to a strong pulsation of the C_{if} voltage and a considerable ripple on the current $i_{if1,i}$ and $i_{if2,i}$. To cope with this problem every module is endowed with a booster converter (see Fig. 1) whose architecture is sketched in Fig. 3. The converter task is twofold:

1. to keep the current $i_{bos,i}$ flowing trough the booster inductance of the i-th module constant in order to comply with the PF specification as explained above;
2. to control the oscillations of the DC-link voltage in order to keep $v_{dc,i}$ bounded within a safe range $[V^*_{min}, V^*_{max}]$. In fact, V^*_{max} cannot be overrun to respect capacitor physical constraints. Moreover, a minimal voltage level is necessary to drive the load current.

With respect to the buck topology, exploited for example in [10], the booster one has a lower voltage level on the rectifier, on the input filter and on the converter itself, thus allowing the adoption of more standard power switches.

The input section, made up of AC/DC rectifier, input filter and booster converter, is modeled as follows. Let v_{r1}, v_{r2} be the voltages and i_{if1}, i_{if1} be the currents at the end of the AC/DC rectifier[1]. The equations of the input filters made by the two inductances $L_{if,i}$ and the capacitor C_{if} are.

$$v_{if,i} = v_{r1,i} - L_{if}\frac{d\,i_{if1,i}}{dt} = v_{r2,i} - L_{if}\frac{d\,i_{if2,i}}{dt}$$

$$i_{if1,i} + i_{if2,i} = C_{if}\frac{d\,v_{if,i}}{dt} + i_{bos,i} \tag{2}$$

or, alternatively, in state space form:

$$\frac{d}{dt}\begin{bmatrix} i_{if1,i} \\ i_{if2,i} \\ v_{if,i} \end{bmatrix} = \begin{bmatrix} 0 & 0 & -\frac{1}{L_{if}} \\ 0 & 0 & -\frac{1}{L_{if}} \\ \frac{1}{C_{if}} & \frac{1}{C_{if}} & 0 \end{bmatrix}\begin{bmatrix} i_{if1,i} \\ i_{if2,i} \\ v_{if,i} \end{bmatrix} + \begin{bmatrix} \frac{1}{L_{if}} & 0 \\ 0 & \frac{1}{L_{if}} \\ 0 & 0 \end{bmatrix}\begin{bmatrix} v_{r1} \\ v_{r2} \end{bmatrix} + \begin{bmatrix} 0 \\ 0 \\ -\frac{i_{bos,i}}{C_{if}} \end{bmatrix} \tag{3}$$

The i-th booster can be modeled as follows:

$$v_{if,i} = L_{bos}\frac{di_{bos,i}}{dt} + (1 - \rho_i)v_{dc,i}$$

$$(1 - \rho_i)i_{bos,i} = C_{dc}\frac{dv_{dc,i}}{dt} + i_{inH,i} \tag{4}$$

Its state space representation is:

$$\frac{d}{dt}\begin{bmatrix} i_{bos,i} \\ v_{dc,i} \end{bmatrix} = \begin{bmatrix} 0 & -\frac{(1-\rho_i)}{L_{bos}} \\ \frac{(1-\rho_i)}{C_{dc}} & 0 \end{bmatrix}\begin{bmatrix} i_{bos,i} \\ v_{dc,i} \end{bmatrix} + \begin{bmatrix} \frac{v_{if,i}}{L_{bos}} \\ -\frac{i_{inH,i}}{C_{dc}} \end{bmatrix} \tag{5}$$

where:

- $i_{bos,i}$, the current running through the booster inductance, is the first state variable;
- $v_{dc,i}$ the DC-link voltage, is the second state variable;
- ρ_i, the modulation index of the switch $S_{bos,i}$, is the input variable;
- $v_{if,i}$ is the voltage delivered by the AC/DC rectifier;
- $i_{inH,i}$ is the current flowing towards the H-bridge.

Output Section

Every module adopts a two quadrant H-bridge (positive and negative voltages, positive currents) in its outer section (see Fig. 3). This kind of implementation

---

[1] The relations between output voltages and currents $v_{r1,2}$, $i_{if1,2}$ and input three-phase voltages and currents are omitted since they follows from standard results on AC/DC converters.

has a drawback: the switching behavior generates a current ripple that has to be damped. This is usually done introducing an output filter after the H-bridge. In this project, besides the filter, a current sharing and optimal interleaving technique have been added to improve overall performances [14].

Let write the output currents (see Fig. 1) as:

$$i_{M,i}(t) = I_{M,i} + \Delta i_{M,i}(t) \quad \text{with } i \in \{1,2,3\}$$
$$i_M(t) = i_{M,1} + i_{M,2} + i_{M,3} = I_M + \Delta i_M(t) \tag{6}$$

where $I_{M,i}$ and I_M represent the mean values while $\Delta i_{M,i}(t)$ and $\Delta i_M(t)$ the current ripples. Using an optimal interleaving among N modules, the module commands are staggered in phase of $2\pi/N$. The resulting equivalent frequency of $\Delta i_M(t)$ is N times the frequency of $\Delta i_{M,i}(t)$ yielding a less stringent output filter dimensioning. Moreover, the split of the total current into N paralleled converters reduces by N times each module current allowing the use of more standard, faster and cheaper switches.

The model of the outer section can be obtained as follows. The voltage and current equations of the i-th module are:

$$v_{M,i} = u'_{M,i} v_{dc,i}$$
$$i_{inH,i} = u'_{M,i} i_{M,i} \quad \text{with } i \in \{1,2,3\} \tag{7}$$

where $u'_{M,i}$ is the modulation index belonging to the set $[-1, 1]$. The input filter voltages and currents are expressed as:

$$v_{M,i} = L_{of} \frac{d\, i_{M,i}}{dt} + v_l \quad \text{with } i \in \{1,2,3\}$$
$$v_l = v_{cof} + R_{of} C_{of} \frac{d\, v_{cof}}{dt} \tag{8}$$

Load Model

Bending effects of the electron beam, focusing and defocusing, are achieved by means of a set of magnets connected through a cable. The electrical model of the load has to capture the different behaviors coming out both at high and low frequencies. A simpler representation is chosen since the current reference has only two components: a continuous component and a sinusoidal one at 5 Hz. The load equivalent circuit Z_l takes into account the load impedance R_{lm} and L_{lm} and the cable characterization R_c and C_c. The final load model is:

$$v_l = R_c i_l + v_{lm}$$
$$v_{lm} = R_{lm} i_{lm} + L_{lm} \frac{di_{lm}}{dt}$$
$$i_l = i_{lm} + C_c \frac{dv_{lm}}{dt} \tag{9}$$

The final state space representation can be obtained coupling the output section equations (8) and the load model relations (9):

$$\dot{\mathbf{x}}_{\text{out}} = \mathbf{A}_{\text{out}}\,\mathbf{x}_{\text{out}} + \mathbf{B}_{\text{out}}\,\mathbf{v_M}$$
$$i_l = \mathbf{C}_{\text{out}}\,\mathbf{x}_{\text{out}} \tag{10}$$

where:

$$\mathbf{x}_{\text{out}} = \begin{bmatrix} i_{M,1} & i_{M,2} & i_{M,3} & i_{lm} & v_{of} & v_c \end{bmatrix}^T$$

$$\mathbf{v_M} = \begin{bmatrix} v_{M,1} & v_{M,2} & v_{M,3} \end{bmatrix}^T$$

$$\mathbf{A}_{\text{out}} = \begin{bmatrix}
\alpha & \alpha & \alpha & 0 & \frac{\alpha}{R_{of}} & \frac{\alpha}{R_c} \\
\alpha & \alpha & \alpha & 0 & \frac{\alpha}{R_{of}} & \frac{\alpha}{R_c} \\
\alpha & \alpha & \alpha & 0 & \frac{\alpha}{R_{of}} & \frac{\alpha}{R_c} \\
0 & 0 & 0 & -\frac{R_{lm}}{L_{lm}} & 0 & \frac{1}{L_{lm}} \\
\beta R_c & \beta R_c & \beta R_c & 0 & -\beta & \beta \\
\kappa R_{of} & \kappa R_{of} & \kappa R_{of} & -\frac{1}{C_c} & \kappa & -\kappa
\end{bmatrix}$$

$$\mathbf{B}_{\text{out}} = \begin{bmatrix}
\frac{1}{L_{of}} & 0 & 0 \\
0 & \frac{1}{L_{of}} & 0 \\
0 & 0 & \frac{1}{L_{of}} \\
0 & 0 & 0 \\
0 & 0 & 0 \\
0 & 0 & 0
\end{bmatrix}$$

$$\mathbf{C}_{\text{out}} = \begin{bmatrix} 1 - R_c\gamma & 1 - R_c\gamma & 1 - R_c\gamma & 0 & \gamma & -\gamma \end{bmatrix}$$

$$\gamma = \frac{1}{R_c + R_{of}}, \qquad \alpha = -\frac{\gamma R_c R_{of}}{L_{of}}$$

$$\beta = \frac{\gamma}{C_{of}}, \qquad \kappa = \frac{\gamma}{C_c}$$

It is worth noting that the current balance is not intrinsically guaranteed due to the asymmetry of the modules. Therefore a suitable control has to be provided.

3   Internal Model Current Control

The choice of an internal model approach for the power supply control is strictly related to the high accuracy requirements and to the requested interleaving coordination of the current sharing topology. In this section, the main features of this controller are deeply analyzed.

The control objective is twofold:

1. the current flowing in the load magnet has to track asymptotically the sinusoidal reference (1) with a steady-state error lower than 50 ppm;
2. currents drawn from each module of the proposed topology have to be equal.

The first control objective can be pursued by means of an high-gain/large-bandwidth controller with sufficiently large gain at the frequencies where the reference harmonic content is relevant (0Hz, 5Hz). This solution is generally realized using an analog hysteresis current controller for each module of the proposed structure with a supervising controller. The second control objective is guaranteed imposing equal references to each module. Anyway, it is well known that hysteresis solutions could generate unpredictable converter switching sequences, weakening the interleaving technique effects and leading to high current ripples [15]. A digital implementation of PID controllers could be exploited as well but, owing to the high gain requirements and unless complicated lag network are added, the resulting controller will have a large bandwidth forcing a very small sampling time.

As a final result, an internal model based solution is clearly the preferable one because

- it is simple (no compensation network is needed) and suitable for digital implementation;
- a small sampling time is not needed since the resulting bandwidth can be kept very narrow (this is admissible because no requirement on the convergence rate is present);
- it guarantees excellent performances in terms of asymptotic tracking.

3.1 System Model and Control Design

The overall output section model represented by equations (10) take into account cable parasitic elements and dynamics related to capacitor C_{of}. However the effects of these elements are not relevant in the control frequency range so a simplified Linear Time-Invariant (LTI) model can be adopted in the control design, since internal model approach guarantees steady-state tracking robustness. The following simplified model represents the basic behavior of the PS combined with the load.

$$\frac{d}{dt} \begin{bmatrix} i_{M,1} \\ i_{M,2} \\ i_{M,3} \end{bmatrix} = -\frac{R_{lm}}{3L_{lm} + L_{of}} \mathbf{A}_{\text{out}}^{\mathbf{R}} \begin{bmatrix} i_{M,1} \\ i_{M,2} \\ i_{M,3} \end{bmatrix} + \mathbf{B}_{\text{out}}^{\mathbf{R}} \begin{bmatrix} v_{M,1} \\ v_{M,2} \\ v_{M,3} \end{bmatrix} \qquad (11)$$

with:

$$\mathbf{A}_{\text{out}}^{\mathbf{R}} = \begin{bmatrix} 1 & 1 & 1 \\ 1 & 1 & 1 \\ 1 & 1 & 1 \end{bmatrix}, \quad \mathbf{B}_{\text{out}}^{\mathbf{R}} = \begin{bmatrix} \delta & \zeta & \zeta \\ \zeta & \delta & \zeta \\ \zeta & \zeta & \delta \end{bmatrix}$$

$$\delta = \frac{2L_{lm} + L_{of}}{3L_{lm}L_{of} + L_{of}^2}, \quad \zeta = -\frac{L_{lm}}{3L_{lm}L_{of} + L_{of}^2} \qquad (12)$$

Let define $u_{M,i}$ as:

$$u_{M,i} = u'_{M,i} \frac{v_{dc,i}}{V_{max}^*} = \frac{v_{M,i}}{V_{max}^*} \qquad (13)$$

where $u'_{M,i} \in [-1, 1]$ is the modulation index of the i-th module.

According to the above equations, the control indexes u_{Mt}, u_{d1} and u_{d2} are designed to control i_{lm}, i_{d1} and i_{d2} respectively by means of a digital implementation of the internal model principle.

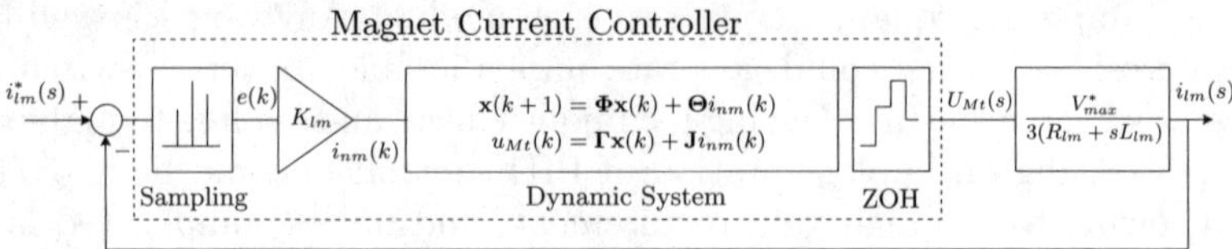

Fig. 4. Load current controller and correspondent plant

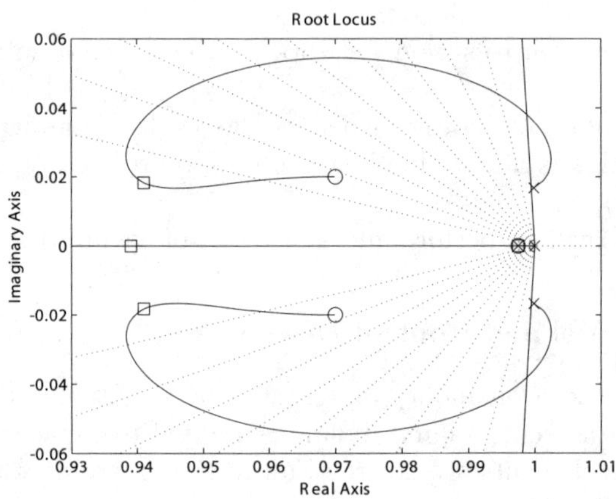

Fig. 5. Root locus of the magnet current controller and its plant

Load Current Controller

The internal model based load current controller is made up of a digital dynamic system and a simple gain K_{lm} as sketched in Fig. 4. The digital dynamic system is designed as follows:

$$\mathbf{x}(k+1) = \mathbf{\Phi x}(k) + \mathbf{\Theta} K_{lm} e(k)$$
$$u_{Mt}(k) = \mathbf{\Gamma x}(k) + \mathbf{J} K_{lm} e(k) \tag{14}$$

with:

$$e(k) = i_{lm}^*(k) - i_{lm}(k)$$

$$\mathbf{\Phi} = \begin{bmatrix} 1 & 0 & 0 \\ 0 & \cos(2\pi 5 T_s) & \sin(2\pi 5 T_s) \\ 0 & -\sin(2\pi 5 T_s) & \cos(2\pi 5 T_s) \end{bmatrix}, \qquad \mathbf{\Theta} = \begin{bmatrix} b_0 \\ b_1 \\ b_2 \end{bmatrix}$$

$$\mathbf{\Gamma} = \begin{bmatrix} -1 & -1 & 0 \end{bmatrix}, \qquad \mathbf{J} = 1 \tag{15}$$

The matrix $\mathbf{\Phi}$ represents the digital internal model of the current reference: the term 1 in the first row is the model of the DC component I_0 while the other not null terms play the role of a digital oscillator with frequency $5Hz$. The value of $\boldsymbol{\Gamma}$ is chosen to guarantee the observability of the couple $(\boldsymbol{\Phi}, \boldsymbol{\Gamma})$ and $\mathbf{J}$ is the proportional part of the controller which ensures robustness.

Assume that the discrete time plant is obtained from the continuous one by means of a zero holder method discretization with sampling time equal to $0.533ms$ ($f_s = 1875Hz = 1/4f_{PWM}$). The zeros of the dynamic system transfer function are chosen to guarantee stability for the closed loop system. The first zero cancels the plant pole while the other two act like attractors for the imaginary poles of the controller ensuring stability. The controller gain is selected as $K_{lm} = 0.176$ and the corresponding poles of the closed loop system are marked with squares. The root locus of Fig. 5 is obtained. Then, the resulting $\boldsymbol{\Theta}$ is:

$$z_{t1,2,3} = \begin{bmatrix} e^{(-\frac{R_{lm}}{L_{lm}}T_s)} \\ 0.97 + 0.02j \\ 0.97 - 0.02j \end{bmatrix} \Rightarrow \boldsymbol{\Theta} = \begin{bmatrix} -0.0117 \\ -0.0506 \\ -0.0692 \end{bmatrix} \tag{16}$$

In conclusion, the load current controller transfer function is:

$$G_c(z) = \frac{U_{Mt}(z)}{I^*(z) - I_{lm}(z)} = \frac{0.176(z - 0.9975)(z^2 - 1.94z + 0.9413)}{(z - 1)(z^2 - 2z + 1)} \tag{17}$$

where $I^*(z)$ is the $\mathcal{Z}$-transform of the sampled magnet current reference.

Difference Current Controllers

The structure of the difference current controllers is the same of (14) with equal values for $\boldsymbol{\Phi_d}$, $\boldsymbol{\Gamma_d}$, $\mathbf{J_d}$ of the correspondent matrices. However, the different plants (Fig. 6) imply a different choice of the zeros of the controller transfer function and, consequently, of $\boldsymbol{\Theta_d}$ and K_{dif}:

$$z_{d1,2,3} = \begin{bmatrix} e^{(-\frac{R_{lm}+R_c}{L_{lm}}T_s)} \\ 0.97 + 0.02j \\ 0.97 - 0.02j \end{bmatrix} \Rightarrow \boldsymbol{\Theta_d} = \begin{bmatrix} -0.0160 \\ -0.0471 \\ -0.0726 \end{bmatrix} \tag{18}$$

The closed loop root locus is depicted in Fig. 7(a) and Fig. 7(b). The squares spot the system poles for the gain selected, $K_{dif} = 0.0039$. In the end, the difference current controller transfer functions are:

$$G_{cd}(z) = \frac{U_{di}(z)}{I_{di}(z)} = -\frac{0.00392(z - 0.9965)(z^2 - 1.94z + 0.9413)}{(z - 1)(z^2 - 2z + 1)}, \quad i = 1,2 \tag{19}$$

Remark 1. The control actions u_{d1} and u_{d2} should be equal to zero in ideal conditions, in fact current balancing control is inserted only to cope with asymmetries of the power modules.

Remark 2. The $u_{M,i}$ commands imposed by the controllers have to be transformed in modulation indexes for the interleaving PWM of the H-bridge switches.

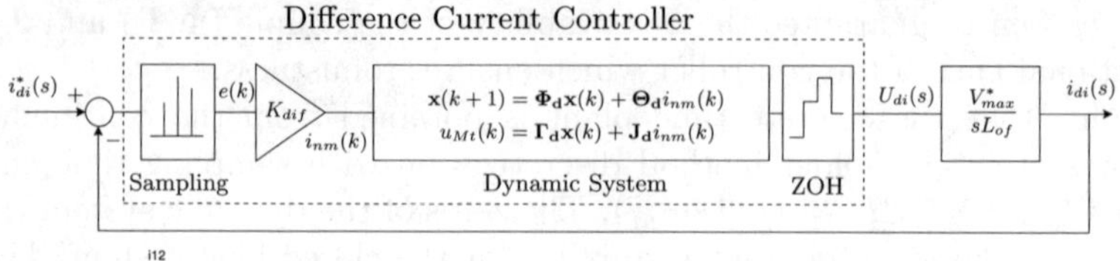

Fig. 6. Difference current controller and correspondent plant

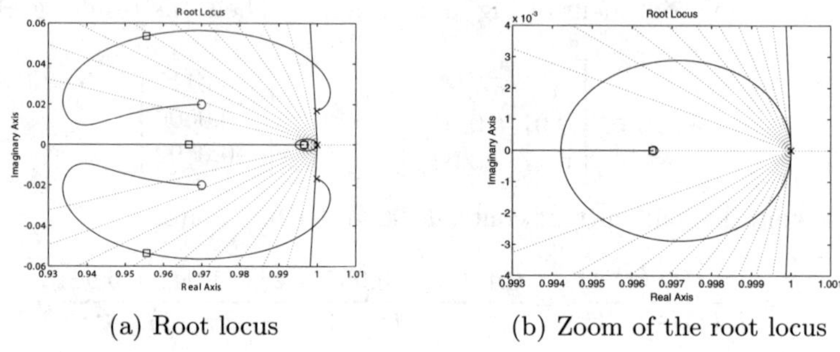

(a) Root locus

(b) Zoom of the root locus

Fig. 7. Root locus of the difference current controller and its plant

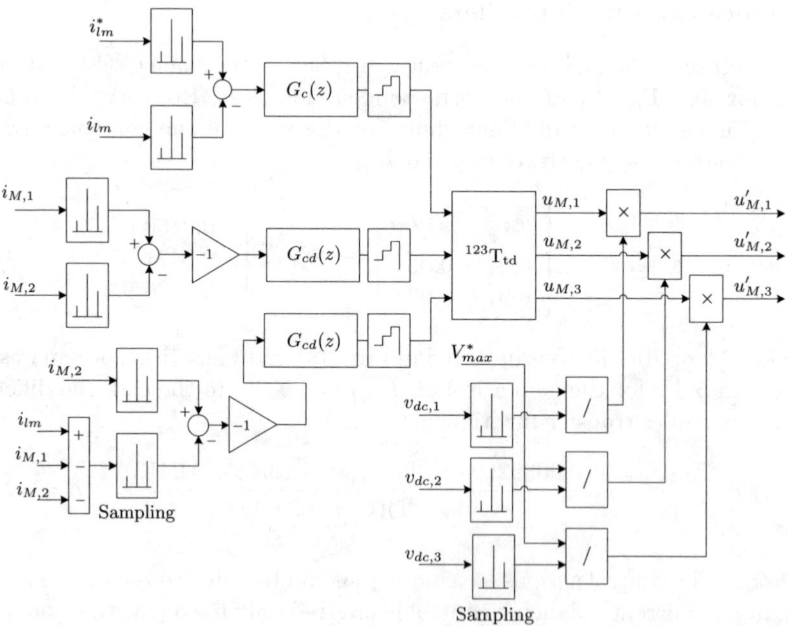

Fig. 8. Current Controller

This task is quite critical since, as stated in section 2, the $v_{dc,i}$ have relevant oscillations owing to exchange of reactive power with the load magnet.

The final version of the internal model controller, implemented in suitable digital cards, is depicted in Fig 8. Its design takes into account that the values directly sensed are i_{lm}, $i_{M,1}$ and $i_{M,2}$ (see Fig. 1) and that the modulation indexes delivered to the PWM modulators have to be $u'_{M,i}$ and not $u_{M,i}$.

4   Cascade Booster Controller

The control systems of the booster converters have to fulfil two main objectives:

- to comply with the requirement on the PF (see Section 2)
- to keep the DC-link voltage oscillations inside a safe range

The booster converter equations (20) that follow are obtained by elaborating (5) and (7):

$$\frac{d\,v_{dc,i}}{dt} = -\left(\frac{i_{M,i}v_{M,i}}{C_{dc}}\right)\frac{1}{v_{dc,i}} + \frac{(1-\rho_i)}{C_{dc}}i_{bos,i}$$

$$\frac{d\,i_{bos,i}}{dt} = -\frac{(1-\rho_i)}{L_{bos}}v_{dc,i} + \frac{v_{if,i}}{L_{bos}} \tag{20}$$

This system is clearly a nonlinear underactuated system. It is a nonlinear system due to the presence of the term $1/v_{dc,i}$ and of the product between the control input ρ_i and the state $[v_{dc,i} \quad i_{bos,i}]^T$; it is underactuated because there are one input, ρ_i, and two control targets, $v_{dc,i}$ and $i_{bos,i}$.

Another important feature of the booster converter is the term $\frac{v_{M,i}i_{M,i}}{C_{dc}}$. As discussed in the previous Section, the internal model-based controller ensures the asymptotic convergence of each module output current $i_{M,i}$ to $i^*_{lm}/3$ and of

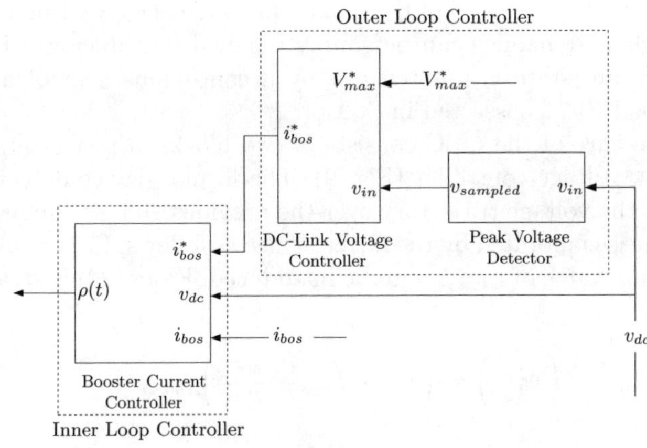

Fig. 9. Architecture of the DC-link controllers

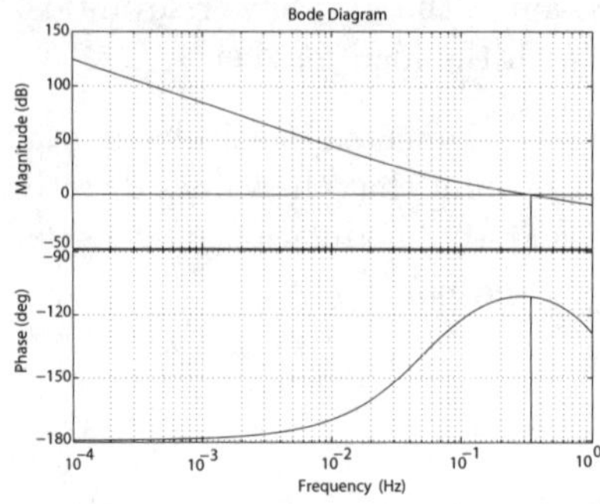

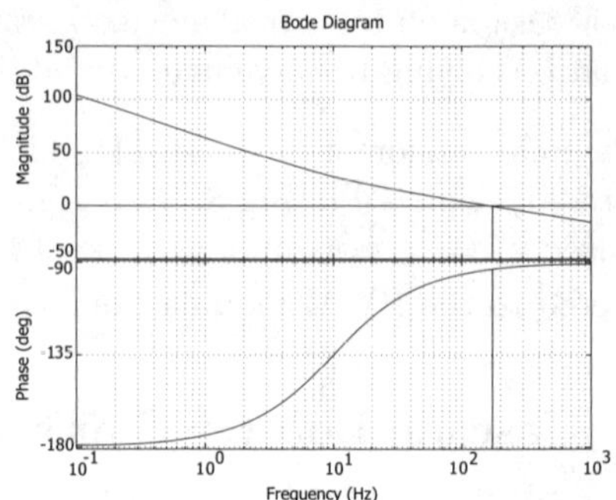

Fig. 10. Bode diagram of the plant controlled by OLC

Fig. 11. Bode Diagram of the plant controlled by ILC

the total current i_{lm} to i_{lm}^*. The single module voltage $v_{M,i}$ can be computed through (11) and therefore $\frac{v_{M,i} i_{M,i}}{C_{dc}}$, although time-variant, is asymptotically known and periodic with period equal to $T_r = 1/f_r$. Bearing in mind all these system features, the booster controllers are designed using a cascade configuration (see Fig. 9).

4.1 Outer Loop Controller

The Outer Loop Controller (OLC) is designed to control the maximum value of DC-link voltage trajectory, meanwhile allowing $v_{dc,i}$ to freewheel under this value.

This fact has two consequences. On the one hand, when the maximum value of the DC-link voltage is under control, a suitable dimensioning of the DC-link capacitors ensures bounded oscillations inside a safe range $[V_{min}^*, V_{max}^*]$ even in the worst case, i.e. when the load draws the maximum current from the DC-link. On the other hand the control of the maximum value of $v_{dc,i}$ without taking into account its whole dynamics can be simply pursued introducing a PI controller whose equilibrium state is pointed out by a continuous control action, $i_{bos,i}$, ensuring a good PF, as asserted in Section 2.

The architecture of the OLC consists of two blocks: a peak voltage detector and a DC-link voltage controller (Fig. 9). The former device detects the maximum value of the voltage trajectory over the previous 200 ms time window. The obtained value is elaborated by the latter device as follows. First of all, the model of the maximum value of $v_{dc,i}$ has to be introduced. From (20) is straightforward to obtain:

$$C_{dc}\frac{1}{2}\frac{d}{dt}\left(v_{dc,i}^2\right) = \left(v_{if} - L_{bos}\frac{d\,i_{bos,i}}{dt}\right) i_{bos,i} - i_{M,i} v_{M,i} \tag{21}$$

This equation represents the power balance of the i-th booster: the left hand is the power on the capacitor C_{dc}, the right hand is the sum of the power flowing inside the booster, the power stored inside the inductance L_{bos} and the power flowing into the H-bridge (a losses free bridge is assumed). Integrating (21) over a time window equal to $T_r = 1/f_r = 200$ ms and assuming that the current running through L_{bos} is constant for the period taken into account, the following relation is obtained:

$$
\begin{aligned}
\frac{C_{dc}}{2} \int_{kT_r}^{(k+1)T_r} \frac{d}{dt}\left(v_{dc,i}^2\right) dt &= \frac{C_{dc}}{2}\left(v_{dc,i}^2(k+1) - v_{dc,i}^2(k)\right) = \\
&= E_c(k+1) - E_c(k) = \\
&= \Delta E_{in}(k+1,k) - \Delta E_{L_{bos}}(k+1,k) - \Delta E_{out}(k+1,k)
\end{aligned}
\tag{22}
$$

where:

$$
\begin{aligned}
\Delta E_{in}(k+1,k) &= \int_{kT_r}^{(k+1)T_r} v_{if} i_{bos}\, dt \simeq T_r \bar{v}_{if} i_{bos,i}(k) \\
\Delta E_{L_{bos}}(k+1,k) &= \int_{kT_r}^{(k+1)T_r} L_{bos}\frac{d\, i_{bos,i}}{dt} i_{bos,i}\, dt = \\
&= \frac{1}{2} L_{bos}\left(i_{bos,i}^2(k+1) - i_{bos,i}^2(k)\right) \\
\Delta E_{out}(k+1,k) &= \int_{kT_r}^{(k+1)T_r} i_{M,i} v_{M,i}\, dt
\end{aligned}
\tag{23}
$$

$\bar{v}_{if}$ is the mean value of v_{if} over a period. Equation (22) is an energy balance and, properly rearranged, yields to the discrete time model of $v_{dc,i}$. The current reference (1) is periodic and therefore the voltage $v_{dc,i}$ oscillates with the same frequency at steady-state. On the other hand, during the transient the time interval between two consecutive peaks varies from a minimum of 0 ms and 400 ms. Since the mean value between these two bounds is 200 ms, the best choice for the sampling time of the discrete model is equal to 200 ms as well. Then, being kT_r and $(k+1)T_r$ the instants when the $v_{dc,i}$ reaches its maximum value, the following expression can be achieved:

$$
\begin{aligned}
(v_{dc,i}^{max}(k+1))^2 = (v_{dc,i}^{max}(k))^2 + \frac{2}{C_{dc}}\Big(& T_r \bar{v}_{if} i_{bos,i}(k) + \\
& - \frac{1}{2} L_{bos}\left(i_{bos,i}^2(k+1) - i_{bos,i}^2(k)\right) + \\
& - \int_{kT_r}^{(k+1)T_r} i_{M,i} v_{M,i}\, dt \Big)
\end{aligned}
\tag{24}
$$

Linearizing the above model with an initial point equal to V_{max}^*, the discrete time model of the maximum value of $v_{dc,i}$ is:

$$v_{dc,i}^{max}(k+1) = v_{dc,i}^{max}(k) + \frac{1}{C_{dc}}\left(T_r \frac{\bar{v}_{if}}{V_{max}^*} i_{bos,i}(k) + \right.$$
$$- \frac{1}{2}\frac{L_{bos}}{V_{max}^*}\left(i_{bos,i}^2(k+1) - i_{bos,i}^2(k) \right) + \qquad (25)$$
$$\left. - \frac{1}{V_{max}^*} \int_{kT_r}^{(k+1)T_r} i_{M,i} v_{M,i}\, dt \right)$$

The term $\frac{1}{2}\frac{L_{bos}}{V_{max}^*}\left(i_{bos,i}^2(k+1) - i_{bos,i}^2(k) \right)$ is negligible with respect to $T_r \frac{\bar{v}_{if}}{V_{max}^*} i_{bos,i}(k)$ since in steady state condition $i_{bos,i}(k) \simeq i_{bos,i}(k+1)$. The last term is a disturbance that has to be rejected.

In the end the discrete time model of the maximum voltage of the DC-link is:

$$v_{dc,i}^{max}(k+1) = v_{dc,i}^{max}(k) + \frac{1}{C_{dc}}\left(T_r \frac{\bar{v}_{if}}{V_{max}^*} i_{bos,i}(k) - d_{i_{inH},i}(k,k+1) \right) \qquad (26)$$

Defining the voltage error:

$$\tilde{v}_{dc,i}(k) = v_{dc,i}^{max}(k) + V_{max}^* \qquad (27)$$

the following plant is obtained:

$$\tilde{v}_{dc,i}(k+1) = \tilde{v}_{dc,i}(k) + \frac{1}{C_{dc}}\left(T_r \frac{\bar{v}_{if}}{V_{max}^*} i_{bos,i}(k) - d_{i_{inH},i}(k,k+1) \right) \qquad (28)$$

To stabilize this system and to reject the mean value of the disturb $d_{i_{inH},i}(k,k+1)$, a simple proportional-integral controller is designed:

$$R_{OLC}(z) = \frac{I_{bos,i}^*(z)}{\tilde{V}_{dc,i}(z)} = -\left(k_p^{dc} + T_r \frac{k_i^{dc}}{z-1} \right) \qquad (29)$$

where the operator z is related to a sample frequency equal to $T_r = 1/f_r = 0.2$ s. The control variable delivered by the regulator is the current reference $i_{bos,i}^*$ that will be tracked by the Intenal Loop Controller. The parameters of Table 1 are considered and the gains of the regulator are set equal to $k_i^{dc} = 16.5 \cdot 10^{-3}$ and $k_p^{dc} = 49.6 \cdot 10^{-3}$. The values of k_i^{dc} and k_p^{dc} are selected to keep the fastest dynamics of the open loop far from $f_r = 5Hz$ and to obtain a satisfactory phase margin of about 70^0 at a frequency near to $0.3Hz$: Fig. 10.

4.2 Inner Loop Controller

The aim of the ILC is to track the desired current reference generated by the OLC. To perform this task a simple PI controller with a PWM modulator is designed as follows.

The inductor behavior is described by the equation:

$$L_{bos}\frac{di_{bos,i}}{dt} = v_{if} - (1-\rho_i)v_{dc,i}$$

where ρ_i is the modulation index of the switch $S_{bos,i}$. Defining:

$$\rho_i = 1 - \frac{v_{if}}{v_{dc,i}} + \frac{1}{v_{dc,i}}\hat{\rho}_i$$

and the current error:

$$\tilde{i}_{bos,i} = i_{bos,i} - i^*_{bos,i}$$

the plant to be controlled is:

$$\frac{d\tilde{i}_{bos,i}}{dt} = \frac{1}{L_{bos}}\hat{\rho}_i - \frac{di^*_{bos,i}}{dt}$$

and the following PI control is exploited:

$$R_{ILC}(s) = \frac{\hat{P}_i(s)}{\tilde{I}_{bos,i}(s)} = -\left(k_p^{idc} + \frac{k_i^{idc}}{s}\right) \tag{30}$$

The corresponding discrete time version is obtained by means of a Forward Euler method with sampling frequency f_s.

The parameters of Table 1 are considered and the modulation index ρ_i is performed by a simple PWM modulator with frequency f_{PWM}. The gains $k_p^{idc} = 5.0329$ and $k_i^{idc} = 316.23$ are tuned to obtain the desired phase margins of 86.4^0 at frequency $161Hz$: Fig. 11.

4.3   Capacitor Design

Key point of the PS design is the dimensioning of the DC-links. Their correct behavior does not depend on the voltage trajectories but only on the boundedness of voltages $v_{dc,i}$ between an upper value V^*_{max} and a lower value V^*_{min}. V^*_{max} cannot be overrun to respect capacitor physical constraints and V^*_{min} has to ensure the possibility of driving the current on the load. Moreover, if $v_{dc,i}$ becomes too small, modulation indexes $u'_{M,i}$ bigger than one could be requested thus introducing saturation phenomena. The formerly designed OLCs and ILCs keep under control the maximum values of $v_{dc,i}$.

The values of C_{dc} are obtained balancing the energies when the maximum current is drawn from the load. In this way, when references with smaller I_{AC} have to be tracked, the energy exchanged between the magnet and the capacitors C_{dc} is reduced and the oscillations of the DC-link voltages are reduced too. So, the minimum $v_{dc,i}$ value is greater than V^*_{min}.

Now assume that losses on the whole outer section are compensated by the power delivered by the booster converters and that the power stored in the active elements of the cable and in the output filter is negligible. So, the energy balance can be done taking into account only the DC-link capacitors and the magnet equivalent inductor. The energy stored in the magnet in the charging half period of the sinusoidal i_{lm} can be calculated as:

$$\Delta E_{L_{lm}} = \int_{-T_r/4}^{T_r/4} i_{lm}(t) L_{lm} \frac{di_{lm}(t)}{dt} dt = \frac{1}{2} L_{lm}(I_{max}^2 - I_{min}^2) \qquad (31)$$

where i_{lm} is approximated with i_{lm}^* with $I_{AC} = 99A$ and $T_r = 1/f_r$. Let $\hat{C}_{dc} = 3C_{dc}$ be the parallel of the three capacitors. In the same time interval the energy delivered by the three modules is:

$$\Delta E_{\hat{C}_{dc}} = \int_{-T_r/4}^{T_r/4} v_{dc}(t) \hat{C}_{dc} \frac{dv_{dc}(t)}{dt} dt = \frac{1}{2} \hat{C}_{dc}(V_{max}^{*2} - V_{min}^{*2}) \qquad (32)$$

Given the values of V_{max}^* and V_{min}^*, the value of $\hat{C}_{dc}$, and therefore of C_{dc}, is straightforward.

The previous procedure yields useful results for the dimensioning of the DC-link capacitors. However it is worth to mention that this type of results is a little rough and should be refined through simulative or experimental tests.

5   Simulation Results

Extensive simulations were carried out to test the adopted control strategies. The overall system has been considered and the parameters of Table 1 were assumed. First of all, the performances of the internal model current controller are discussed and its effectiveness demonstrated. Then, the cascade booster controller is analyzed.

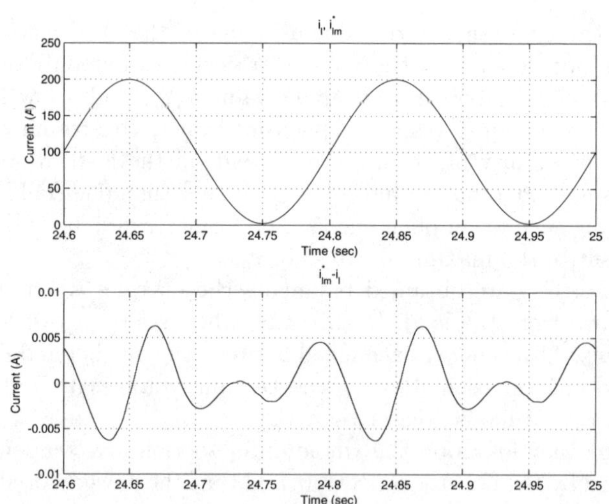

Fig. 12. Magnet current, reference current and current error

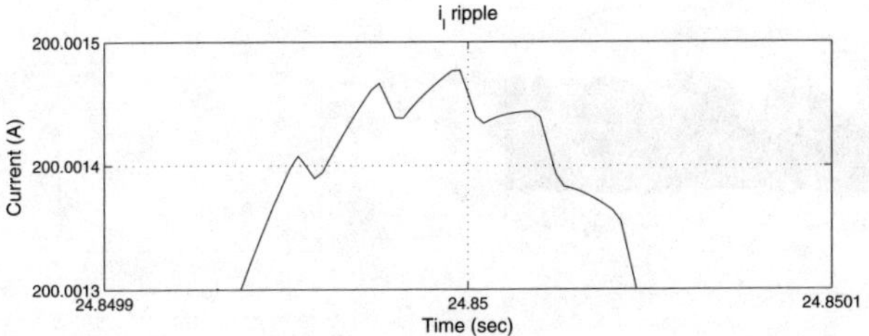

Fig. 13. Current ripple

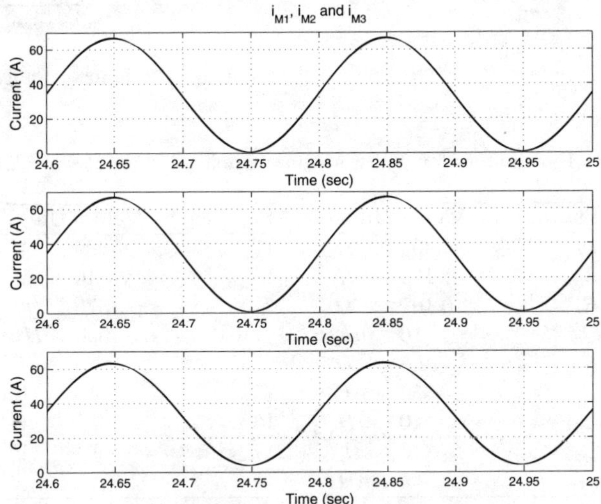

Fig. 14. Currents of the modules

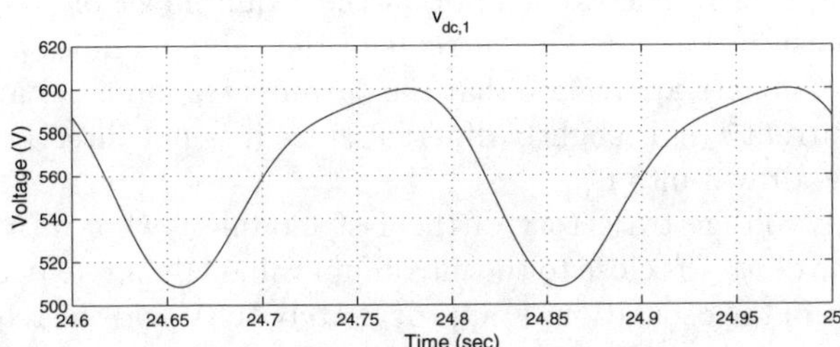

Fig. 15. Trajectory of the $v_{dc,1}$

The proposed results refer to a simulation in full output power (i.e. the reference current is the maximum allowable, $I_{AC} = 99$A). The current reference, the load current i_l and the current tracking error are shown in Fig. 12. Thanks to the internal model based control the tracking of the current is very good.

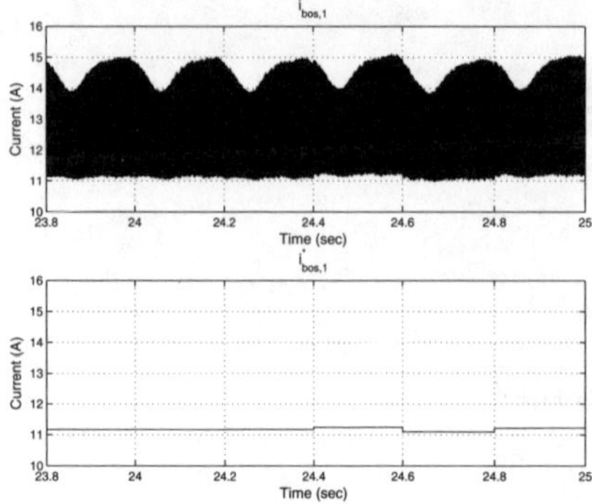
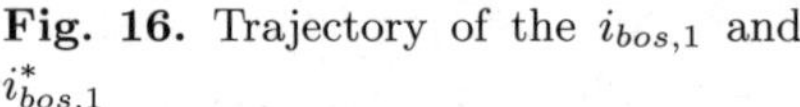

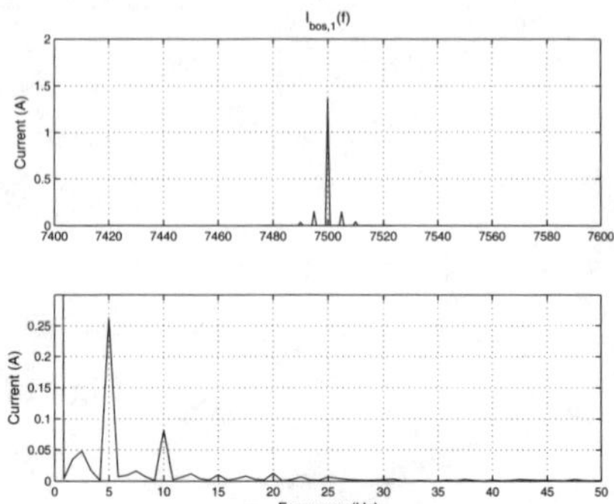

Fig. 16. Trajectory of the $i_{bos,1}$ and $i^*_{bos,1}$

Fig. 17. Fourier analysis of $i_{bos,1}$

Table 1. Parameters for Booster Quadrupole Magnet Power Converter

Parameters	Values	Units	Parameters	Values	Units
R_{lm}	0.496	Ω	V^*_{max}	600	V
L_{lm}	105	mH	V^*_{min}	510	V
R_c	0.187	Ω	f_s	1875	Hz
C_c	16	nF	f_{PWM}	7500	Hz
R_{of}	12.5	Ω	C_{if}	5	mF
C_{of}	350	μF	L_{if}	10	mH
L_{of}	10	mH	V_{FD}	0.9	V
C_{dc}	16	mF	V_{line}	294	V
L_{bos}	5	mH			

Error is kept below the admissible limit of 10mA as requested by the control specification 1), in 2.1. The satisfaction of the requirement of a ripple equal or less 2mA is shown in Fig. 13. The currents of the three modules are depicted in Fig. 14. It is possible to appreciate that the currents i_{M1} and i_{M2} are definitively similar. The current i_{M3} is slightly different since it is not directly sensed and a small current is drawn into C_{of}.

The DC-link voltage trajectory of the 1-st module is shown in Fig. 15. The maximum value of $v_{dc,1}$ is close to the maximum value V^*_{max} as expected while the minimum value of the oscillations is approximately 508V and the requirements on the upper and lower bounds of the safe voltage range are substantially satisfied.

The value of the DC-link capacitance C_{dc} needs further analysis. As asserted in 4.3 the algorithm for the dimensioning of the DC-link is a little rough and has to be tuned by means of simulations. Considering the specifications stated in 2.1, the value of I_{max} and I_{min} are respectively $200A$ and $2A$ in full output power conditions. Adopting the values of V^*_{max} and V^*_{min} of Table 1, a C_{dc} value of $14mF$ is computed through (31) and (32). Simulations highlighted that C_{dc} value has to be increased to take into account the reactive power of output

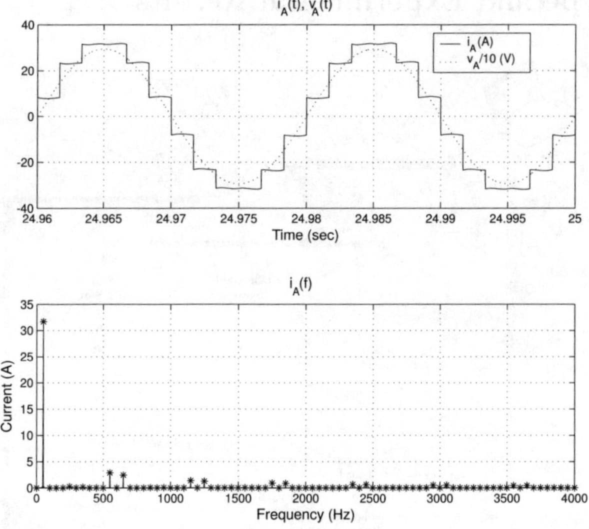

Fig. 18. Mains current and voltage. Fourier analysis of i_A.

filter and cables. A slightly larger value of $16mF$ are selected. Similar results are obtained for the 2-nd and 3-rd module.

The $i_{bos,1}$ and its reference $i^*_{bos,1}$, delivered by the OLC, are sketched in Fig. 16. The OLC control objective of null error and constant output is not reached because of the approximations introduced: the $i^*_{bos,1}$ reference denotes a tiny residual oscillation less of 0.4%. Anyway, this oscillation can be tolerated. The Fourier analysis of $i_{bos,1}$ (Fig 17) denotes a main continuous component of $12.86A$ and two spurious harmonics.

Finally, the analysis of the PF is reported. The mains voltage and current of phase A are depicted in Fig. 18. Analogous results can be shown for the phases B and C. The dominant component of the mains current is the 5Hz fundamental (second picture of Fig. 18). The values of the mains voltages and currents yield the following PF for the connection between the PS and the line:

$$PF = \frac{P_{in}}{\left(\mathbf{v}^{rms}\right)^{\mathbf{T}}\left(\mathbf{i}^{rms}\right)} = 0.988 \tag{33}$$

where:

$$\mathbf{v}^{rms} = \begin{bmatrix} v_A^{rms} \\ v_B^{rms} \\ v_C^{rms} \end{bmatrix}, \quad \mathbf{i}^{rms} = \begin{bmatrix} i_B^{rms} \\ i_B^{rms} \\ i_C^{rms} \end{bmatrix} \tag{34}$$

Hence, requirement 3) of 2.1 is substantially fulfilled and a PF close to the unit is achieved.

6   Prototype and Experimental Results

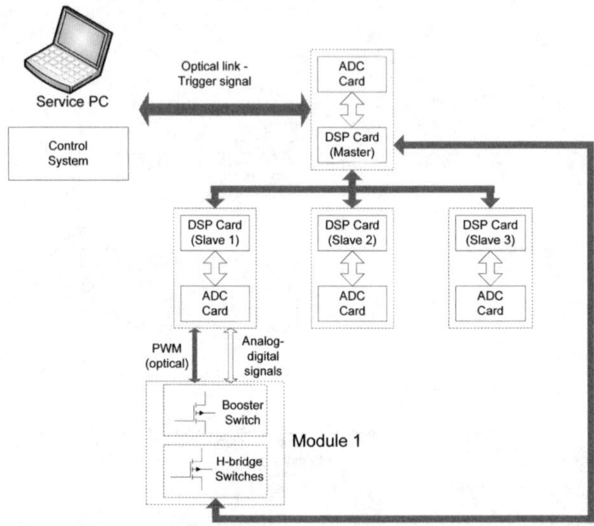

Fig. 19. Architecture of the control card

6.1   Prototype Digital Control

A prototype of the whole power supplier of Fig. 1 was built in collaboration with O.C.E.M. S.p.A, a company located in San Giorgio di Piano, Bologna. In particular the control above described was implemented by means of suitable DSP- and ADC-cards developed for this type of applications and furnished by Diamond Light Source Ltd. The main advantages of using these cards are:

- they are fully developed and have been shown to meet the required performance;
- they provide a common interface for all power supplies to the Control System, and can make use of the EPICS drivers that have already been developed.

DSP-cards exploit an Analog Devices SHARC digital signal processor with floating point capability, 60MHz clock frequency. The flexibility of the card is ensured by a *Field Programmable Gate Array* (FPGA) which performs all the communications and act as a really precise PWM modulator. *Pulse Repetition Modulation* technique is coupled with PWM to comply with the requested high accuracy. The communications with the control system and PWM signals are delivered by means of optical fibres. Two fast serial links are provided: one for communications between DSP-card and ADC-card, the other one for communications between more DSP-cards to get multiprocessor capability.

The ADC-cards are the key parts to obtain the precision requested by the control specifications. The requirement of a precision greater than 10 ppm asks for 17 bits plus sign for the current measurement: this resolution could be provided

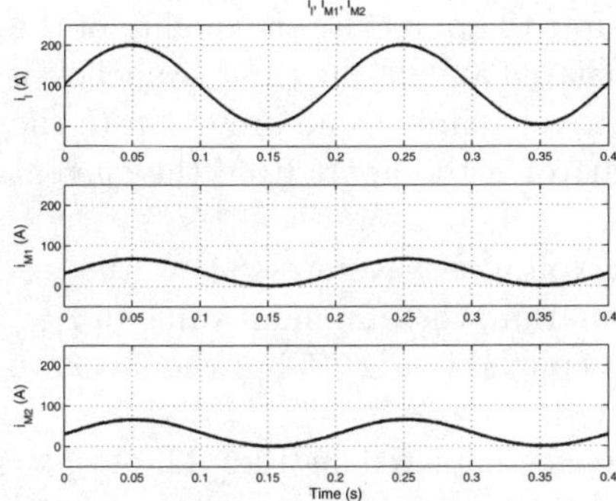

Fig. 20. Experimental results: i_l, $i_{M,1}$ and $i_{M,2}$

Fig. 21. Experimental results: i_l, $v_{dc,1}$ and $v_{dc,2}$

by two 16 bit two-channel ADCs. Another 16 bit four-channel ADC is provided for DC-link voltage measurement. Finally, two 14 bit DAC-channels are used to monitor any internal variable of the controller as well as for maintenance and troubleshooting. For more details see [9].

The cards are organized in the master-slave architecture depicted in Fig. 19. The master controls the load current and delivers the H-bridge switch commands performing the internal model control described in Section 3 while the slave cards controls the DC-link voltage (29) and (30).

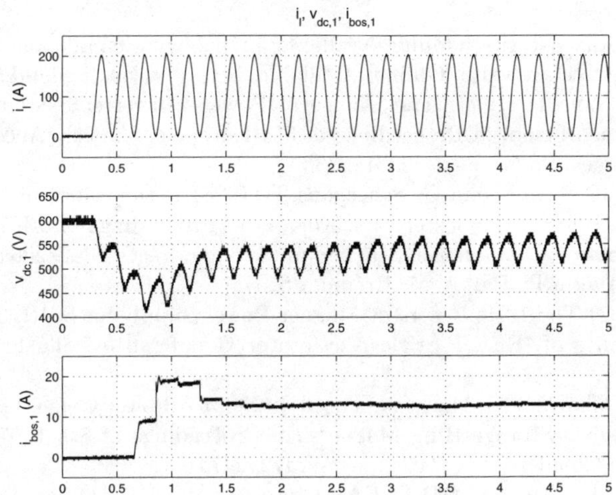

Fig. 22. Experimental results: i_l, $v_{dc,1}$ and $i_{bos,1}$

6.2 Experimental Results

The experimental tests obtained both in transitory and steady state confirm the quality of the control design. In Fig. 20 the load current i_l and the module currents $i_{M,1}$ and $i_{M,2}$ are reported. It is possible to appreciate the quality of the load current tracking. The burden is equally shared among the three branches of the PS. In particular $i_{M,1}$ and $i_{M,2}$ are definitively similar and are equal to the third part of i_l. $i_{M,3}$ is equal to the third part of i_l too apart from the current flowing in C_{of}.

In Fig. 21 the load current and DC-link voltages waveforms show the expected trend. Due to the introduced approximation, the minimum value of $v_{dc,i}$ is slightly smaller than the one expected and depicted in Fig. 15 but the ultimate goal of a controlled voltage pulsation is ensured.

The transient performances of the device are depicted in Fig. 22. At $t = 0.26sec$ the load is switched on. The DC-link voltage reaches a value of $400V$ but the OLC is able to drive the voltage into the desired set $[500V, 600V]$ and keeps it under control. The corresponding booster current is depicted in the third picture. After a first phase in which the value of $20A$ is reached, the current running through L_{bos} decreases to a constant value.

Acknowledgments

The authors wish to thank Mr. J. A. Dobbing and Mr. R. J. Rushton from Diamond Light Source Ltd and Mr. G. Taddia and Mr. M. Pretelli from OCEM S.p.A. for their support, suggestions and the helpful discussions.

References

1. The CANDLE web site [Online] Available: http://www.candle.am/~TDA/
2. The DIAMOND web site [Online] Available: http://www.diamond.ac.uk/
3. Marks N, Poole D (1996) The Choice of Power Converter Systems for a 3GeV Booster Synchrotron, in Proceedings of 5th European Particle Accelerator Coonference , Sitges, Spain, pages 2331–2333
4. Burkmann K, Schindhelm G, Scheegans T (1998) Performance of the White Circuits of the BESSY II Booster Synchrotron, in Proceedings of 6th European Particle Accelerator Coonference, Stockholm, Sweden, pages 2062–2064
5. Hettel R, Averill R, Baltay M, Brennan S, Harris C, Horton M, Jach C, Sebek J, Voss J (1991) The 10Hz Resonant Magnet Power Supply for SSRL 3GeV Injector, in Proceedings of IEEE Particle Accelerator Coonference , San Francisco, pages 926–928
6. Griffiths S, Charnley G, Marks N, Theed J (2002) A Power Converter Overview for the DIAMOND Storage Ring Magnets, in Proceedings of 8th European Particle Accelerator Coonference , Paris, pages 2472–2474
7. Bellomo P, de Lira A (2004) SPEAR3 Intermediate DC Magnet Power Supplies, in Proceedings of 9th European Particle Accelerator Coonference , Lucerne, pages 1798–1800
8. Irminger G, Horvat M, Jenni F, Boksberger H (1998) A 3Hz, 1MWpeak Bending Magnet Power Supply for the Swiss Light Source (SLS), Paul Scherrer Institut
9. Jenni F, Tanner L, Horvat M (2002) A Novel Control Concept for Highest Precision Accelerator Power Supplies, in Proceedings of 10th International Power Electronics and Motion Control Conference , Cavtat & Dubrovnik

10. Jenni F, Boksberger H, Irminger G (1999) DC-Link Control for a 1MVA-3Hz Single Phase Power Supply, in Proceedings of 30th Annual IEEE Power Electronics Specialists Conference, Charleston, vo2 l, pages 1172–1176
11. Pett J, Barnett I, Fernqvist G, Hundzinger D, Perreard J-C (1996) A Strategy for Controlling the LHC Magnet Currents, in Proceedings 5th European Particle Accelerator Coonference , Sitges, Spain, pages 2317–2319
12. King Q, Barnett I, Hundzinger D, Pett J (1999) Developments in the High Precision Control of Magnet Currents for LHC, in Proceedings of IEEE Particle Accelerator Coonference , New York, pages 3743–3745
13. Dobbing J A, Abraham C A, Rushton R J, Cagnolati F, Pretelli M P C, Sita L, Facchini G, Rossi C (2006) Diamond booster magnet power converters, in Proceedings of 10th European Particle Accelerator Conference, Edinburgh, UK, pages 2664–2666
14. Chang C (1995) Current ripple bounds in interleaved dc-dc power converters, in Proceedings of International Conference on Power Electronics and Drive Systems, vo2 l, pages 738–743
15. U-97, Modelling, analysis and compensation of the current-mode converter, pages 278–291

Distributed PCHD-Systems, from the Lumped to the Distributed Parameter Case

Kurt Schlacher

Institute of Automatic Control and Control Systems Technology, Johannes Kepler University Linz
kurt.schlacher@jku.at

Summary. The Hamiltonian approach has turned out to be an effective tool for modeling, system analysis and controller design in the lumped parameter case. There exist also several extensions to the distributed parameter case. This contribution presents a class of extended distributed parameter Hamiltonian systems, which preserves some useful properties of the well known class of Port Controlled Hamiltonian systems with Dissipation. In addition, special ports are introduced to take the boundary conditions into account. Finally, an introductory example and the example of a piezoelectric structure, a problem with two physical domains, show, how one can use the presented approach for modeling and design.

Keywords: Distributed Parameter Systems, Hamiltonian Systems with Input and Dissipation.

1 Introduction

Modern model based control requires accurate mathematical models of the plant to be controlled. Since the modern theory can deal with many different classes of models, we restrict ourselves here to a certain class of models, which appears often in classical physics. These models have a rich structure, they are much more than a set of arbitrary differential equations. Roughly speaking, these models are derived from two sets of laws. The first one encompasses the so called balance equations and/or conservation laws. Typical representatives are the balance of linear momentum, of energy, the conservation of mass, of charge etc. Such a law consists of a storage and a flow to the storage. A discrete storage is described by ordinary differential equations, a distributed storage leads to partial differential equations. In the ODE case the flows are also discrete quantities, whereas we have to deal with flows over surfaces in the PDE case. The second set of laws contains the so called constitutive equations, which describe the flows in dependence of the behavior of the matter. If one looks back to the models, which describe dynamic systems of classical physics, then it is no wonder, why Lagrangian and Hamiltonian methods are so successful in this field.

This contribution deals with an Hamiltonian approach to lumped and distributed parameter systems. In Section 2 we give a short overview of the lumped

C. Bonivento et al. (Eds.): Adv. in Control Theory and Applications, LNCIS 353, pp. 239–255, 2007.
springerlink.com

parameter case, where the main ingredients and tools for the construction of a **p**ort **c**ontrolled **H**amiltonian system with **d**issipation or PCHD system are presented. In Section 3 we extend this approach to distributed parameter systems. This extension is not unique at all, but the presented approach has the capability to cover several physical systems of interest. We use differential geometry as the basic mathematical tool. Since the methods for PDEs are not so well known like the methods for ODEs, the reader will find a short summary of these methods in the Appendix. It is worth mentioning that the presented approach is a formal one, where we assume the existence of several maps and the correctness of our manipulations, but we show neither their existence nor we prove the correctness from the functional analysis point of view. We will use the tensor notation with Einstein's convention for sums to keep the formulas as short as possible. In addition, we confine ourselves to the time invariant case.

2 The Finite Dimensional Case

Let us consider the following system

$$\dot{x}^{\alpha} = \left(J^{\alpha\beta}\left(x\right) - R^{\alpha\beta}\left(x\right) \right) \partial_{\beta} H\left(x\right) + B_{\varsigma}^{\alpha}\left(x\right) u^{\varsigma} \,, \tag{1}$$

$\alpha, \beta = 1, \ldots, q$, $\varsigma = 1, \ldots, r$, also called a PCHD-system, with the q-dimensional state $x \in \mathbb{R}^{q}$ and the r-dimensional input $u \in \mathbb{R}^{r}$. The skew symmetric matrix $\left[J^{\alpha\beta} \right]$ is the structure matrix, the positive semi definite matrix $\left[R^{\alpha\beta} \right]$ is the dissipation matrix, and H is called the Hamiltonian, see also [11] and the citations therein. From the relations

$$\frac{\mathrm{d}}{\mathrm{d}t} H = \underbrace{\partial_{\alpha} H \, J^{\alpha\beta} \partial_{\beta} H}_{=0} - \underbrace{\partial_{\alpha} H \, R^{\alpha\beta} \partial_{\beta} H}_{\geq 0} + \underbrace{\partial_{\alpha} H \, B_{\varsigma}^{\alpha} \, u^{\varsigma}}_{y_{\varsigma}} \tag{2}$$

we see that H is a constant of motion for vanishing dissipation $R^{\alpha\beta} = 0$ and no input. If the system is completed by the outputs

$$y_{\varsigma} = \partial_{\alpha} H \, B_{\varsigma}^{\alpha} \,, \tag{3}$$

then the pair $(y_{\varsigma}, u^{\varsigma})$, also called a port, describes a discrete flow to the system, which changes the stored quantity H. One uses R to cover internal losses of H.

Since the equation (1) describes the dynamical system in special coordinates, we work out the coordinate free description first. We assume, the state x lives in a q-dimensional smooth manifold $\mathcal{X}$ equipped locally with coordinates (x^{α}), $\alpha = 1, \ldots, q$. Let $\mathcal{T}(\mathcal{X})$, $\mathcal{T}^{*}(\mathcal{X})$ denote the tangent and cotangent bundle of $\mathcal{X}$ equipped with the coordinates $(\dot{x}^{\alpha})$ and $(\dot{x}_{\alpha})$ according to the bases $\{\partial_{\alpha}\}$, $\{\mathrm{d}x^{\alpha}\}$ respectively. Given a function $H \in C^{\infty}(\mathcal{X})$ and a tangent vector field $v = v^{\alpha} \partial_{\alpha} \in \Gamma(\mathcal{T}(\mathcal{X}))$, then the change of H into the direction v at the point x follows as

$$v\left(H\right) = v \rfloor \mathrm{d}H = v^{\alpha} \partial_{\alpha} \rfloor \partial_{\beta} H \mathrm{d}x^{\beta} = v^{\alpha} \partial_{\alpha} H \tag{4}$$

Obviously, the one form $\mathrm{d}H \in \Gamma\left(T^*\left(\mathcal{X}\right)\right)$ is the image of H by the exterior derivative d, and the contraction $\rfloor$ describes the canonical product $T\left(\mathcal{X}\right) \times T^*\left(\mathcal{X}\right) \to C^\infty\left(\mathcal{X}\right)$. The matrices $\left[J^{\alpha\beta}\right]$, $\left[R^{\alpha\beta}\right]$ represent linear maps of the type

$$J, R : T^*\left(\mathcal{X}\right) \to T\left(\mathcal{X}\right) \tag{5}$$

with J skew symmetric and R positive semi definite. Obviously, the construction of the autonomous system is based on the sequence

$$C^\infty\left(\mathcal{X}\right) \xrightarrow{\mathrm{d}} T^*\left(\mathcal{X}\right) \xrightarrow{J,R} T\left(\mathcal{X}\right) . \tag{6}$$

It is worth mentioning that $v \in \Gamma\left(T\left(\mathcal{X}\right)\right)$ with

$$v^\alpha = \left(J^{\alpha\beta} - R^{\alpha\beta}\right) \partial_\beta H \tag{7}$$

is met.

To incorporate the input u and output y, we choose two linear spaces $\mathcal{U}$, $\mathcal{Y} = \mathcal{U}^*$ dual to each other, equipped with coordinates (u^ς), (y_ς), $\varsigma = 1, \dots, r$. Obviously, the natural product $u^\varsigma y_\varsigma$ describes the total flow to the system for the choice of (3). It is also worth mentioning that v

$$v^\alpha = \left(J^{\alpha\beta} - R^{\alpha\beta}\right) \partial_\beta H + B_\varsigma^\alpha u^\varsigma \tag{8}$$

is not a tangent vector field on $\mathcal{X}$ in contrast to (7). Let us introduce the bundle $\mathcal{X} \times \mathcal{U} \xrightarrow{\rho} \mathcal{X}$, where $\rho = \mathrm{pr}_1$ is the projection on the first factor, then a section of this bundle is state feedback. Therefore, $v \in \Gamma\left(\rho^*\left(T\left(\mathcal{X}\right)\right)\right)$ is met, where $\rho^*\left(T\left(\mathcal{X}\right)\right)$ is the pullback of $T\left(\mathcal{X}\right)$ by ρ^*, see the Appendix. Now, it is easy to show that v is a true tensor object. Obviously, the matrix $\left[B_\varsigma^\alpha\right]$ describes the input map $B : \mathcal{U} \to T\left(\mathcal{X}\right)$, as well as its dual $B^* : T^*\left(\mathcal{X}\right) \to \mathcal{Y}$.

Of special physical interest is the case of the existence of functions $H_\varsigma \in C^\infty\left(\mathcal{X}\right)$ such that

$$B_\varsigma^\alpha = -J^{\alpha\beta}\partial_\beta H_\varsigma \quad , \quad R^{\alpha\beta}\partial_\beta H_\varsigma = 0 \quad , \quad \partial_\alpha H_\varsigma J^{\alpha\beta}\partial_\beta H_\tau = 0 \tag{9}$$

is met, because then one can rewrite the system (1) as

$$\begin{aligned}
\dot{x}^\alpha &= v^\alpha = \left(J^{\alpha\beta} - R^{\alpha\beta}\right) \partial_\beta \left(H - H_\varsigma u^\varsigma\right) = \left(J^{\alpha\beta} - R^{\alpha\beta}\right) \partial_\beta H + B_\varsigma^\alpha u^\varsigma \quad , \\
y_\varsigma &= v\left(H_\varsigma\right) = B_\varsigma^\alpha \partial_\alpha H
\end{aligned} \tag{10}$$

with the extended Hamiltonian H_e,

$$H_e = H - H_\varsigma u^\varsigma . \tag{11}$$

If one chooses $Y_\varsigma = H_\varsigma$ for the output, then y_ς follows from $y_\varsigma = \frac{\mathrm{d}}{\mathrm{d}t}Y_\varsigma = v\left(Y_\varsigma\right)$. In many cases one can measure Y_ς and y_ς simultaneously. Therefore, a control law of the type

$$u^\varsigma = -P^{\varsigma\tau}Y_\tau - D^{\varsigma\tau}y_\tau , \quad \tau = 1, \dots, r \tag{12}$$

with positive (semi) definite matrices $[P^{\varsigma\tau}]$, $[D^{\varsigma\tau}]$, $P^{\varsigma\tau}, D^{\varsigma\tau} \in \mathbb{R}$ is easy to implement. It is straightforward to show, that the equations of motion of the closed loop follow from

$$H_c = H + \frac{1}{2} H_\varsigma P^{\varsigma\tau} H_\tau \,, \quad J_c^{\alpha\beta} = J \,, \quad R_c^{\alpha\beta} = R^{\alpha\beta} + B_\varsigma^\alpha D^{\varsigma\tau} B_\tau^\beta \,, \qquad (13)$$

where the index c points to quantities of the closed loop. It is worth mentioning that the PD control law (12) is very often applied to mechanical systems.

We will not discuss further control strategies for lumped PCHD systems, the reader is kindly asked to consult the very rich literature, see e.g. [11] and the citations therein.

3   The Infinite Dimensional Case

To extend the consideration from the finite dimensional case to the infinite dimensional one, we have to find counterparts for the state and the state manifold, for the Hamiltonian and for the relations (4), (8) and the maps (5), (6). If we think of physical examples like beams, strings, etc., then it will be clear, that a state of a distributed parameter system is given by a certain set of functions defined on a domain $\mathcal{D}$. Therefore, we introduce a compact set $\mathcal{D}$ with coordinates (X^i), $i = 1,\ldots,p$, which is equipped with the global volume form $dX = \bigwedge_{i=1}^{p} dX^i$. Since $\mathcal{D}$ is orientable, $\partial\mathcal{D}$ denotes the coherently oriented boundary with coordinates $\left(\bar{X}^{\bar{i}}\right)$, $\bar{i} = 1,\ldots,p-1$. We call the coordinates adapted to the boundary, if $\overline{X}^{\bar{i}} = X^{\bar{i}}$ is met. Since a state is given by a set of functions on $\mathcal{D}$, we introduce the state bundle $\mathcal{X} \xrightarrow{\pi} \mathcal{D}$, see the Appendix, with coordinates $\left(X^i, x^\alpha\right)$, $\alpha = 1,\ldots,q$, such that a section σ of $\mathcal{X}$ defines a state x by $x^\alpha = \sigma^\alpha(X)$, $X \in \mathcal{D}$. Let $i : \partial\mathcal{D} \to \mathcal{D}$ denote the inclusion map, then $i^*(\mathcal{X})$ is the pull back of $\mathcal{X}$ to the boundary $\partial\mathcal{D}$.

To proceed with our program, we have to find a counterpart for the Hamiltonian. Again from physics it is evident, that we introduce the Hamiltonian functional $\mathfrak{H} : \Gamma(\mathcal{X}) \to \mathbb{R}$ by

$$\mathfrak{H}(\sigma) = \int_{\mathcal{D}} H \circ j^m(\sigma)\,dX \,, \quad \sigma \in \Gamma(\mathcal{X}) \,, \quad H \in C^\infty(J^m(\mathcal{X})) \,, \qquad (14)$$

where the Hamiltonian density depends not only on x^α but also on the jet coordinates x_I^α, $\#I \le m > 0$ in general, see also the Appendix.

To define a counterpart for (4), we replace the tangent vector field by a so called evolutionary vector field $v = v^\alpha \partial_\alpha \in \Gamma\left(\pi_0^{n,*}(V(\mathcal{X}))\right)$, $v^\alpha \in C^\infty(J^n(\mathcal{X}))$, see also the Appendix, which corresponds to the set of partial differential equations

$$\dot{X} = 0, \quad \dot{x}^\alpha = v^\alpha \,. \qquad (15)$$

Now, one can show, see the Appendix, that the change $v(\mathfrak{H})$ of $\mathfrak{H}$ into the direction of v at σ is given by the formula

$$v\left(\mathfrak{H}\right)(\sigma) = \int_{\mathcal{D}} j^{m+n}\left(\sigma\right)^{*}\left(j^{m}\left(v\right)\left(H\mathrm{d}X\right)\right), \tag{16}$$

which is the wanted counterpart of (4). Here, we confine ourselves to the case $m = 1$, since it can be solved by the straightforward application of the integration by part technique. The general case $m \geq 1$ can be found in [3].

3.1 First Order Hamiltonian

To find a manageable expression for (16) with $H \in C^{\infty}\left(J\left(\mathcal{X}\right)\right)$, we look at the relations

$$j\left(v\right)\left(H\mathrm{d}X\right) = j\left(v\right)\rfloor\mathrm{d}H \wedge \mathrm{d}X = \left(v^{\alpha}\partial_{\alpha}H + d_{i}\left(v^{\alpha}\right)\partial_{\alpha}^{1_{i}}H\right)\mathrm{d}X$$

and derive by integration by parts, see the Appendix for the horizontal differential d_{H}, the expression

$$\begin{aligned}
j\left(v\right)\left(H\mathrm{d}X\right) &= \left(\left(v^{\alpha}\partial_{\alpha}H - v^{\alpha}d_{i}\partial_{\alpha}^{1_{i}}H\right) + d_{i}\left(v^{\alpha}\partial_{\alpha}^{1_{i}}H\right)\right)\mathrm{d}X \\
&= v\rfloor\left(\partial_{\alpha}H - d_{i}\partial_{\alpha}^{1_{i}}H\right)\mathrm{d}x^{\alpha} \wedge \mathrm{d}X \\
&\quad + d_{H}\left(v\rfloor\partial_{\alpha}^{1_{i}}H\mathrm{d}x^{\alpha} \wedge \partial_{i}\rfloor\mathrm{d}X\right).
\end{aligned}$$

From this relation we extract two new maps. The variational derivative δ is the map

$$\delta : \pi_{0}^{1,*}\left(\bigwedge_{p}^{p}\left(\mathcal{T}^{*}\left(\mathcal{X}\right)\right)\right) \to \pi_{0}^{2,*}\left(\bigwedge_{p}^{p+1}\left(\mathcal{T}^{*}\left(\mathcal{X}\right)\right)\right) \tag{17}$$

given in coordinates by

$$\delta_{\alpha}\left(H\right) = \left(\partial_{\alpha}H - d_{i}(\partial_{\alpha}^{1_{i}})\right)H. \tag{18}$$

The second map $_{\partial}\delta$,

$$_{\partial}\delta : \pi_{0}^{1,*}\left(\bigwedge_{p}^{p}\left(\mathcal{T}^{*}\left(\mathcal{X}\right)\right)\right) \to i^{*}\left(\pi_{0}^{1,*}\left(\bigwedge_{p-1}^{p}\left(\mathcal{T}^{*}\left(\mathcal{X}\right)\right)\right)\right) \tag{19}$$

follows in coordinates as

$$i^{*}\left(\partial_{\alpha}^{1_{i}}H\mathrm{d}x^{\alpha} \wedge \partial_{i}\rfloor\mathrm{d}X\right) = _{\partial}\delta_{\alpha}\left(H\right)\mathrm{d}x^{\alpha} \wedge \mathrm{d}\overline{X}. \tag{20}$$

Since coordinates adapted to the boundary simplifies several expressions, we use them from now on. E.g. expression (20) takes the simpler form

$$i^{*}\left(\partial_{\alpha}^{i}H\mathrm{d}x^{\alpha} \wedge \partial_{i}\rfloor\mathrm{d}X\right) = \partial_{\alpha}^{1_{p}}H\mathrm{d}x^{\alpha} \wedge \mathrm{d}\overline{X}.$$

Two interesting facts are worth mentioning. If we compare the bundles $i^{*}\left(J\left(\mathcal{X}\right)\right)$, $J\left(i^{*}\left(\mathcal{X}\right)\right)$ with coordinates adapted to the boundary $\left(X^{\bar{i}}, x^{\alpha}, x_{1_{i}}^{\alpha}\right)$, $\left(X^{\bar{i}}, x^{\alpha}, x_{1_{\bar{i}}}^{\alpha}\right)$, then they differ by the variables $x_{1_{p}}^{\alpha}$. Just the derivatives of H with respect to $x_{1_{p}}^{\alpha}$ enter the map $_{\partial}\delta$.

With the help of (17, 18) and (19, 20) we obtain the required expression for (16) for first order Hamiltonians as

$$v\left(\mathfrak{H}\right)(\sigma) = \int_{\mathcal{D}} j^{n+2}\left(\sigma\right)^{*}\left(v\rfloor\delta\left(H\mathrm{d}X\right)\right) + \int_{\partial\mathcal{D}} j^{n+1}\left(\sigma\right)^{*}\left(v\rfloor\delta_{\partial}\left(H\mathrm{d}X\right)\right) \qquad (21)$$

given in coordinates by

$$v\left(\mathfrak{H}\right)(\sigma) = \int_{\mathcal{D}} j^{n+2}\left(\sigma\right)^{*}\left(v^{\alpha}\delta_{\alpha}\left(H\right)\mathrm{d}X\right) + \int_{\partial\mathcal{D}} j^{n+1}\left(\sigma\right)^{*}\left(v^{\alpha}\partial_{\alpha}^{1_{p}}\left(H\right)\mathrm{d}\overline{X}\right) \ . \qquad (22)$$

These relations, they are the counterpart of (4), are the basis for the following investigations.

3.2 Evolutionary Equations for First Order Hamiltonian

Here, we propose the following set of equations

$$\dot{X}^{i} = 0 \ , \quad \dot{x}^{\alpha} = v^{\alpha} = \left(J^{\alpha\beta} - R^{\alpha\beta}\right)\delta_{\beta}H + B_{\varsigma}^{\alpha}u^{\varsigma} \qquad (23)$$

as generalization of (1) to the distributed parameter case. The matrices J, R describe a skew symmetric and a positive semi definite map

$$J, R : \pi_{0}^{2,*}\left(\bigwedge_{p}^{p+1}\left(\mathcal{T}^{*}\left(\mathcal{X}\right)\right)\right) \to \pi_{0}^{2,*}\left(\mathcal{V}\left(\mathcal{X}\right)\right) \ ,$$

which are the counterpart of (5). In addition, there is a natural product given by the contraction $\rfloor$,

$$\rfloor : \pi_{0}^{2,*}\left(\mathcal{V}\left(\mathcal{X}\right)\right) \times \pi_{0}^{2,*}\left(\bigwedge_{p}^{p+1}\left(\mathcal{T}^{*}\left(\mathcal{X}\right)\right)\right) \to \pi_{0}^{2,*}\left(\bigwedge_{p}^{p}\left(\mathcal{T}^{*}\left(\mathcal{X}\right)\right)\right) \ .$$

The evaluation of the volume integral of (22) for v from (23) leads to

$$\int_{\mathcal{D}} v^{\alpha}\delta_{\alpha}\left(H\right)\mathrm{d}X = \int_{\mathcal{D}}\big(\underbrace{\delta_{\alpha}\left(H\right)J^{\alpha\beta}\delta_{\beta}H}_{=0} - \underbrace{\delta_{\alpha}\left(H\right)R^{\alpha\beta}\delta_{\beta}H}_{\geq 0} + \delta_{\alpha}\left(H\right)B_{\varsigma}^{\alpha}u^{\varsigma}\big)\mathrm{d}X \ ,$$

which is the counterpart of (2). Again, there is a canonical output y_{ς},

$$y_{\varsigma} = B_{\varsigma}^{\alpha}\delta_{\alpha}\left(H\right) \ , \qquad (24)$$

see (3) for the lumped parameter case. Following these considerations we choose the vector bundle $\mathcal{U} \xrightarrow{\rho} \mathcal{D}$ with coordinates $\left(X^{i}, u^{\varsigma}\right)$, $\varsigma = 1, \ldots, r$ for the input space. Then the output space $\mathcal{Y}$ is the bundle $\mathcal{Y} \xrightarrow{\rho^{*}} \mathcal{D}$ with coordinates $\left(X^{i}, y_{\varsigma}\right)$ dual to $\mathcal{U}$ with respect to the product $\langle\cdot,\cdot\rangle : \mathcal{Y} \times \mathcal{U} \to \bigwedge^{p}\left(\mathcal{T}^{*}\left(\mathcal{D}\right)\right)$ given in coordinates by $y_{\varsigma}u^{\varsigma}\mathrm{d}X$. Obviously, the pair $\left(y_{\varsigma}, u^{\varsigma}\right)$ is nothing else than a port distributed over $\mathcal{D}$. Here, the matrix $[B_\varsigma^\alpha]$ describes the input map $B : \mathcal{U} \to \pi_0^{2,*}(\mathcal{V}(\mathcal{X}))$, as well as its dual $B^* : \pi_0^{2,*}\left(\bigwedge_p^{p+1}(\mathcal{T}^*(\mathcal{X}))\right) \to \mathcal{Y}$.

It must be mentioned here, that there are other generalizations of Hamiltonian systems to the distributed parameter case than (23), see [5], [9], [12]. A straightforward extension would be the replacement of the maps J, R, B by suitable vector valued differential operators.

The introduction of ports distributed over $\partial\mathcal{D}$ is more tricky than the previous case and it is based on the following considerations. We have to add initial and boundary conditions to (23) to complete the problem. The initial conditions are simply given by $x^\alpha = \sigma_0^\alpha$, $\sigma_0 \in \Gamma(\mathcal{X})$ for $t = 0$. To proceed, we assume the existence of a special subset of the dependent variables. Let $\{\bar{x}^\varsigma\}$ be a subset of the variables $\{x^\alpha\}$ with the following properties. From the relations $\dot{\bar{x}}^\varsigma = f^{\bar{\varsigma}}\left(t, X^{\bar{i}}\right)$ met on $\partial\mathcal{D}$, one is able to determine $\dot{x}^\alpha = f^\alpha\left(t, X^{\bar{i}}\right)$ by differentiation and elimination with the help of (23). The functions $f^{\bar{\varsigma}}$ can be chosen freely, there do not exist hidden constraints between them and their derivatives and/or the initial conditions and their derivatives.[1] Finally, the surface integral of (22) can be rewritten as

$$\int_{\partial\mathcal{D}} v^\alpha \partial_\alpha^{1_p}(H)\,\mathrm{d}\overline{X} = \int_{\partial\mathcal{D}} \dot{\bar{x}}^\varsigma \partial_\varsigma^{1_p}(H)\,\mathrm{d}\overline{X}\,. \tag{25}$$

Now, one possibility for the choice of ports on $\partial\mathcal{D}$ is given by

$$\bar{u}^\varsigma = \dot{\bar{x}}^\varsigma\,, \quad \bar{y}_{\bar{\varsigma}} = -\partial_{\bar{\varsigma}}^{1_p}(H)\,, \tag{26}$$

since it meets

$$\int_{\partial\mathcal{D}} \left(v^{\bar{\varsigma}} \partial_{\bar{\varsigma}}^{1_p}(H) + \bar{y}_{\bar{\varsigma}} \bar{u}^\varsigma\right)\mathrm{d}\overline{X} = 0\,.$$

But the choice

$$\bar{u}^\varsigma = -\partial_{\bar{\varsigma}}^{1_p}(H)\,, \quad \bar{y}_{\bar{\varsigma}} = \dot{\bar{x}}^\varsigma \tag{27}$$

is also possible. One has to decide in the light of the actual problem, whether (26), (27) or a combination of them is the correct choice. Analogously to above we choose the vector bundle $\bar{\mathcal{U}} \xrightarrow{\bar{\rho}} \partial\mathcal{D}$ with coordinates $\left(X^{\bar{i}}, \bar{u}^\varsigma\right)$, $\bar{\varsigma} = 1, \dots, \bar{r}$ for the input space. The output space $\mathcal{Y}$ is the bundle $\bar{\mathcal{Y}} \xrightarrow{\bar{\rho}^*} \partial\mathcal{D}$ with coordinates $\left(X^{\bar{i}}, \bar{y}_{\bar{\varsigma}}\right)$ dual to $\bar{\mathcal{U}}$ with respect to the product $\langle\cdot,\cdot\rangle : \bar{\mathcal{Y}} \times \bar{\mathcal{U}} \to \bigwedge^p(\mathcal{T}^*(\partial\mathcal{D}))$ given in coordinates by $\bar{y}_{\bar{\varsigma}} \bar{u}^\varsigma \mathrm{d}\overline{X}$. Further constructions of ports on $\partial\mathcal{D}$ following physical considerations can be found in [4].

Before we continue, let us summarize the proposed construction of an autonomous Hamiltonian system with dissipation. The lumped parameter case was based on the sequence (6). Since we have the two maps δ, $_\partial\delta$, see (17), (19), now, the distributed parameter counterpart of (6) splits in the following manner

---

[1] This may be violated on a set of measure zero.

$$\pi_0^{2,*}\left(\Lambda_p^{p+1}\left(\mathcal{T}^*\left(\mathcal{X}\right)\right)\right) \quad\xrightarrow{\ J,R\ }\quad \pi_0^{2,*}\left(\mathcal{V}\left(\mathcal{X}\right)\right)$$

$$\pi_0^{1,*}\left(\Lambda_p^{p}\left(\mathcal{T}^*\left(\mathcal{X}\right)\right)\right)\quad\overset{\delta}{\nearrow}$$

$$\underset{\partial\delta}{\searrow}$$

$$i^*\left(\pi_0^{1,*}\left(\Lambda_{p-1}^{p}\left(\mathcal{T}^*\left(\mathcal{X}\right)\right)\right)\right) \tag{28}$$

In addition, one has to add suitable boundary conditions to complete the system.

Also in the distributed parameter case there exists a more special choice for the inputs than (23). If there exist functions $H_\varsigma \in C^\infty\left(\mathcal{X}\right)$, such that

$$B_\varsigma^\alpha = -J^{\alpha\beta}\delta_\beta H_\varsigma \quad,\quad R^{\alpha\beta}\delta_\beta H_\varsigma = 0 \quad,\quad \delta_\alpha\left(H_\varsigma\right)J^{\alpha\beta}\delta_\beta H_\varsigma = 0$$

is met, then one can rewrite the system as

$$\begin{aligned}
\dot{x}^\alpha &= v^\alpha = \left(J^{\alpha\beta}-R^{\alpha\beta}\right)\delta_\beta\left(H - H_\varsigma u^\varsigma\right) = \left(J^{\alpha\beta}-R^{\alpha\beta}\right)\delta_\beta H + B_\varsigma^\alpha u^\varsigma\\
y_\varsigma &= v\left(H_\varsigma\right) = B_\varsigma^\alpha \delta_\alpha H\ .
\end{aligned} \tag{29}$$

with the extended Hamiltonian density H_e,

$$H_e = H - H_\varsigma u^\varsigma\ . \tag{30}$$

Obviously, this is the distributed parameter counterpart of (9), (10). Again, if one chooses $Y_\varsigma = H_\varsigma$ for the output, then y_ς follows from $y_\varsigma = \frac{\mathrm{d}}{\mathrm{dt}}Y_\varsigma = v\left(Y_\varsigma\right)$.

Finally, an interesting case is worth mentioning. If the functions H_ς meet $H_\varsigma \in C^\infty\left(J\left(\mathcal{X}\right)\right)$ only, then the input map B contains already differential operators because of

$$J^{\alpha\beta}\delta_\beta\left(H_\varsigma u^\varsigma\right) = J^{\alpha\beta}\left(\partial_\beta\left(H_\varsigma\right)u^\varsigma - d_i\left(\partial_\beta^{1i}\left(H_\varsigma\right)u^\varsigma\right)\right)\ , \tag{31}$$

where d_i denotes a total derivative of the jet manifold $J^2\left(\mathcal{X}\times_\mathcal{D}\mathcal{U}\right)$. Because of the complexity of this problem, we will not dicuss this problem further, but we will present an example.

Also here we discuss simple PD control laws for the system (29) only. An introduction to more complex approaches can be found in [10]. Let us assume, we can measure Y_ς and y_ς simultaneously, and it is possible to implement the distributed control law

$$u^\varsigma = -P^{\varsigma\tau}Y_\tau - D^{\varsigma\tau}y_\tau\ , \quad P^{\varsigma\tau}, D^{\varsigma\tau} \in C^\infty\left(\mathcal{D}\right)\ , \tag{32}$$

with positive (semi) definite matrices $[P^{\varsigma\tau}]$, $[D^{\varsigma\tau}]$. A short calculation shows, that the equations of motion of the closed loop follow from

$$H_c = H + \frac{1}{2}H_\varsigma P^{\varsigma\tau}H_\tau\ , \quad J_c^{\alpha\beta} = J\ , \quad R_c^{\alpha\beta} = R^{\alpha\beta} + B_\varsigma^\alpha D^{\varsigma\tau}B_\tau^\beta\ . \tag{33}$$

Obviously, the equations (32), (33) are the counterpart of (12), (13). To complete the problem, we connect the systems on $\partial\mathcal{D}$ by ports of the type (27) to a PD controller

$$\bar{u}^{\bar{s}} = \bar{P}^{\bar{s}\bar{\tau}}\bar{Y}_{\bar{\tau}} + \bar{D}^{\bar{s}\bar{\tau}}\bar{y}_{\bar{\tau}} \tag{34}$$

with $\bar{y}_{\bar{s}} = \dot{\bar{x}}^{\bar{s}}$, $\bar{Y}_{\bar{s}} = \bar{x}^{\bar{s}}$ and positive (semi) definite matrices $\left[\bar{P}^{\bar{s}\bar{\tau}}\right]$, $\left[\bar{D}^{\bar{s}\bar{\tau}}\right]$, $\bar{P}^{\bar{s}\bar{\tau}}, \bar{D}^{\bar{s}\bar{\tau}} \in C^{\infty}(\partial\mathcal{D})$. Of course, it must be possible to measure $\bar{y}_{\bar{s}}$ and $\bar{Y}_{\bar{s}}$. Finally, the Hamiltonian functional $\mathfrak{H}$ of the closed loop follows as

$$\mathfrak{H} = \int_{\mathcal{D}} H_c \mathrm{d}X + \int_{\partial\mathcal{D}} \frac{1}{2}\bar{Y}_{\bar{s}}\bar{P}^{\bar{s}\bar{\tau}}\bar{Y}_{\bar{\tau}}\mathrm{d}\overline{X}$$

after a short calculation. In a similar way one derives $v(\mathfrak{H})$,

$$v(\mathfrak{H}) = -\int_{\mathcal{D}} \delta_{\alpha}(H_c) R_c^{\alpha\beta}\delta_{\beta}H_c\mathrm{d}X - \int_{\partial\mathcal{D}} \bar{y}_{\bar{s}}\bar{D}^{\bar{s}\bar{\tau}}\bar{y}_{\bar{\tau}}\mathrm{d}\overline{X} \,,$$

where $v^{\alpha} = \dot{x}^{\alpha}$ denotes the evolutionary field generated by the equations of motion. From these relations it follows, that $\mathfrak{H}$ is a candidate for a Liapunov function, provided $\mathfrak{H}$ is a positive definite functional.

4   Examples

In this section we present two examples, the first one is the one dimensional wave equation, the second one deals with a piezoelectric mechanical structure. The purpose of the first example, which is quite simple, is to show the application of the previous introduced mathematical machinery, whereas the second one is supposed to prove the applicability of the proposed approach to problems with at least two physical domains.

4.1   Wave Equation

Let us consider a rod of length L. Therefore, we choose $\mathcal{D} = [0, L]$ with coordinate X. The dependent variables are the displacement u and the linear momentum p. Therefore, (X, u, p) are the coordinates of the state bundle $\mathcal{X}$. The Hamiltonian functional is given by the total energy of the rod $\mathfrak{H}$

$$\mathfrak{H} = \int_0^L \underbrace{\left(\frac{1}{2\rho}(p)^2 + \frac{E}{2}(u_{1_1})^2\right)}_{=H} \mathrm{d}X \,, \quad H \in C^{\infty}(J(\mathcal{X})) \,,$$

where $\rho \in \mathbb{R}^+$ is the mass density and E Young's modulus. The structure matrix J is

$$J = \begin{bmatrix} 0 & 1 \\ -1 & 0 \end{bmatrix} \,.$$

With the variational derivatives for this problem

$$\delta_u = \partial_u - d_1\partial_u^1 \,, \quad \delta_p = \partial_p \,, \quad d_1 = \partial_x + u_{1_1}\partial_u + u_{1_1+1_1}\partial_u^{1_1}$$

the evolutionary equations of the free system without dissipation ($R = 0$) are

$$\begin{bmatrix} \dot{u} \\ \dot{p} \end{bmatrix} = \begin{bmatrix} 0 & 1 \\ -1 & 0 \end{bmatrix} \begin{bmatrix} \delta_u H \\ \delta_p H \end{bmatrix} = \begin{bmatrix} p/\rho \\ E u_{1_1+1_1} \end{bmatrix} . \tag{35}$$

With the boundary maps

$$_\partial \delta_u = \partial_u^{1_1} , \quad _\partial \delta_p = \partial_p^{1_1}$$

one derives the pair $\dot{u}$, Eu_{1_1} according to (25). Please note, that $\dot{p}$ can be derived form $\dot{u}$ by differentiation and that $_\partial \delta_p H = 0$ is met. Since $\dot{u}$ need not to be continuous, one can assign any L_2 function to $\dot{u}$ at $\partial \mathcal{D}$. Here, we choose the boundary condition

$$\dot{u} = 0 \quad \text{for} \quad X = 0 \quad \text{and} \quad Eu_{1_1} = 0 \quad \text{for} \quad X = L \tag{36}$$

to prohibit energy exchange over the boundary.

Now we assume, there exists a distributed input F, such that the equation (23) takes the form

$$\begin{bmatrix} \dot{u} \\ \dot{p} \end{bmatrix} = \begin{bmatrix} p/\rho \\ Eu_{1_1+1_1} \end{bmatrix} + \begin{bmatrix} 0 \\ 1 \end{bmatrix} F . \tag{37}$$

The input can also be absorbed into the Hamiltonian density $H_e = H - uF$, see (29). A short calculation shows that the collocated output is given by $y = \dot{u} = p/\rho$, see (24). Let us choose the control law, see (32),

$$F = -Ku - D\dot{u} , \quad K, D \in \mathbb{R}^+ , \tag{38}$$

then the closed loop has again the structure of a PCHD system with Hamiltonian functional

$$\mathfrak{H} = \int_0^L \underbrace{\left(\frac{1}{2\rho} (p)^2 + \frac{K}{2} (u)^2 + \frac{E}{2} (u_{1_1})^2 \right)}_{H_c} dX$$

and evolutionary equations

$$\begin{bmatrix} \dot{u} \\ \dot{p} \end{bmatrix} = \left(\begin{bmatrix} 0 & 1 \\ -1 & 0 \end{bmatrix} - \begin{bmatrix} 0 & 0 \\ 0 & D \end{bmatrix} \right) \begin{bmatrix} \delta_u H_c \\ \delta_p H_c \end{bmatrix} .$$

With the boundary conditions from above, we derive the relation

$$v(\mathfrak{H}) = - \int_0^L D \left(\frac{p}{\rho} \right)^2 dX \le 0 .$$

Therefore, $\mathfrak{H}$ is a candidate for a Liapunov functional.

Often, a distributed input like (37) is impossible to implement, but boundary control is possible. Therefore, we change the boundary conditions (36) at $X = L$, see (34), to

$$E\, u_{1_1}|_{X=L} = -\bar{K}\, u|_{X=L} - \bar{D}\, \dot{u}|_{X=L} , \quad \bar{K}, \bar{D} \in \mathbb{R}^+ , \tag{39}$$

and introduce the new Hamiltonian functional $\mathfrak{H}$,

$$\mathfrak{H} = \int_0^L \left(\frac{1}{2\rho} \left(p\right)^2 + \frac{E}{2} \left(u_{1_1}\right)^2 \right) \mathrm{d}X + \frac{\bar{K}}{2} \left(u|_{X=L}\right)^2 .$$

The equations of motion are the same like (35). Again, we conclude from the relation

$$v\left(\mathfrak{H}\right) = -\bar{D} \left(\dot{u}|_{X=L}\right)^2 \leq 0 ,$$

that $\mathfrak{H}$ is a candidate for a Liapunov functional. Of course, one can combine the control laws (38) and (39).

4.2 Piezoelectric Structures

Although we confine ourselves to models of linearized elasticity, linearized quasi static electrodynamics combined with linear constitutive relations, the models are already quite complex, see [8]. Therefore, we present two approaches only, how one can set up such a model from a Hamiltonian point of view, but we will not explain the fundamental basics.

Let $\mathcal{D}$ denote the domain of the 3-dimensional mechanical structure equipped with the Euclidean coordinates $\left(X^i\right)$, $i = 1, 2, 3$, which are used to mark the positions of the mass points. The actual position of a mass point X is given by $u^\alpha + \delta_i^\alpha X^i$, where u^α, $\alpha = 1, 2, 3$ are the displacements. The state of the elastic structure, is given by the positions, or equivalently by the displacements u^α, and linear momenta $p = \rho \dot{u}$ with the mass density $0 < \rho \in C^\infty\left(\mathcal{D}\right)$. Therefore, we choose the state bundle $(\mathcal{X}, \pi, \mathcal{D})$ equipped with local coordinates $\left(X^i, u^\alpha, p_\alpha\right)$. Let us assume, the piezoelectric material admits a stored energy density e_E,

$$\int_{\mathcal{S}} \mathrm{d}e_E \wedge \mathrm{d}X = \int_{\mathcal{S}} \left(\sigma^{ij} \mathrm{d}\varepsilon_{ij} + E_\varsigma \mathrm{d}D^\varsigma\right) \wedge \mathrm{d}X ,$$

which holds for any nice subset $\mathcal{S} \subset \mathcal{D}$. Here, $E_\varsigma, \varsigma = 1, 2, 3$ are the components of electrical field strength E, D^ς are the components of the electrical flux density D, σ^{ij}, $j = 1, 2, 3$ are the components of the stress tensor σ and ε_{ij}, $2\varepsilon_{ij} = \delta_{j\alpha} u_i^\alpha + \delta_{i\beta} u_j^\beta$, are the components of the strain tensor ε. Since we choose the electrical field strength E for the plant input, we rewrite the relation from above as

$$\int_{\mathcal{S}} \mathrm{d} \underbrace{\left(e_E - D^i E_i\right)}_{\hat{e}_E} \wedge \mathrm{d}X = \int_{\mathcal{S}} \left(\sigma^{ij} \mathrm{d}\varepsilon_{\alpha\beta} - D^i \mathrm{d}E_i\right) \wedge \mathrm{d}X$$

and derive the modified stored energy density $\hat{e}_E$. The linearized constitutive equations

$$\sigma^{ij} = C^{ijkl} \varepsilon_{kl} - G^{ij\varsigma} E_\varsigma , \qquad D^\varsigma = G^{ij\varsigma} \varepsilon_{ij} + F^{\varsigma\tau} E_\tau$$

with $k, l, \tau = 1, 2, 3$ and $C^{ijkl}, G^{ij\varsigma}, F^{\varsigma\tau} \in C^\infty\left(\mathcal{D}\right)$ follow from the density $\hat{e}_E$,

$$\hat{e}_E = \frac{1}{2} \varepsilon_{ij} C^{ijkl} \varepsilon_{kl} - \varepsilon_{ij} G^{ij\varsigma} E_\varsigma - \frac{1}{2} E_\varsigma F^{\varsigma\tau} E_\tau ,$$

because the integrability conditions $C^{ijkl} = C^{jikl} = C^{ijlk} = C^{klij}$, $G^{ij\varsigma} = G^{ij\varsigma}$, $F^{\varsigma\tau} = F^{\tau\varsigma}$ are met.

To derive the Hamiltonian density H of the free system, we combine the kinetic energy density e_K,

$$e_K = \frac{1}{2\rho} p_\alpha \delta^{\alpha\beta} p_\beta$$

with $\hat{e}_E$, for $E_\varsigma = 0$ and get

$$H = \frac{1}{2\rho} p_\alpha \delta^{\alpha\beta} p_\beta + \frac{1}{2} \varepsilon_{ij} C^{ijkl} \varepsilon_{kl} \quad . \tag{40}$$

To take the input E into account, we choose the input bundle $\mathcal{U}$ with coordinates $\left(X^i, E_\varsigma\right)$ and introduce the extended Hamiltonian density H_e, see (30),

$$H_e = H - H^\varsigma E_\varsigma = H - G^{ij\varsigma} \varepsilon_{ij} E_\varsigma \; .$$

After a short calculation the equations of motion, see (29), follow in the form

$$\dot{u}^\alpha = \delta^\alpha H_e = \frac{1}{\rho} \delta^{\alpha\beta} p_\beta \quad , \quad \delta^\alpha = \frac{\partial}{\partial p_\alpha}$$

$$\dot{p}_\alpha = -\delta_\alpha H_e = d_i \left(\delta_{\alpha j} \left(\varepsilon_{kl} C^{ijkl} - \delta_{\alpha j} G^{ij\varsigma} E_\varsigma\right)\right) \quad , \quad \delta_\alpha = \frac{\partial}{\partial u^\alpha} - d_i \frac{\partial}{\partial u^\alpha_{1_i}} \;,$$

Here, J takes the canonical form

$$J = \begin{bmatrix} 0 & I_{3\times 3} \\ -I_{3\times 3} & 0 \end{bmatrix}$$

the matrix R vanishes and d_i is a total derivative of $J^2 \left(\mathcal{X} \times_{\mathcal{D}} \mathcal{U}\right)$.

Several facts are worth mentioning. This example shows, how total derivatives d_i may appear in the input map. Since the basic equations belong also to quasi static electrodynamics, the electrical field E admits a potential P with $E_i = -\partial_i P$. Obviously, the input E must meet certain integrability conditions, and a better choice for the input space starts is a bundle $\mathcal{U}$ with coordinates $\left(X^i, P\right)$. Its first jet manifold $J(\mathcal{U})$ with coordinates $\left(X^i, P, P_{1_i}\right)$ contains the coordinates E_ς in a natural manner because of $E_i = -P_{1_i}$. If the piezoelectric material is a non insulator, then the volume charge density must vanish. Therefore, the additional equation $d_\varsigma D^\varsigma = d_\varsigma \left(G^{ij\varsigma} \varepsilon_{ij} + F^{\varsigma\tau} E_\tau\right) = 0$ must be fulfilled. This relation has been omitted.

In the model from above we have taken into account the influence of the electrical field on the mechanical field, but we neglected the opposite influence. In addition, the choice of P or E for the input is of little practical interest. A more adequate choice for the input are the electrical voltages U^ς, $\varsigma = 1,\ldots,r$ applied to the electrodes, which are embedded in the structure. A special case of practical interest is the case

$$P = \Phi_\varsigma U^\varsigma \quad , \quad \Phi_\varsigma \in C^\infty\left(\mathcal{X}\right) \; .$$

with the extended Hamiltonian density H_e,

$$H_e = H - H^\varsigma E_\varsigma = H + G^{ijk}\varepsilon_{ij}d_j\left(\Phi_\varsigma\right)U^\varsigma .$$

Fortunately, it is often possible to design sensors, which measure $Y^\varsigma = H^\varsigma$ and $y^\varsigma = \frac{\mathrm{d}}{\mathrm{d}t}Y^\varsigma$. In this case the implementation of a control law of the type

$$U^\varsigma = -P^{\varsigma\tau}Y_\tau - D^{\varsigma\tau}y_\tau \quad , \quad \tau = 1,\ldots,r ,$$

is a simple task. But it must be mentioned, that the expensive part of the modeling and design problem is the determination of "optimal" functions Φ_ς.

5 Conclusions

The main goal of this contribution is to present an extension of port controlled description of Hamiltonian with dissipation from the lumped to the distributed parameter case. There exist several extensions, see [5], [9], [12], which start from different descriptions of lumped parameter Hamiltonian systems. The presented approach is aimed to be as close as possible to the lumped parameter class given in coordinates by (1). Roughly speaking, this problem is solved by the equations (23). The construction is mainly based on the diagram (28), which replaces the diagram (6) of the lumped parameter case. This diagram also shows how boundary conditions come up in a natural manner. In addition, it shows also that one can replace the linear maps J, R by suitable vector valued differential operators. The same applies to the input map B. Since this extension increases the complexity of the problem by far, only an example, a piezoelectric structure, has been presented.

The required mathematical machinery, specific for distributed parameter systems, has been summed up in the Appendix. Since the presented approach is a formal one, based on differential geometric considerations, several aspects from functional analysis are missing. E.g. Sobolev norms on linear spaces and manifolds have not been introduced, see e.g. [6], [13], also the existence of solutions and their uniqueness have not been discussed. Nevertheless, the author thinks that the presented approach is an interesting field of research for modeling, system analysis and controller design from a Hamiltonian point of view.

6 Appendix

An introduction to the basics of differential geometry can be found in many textbooks, e.g. in [1], or in books about non linear systems and control, like [7]. Here, we summarize after an introduction some facts about bundles, jet manifolds and semi groups only. The reader is kindly asked to consult the books [5], [9] and [2], to find out more about these topics.

The tangent and cotangent bundle of a q-dimensional manifold $\mathcal{M}$ with coordinates $\left(x^i\right)$, $i = 1,\ldots,q$ are denoted by $\mathcal{T}\left(\mathcal{M}\right)$, $\mathcal{T}^*\left(\mathcal{M}\right)$, where we use the standard coordinates $(x^i, \dot{x}^i)$, $(x^i, \dot{x}_i)$ according to the bases span($\{\partial_i\}$), span($\{\mathrm{d}x^i\}$). The symbol $C^\infty(\mathcal{M})$ stands for the set of smooth functions on $\mathcal{M}$. The bundles of p-forms on $\mathcal{M}$ are denoted by $\bigwedge^p(\mathcal{T}^*(\mathcal{M}))$ and we set $\bigwedge^0(\mathcal{T}^*(\mathcal{M})) = C^\infty(\mathcal{M})$, $\bigwedge^1(\mathcal{T}^*(\mathcal{M})) = \mathcal{T}^*(\mathcal{M})$ and $\bigwedge(\mathcal{T}^*(\mathcal{M}))$ for the exterior algebra generated by $\mathcal{T}^*(\mathcal{M})$. We use the symbol d for the exterior derivative $\mathrm{d} : \bigwedge^r(\mathcal{T}^*(\mathcal{M})) \to \bigwedge^{r+1}(\mathcal{T}^*(\mathcal{M}))$, $r \geq 0$ and denote the contraction of a form along a field by $\rfloor : \mathcal{T}(\mathcal{M}) \times \bigwedge^r(\mathcal{T}^*(\mathcal{M})) \to \bigwedge^{r-1}(\mathcal{T}^*(\mathcal{M}))$, $r \geq 1$.

6.1 Bundles

A bundle is the triple $\mathcal{E} \xrightarrow{\pi} \mathcal{B}$ with the p-dimensional base manifold $\mathcal{B}$, the $(p+q)$-dimensional total manifold $\mathcal{E}$ and a surjective submersion $\pi : \mathcal{E} \to \mathcal{B}$. All fibers $\mathcal{F}_X = \pi^{-1}(X)$, $X \in \mathcal{B}$ are isomorphic to the typical fiber $\mathcal{F}$. We use so called adapted coordinates, where (X^i, x^α), $i = 1, \ldots, p$, $\alpha = 1, \ldots, q$ are coordinates for $\mathcal{E}$ and (X^i) are coordinates for $\mathcal{B}$. A section[2] of $\mathcal{E}$ is a map $\sigma : \mathcal{B} \to \mathcal{E}$ with $\pi \circ \sigma = \mathrm{id}_\mathcal{B}$ with the identity map $\mathrm{id}_\mathcal{B}$ on $\mathcal{B}$. The set of all sections of $\mathcal{E}$ is denoted by $\Gamma(\mathcal{E})$.

The vertical bundle $\mathcal{V}(\mathcal{E})$ of $\mathcal{E}$ is the subbundle of $\mathcal{T}(\mathcal{E})$, which meets, $v \in \mathcal{V}(\mathcal{E})$ implies $\pi_*(v) = 0$. Obviously, $\mathcal{V}(\mathcal{E})$ is a vector bundle. From the cotangent bundle $\mathcal{T}^*(\mathcal{E})$ we derive the subbundles $\bigwedge_r^{s+r}(\mathcal{T}^*(\mathcal{E}))$ with bases $\mathrm{d}x^{\alpha(1)} \wedge \cdots \wedge \mathrm{d}x^{\alpha(s)} \wedge \mathrm{d}X^{i(1)} \wedge \cdots \wedge \mathrm{d}X^{i(r)}$. A short calculation shows that $\bigwedge_r^r(\mathcal{T}^*(\mathcal{E}))$, $1 \leq r \leq p$ and $\bigwedge_p^{p+s}(\mathcal{T}^*(\mathcal{E}))$, $1 \leq s \leq q$ are vector bundles. Of special interest are the annihilator of $\mathcal{V}(\mathcal{E})$ given by $\bigwedge_1^1(\mathcal{T}^*(\mathcal{E}))$, the space of densities given by $\bigwedge_p^p(\mathcal{T}^*(\mathcal{E}))$ with basis $\{\mathrm{d}X\}$, $\mathrm{d}X = \bigwedge_{i=1}^p \mathrm{d}X^i$ and the space of densities with direction given by $\bigwedge_p^{p+1}(\mathcal{T}^*(\mathcal{E}))$ and basis $\{\mathrm{d}x^\alpha \wedge \mathrm{d}X\}$.

6.2 Jet Manifolds

Since we have to determine several partial derivatives of order n, we use the notation of an ordered multi index I with

$$\partial_I = (\partial_1)^{I(1)} \cdots (\partial_p)^{I(p)} \,, \qquad \frac{\partial}{\partial X^i} = \partial_i$$

and $I = I(1), \ldots, I(p)$ and $\#I = \sum_{i=1}^p I(i)$. The special index I, $I(i) = \delta_{ij}$[3] is denoted by 1_j. Finally, the sum $I + J$ of two indices is the sum of their components. Given a bundle $\mathcal{E} \xrightarrow{\pi} \mathcal{B}$ and a section σ then its n-order prolongation $j^n(\sigma)$ of σ is given by

$$x_I^\alpha = \partial_I \sigma^\alpha \,, \quad 1 \leq \#I \leq n \,,$$

where we already used the jet coordinates x_I^α. One can show that the space of all prolongations of n-th order is a manifold $J^n(\mathcal{E})$, called the n-th order jet

---

[2] We use the abbreviation $\mathcal{E}$ for the bundle $\mathcal{E} \xrightarrow{\pi} \mathcal{B}$, whenever the base manifold $\mathcal{B}$ and the projection π are clear from the context.

[3] δ_{ij} denotes the Kronecker symbol with $\delta_{ij} = 0$ for $i \neq j$ and otherweise $\delta_{ii} = 1$.

manifold. Also here we use the adapted coordinates $\left(X^i, x_I^\alpha\right)$, $0 \leq \#I \leq n$, where we set $x_I^\alpha = x^\alpha$ for $\#I = 0$. These jet manifolds are connected by the following sequence

$$J^n\left(\mathcal{E}\right) \overset{\pi_{n-1}^n}{\rightarrow} J^{n-1}\left(\mathcal{E}\right) \rightarrow \cdots \rightarrow J^1\left(\mathcal{E}\right) = J\left(\mathcal{E}\right) \overset{\pi_0^1}{\rightarrow} J^0\left(\mathcal{E}\right) = \mathcal{E} \overset{\pi}{\rightarrow} \mathcal{B} .$$

We also use the abbreviation $\pi_l^k = \pi_{k-1}^k \circ \cdots \circ \pi_l^{l-1}$ and $\pi^k = \pi_{k-1}^k \circ \cdots \circ \pi$. These maps will be used to pull back several bundles. Given a bundle $\mathcal{E} \to \mathcal{B}$ with adapted coordinates (X, x) and a manifold $\mathcal{M}$ with coordinates (z) together with a map $f : \mathcal{M} \to \mathcal{B}$, we call the bundle $f^*\left(\mathcal{E}\right) \overset{\mathrm{pr}_1}{\rightarrow} \mathcal{M}$, $f^*\left(\mathcal{E}\right) = \{(z, (X, x)) \in \mathcal{M} \times \mathcal{E}, \pi\left((X, x)\right) = X = f(z)\}$ the pull back of $\mathcal{E}$ by f. The diagram

$$
\begin{array}{ccc}
f^*\left(\mathcal{E}\right) & \overset{\mathrm{pr}_2}{\rightarrow} & \mathcal{E} \\
\mathrm{pr}_1 \downarrow & & \downarrow \pi \\
\mathcal{M} & \overset{f}{\rightarrow} & \mathcal{B}
\end{array} ,
$$

visualizes this construction. If $\mathcal{M}$ is a bundle $\overline{\mathcal{E}} \overset{\bar{\rho}}{\rightarrow} \mathcal{B}$ and $f = \bar{\rho}$ is met, then we write $\overline{\mathcal{E}} \times_\mathcal{B} \mathcal{E}$ instead of $\bar{\rho}^*\left(\mathcal{E}\right)$.

It is an important fact, that span$\{\mathrm{d}X^i\}$ and the contact forms

$$\theta_I^\alpha = \mathrm{d}x_I^\alpha - x_{I+1_i}^\alpha \mathrm{d}X^i$$

form a basis of $\pi_n^{n+1,*}\left(\bigwedge \mathcal{T}^*\left(J^n\left(\mathcal{E}\right)\right)\right)$. In addition, the n-th order prolongation of $\sigma \in \Gamma\left(\mathcal{E}\right)$ meets

$$j^n\left(\sigma\right)^*\left(\theta_I^\alpha\right) = 0 , \quad 0 \leq \#I < n . \tag{41}$$

Elements of the annihilator of span$\{\theta_I^\alpha\}$ are called total derivatives, a basis is given by span$\{d_i\}$ with

$$d_i = \partial_i + x_{I+1_i}^\alpha \partial_\alpha^I , \quad \partial_\alpha^I = \frac{\partial}{\partial x_I^\alpha} , \tag{42}$$

where d_i is called the total derivative into the direction of X^i. Obviously, d_i is also a map $d_i : C^\infty\left(J^n\left(\mathcal{E}\right)\right) \to C^\infty\left(J^{n+1}\left(\mathcal{E}\right)\right)$, which meets

$$j^{n+1}\left(\sigma\right)^*\left(d_i(f)\right) = \partial_i\left(j^n\left(\sigma\right)^* f\right) , \quad f \in C^\infty\left(J^n\left(\mathcal{E}\right)\right) .$$

With the help of d_i one derives the horizontal differential d_H given by

$$\mathrm{d}_H\left(\omega\right) = \mathrm{d}X^i \wedge d_i\left(\omega\right) \tag{43}$$

in coordinates. It meets for $\omega = h^i \partial_i \rfloor \mathrm{d}X \in \pi_0^{n,*}\left(\bigwedge_{p-1}^{p-1}\left(\mathcal{T}\left(\mathcal{E}\right)\right)\right)$

$$\int_{\mathcal{D}} j^{n+1}\left(\sigma\right)^*\left(\mathrm{d}_H\omega\right) = \int_{\partial\mathcal{D}} \mathrm{d}\left(j^n\left(\sigma\right)^*\left(\omega\right)\right) ,$$

which is nothing else than Stokes' theorem adapted to bundles.

6.3   Semi Groups on Bundles

Let us consider a semi group $\phi_\tau : \mathbb{R}^+ \times \Gamma(\mathcal{E}) \to \Gamma(\mathcal{E})$ that maps sections to sections of the bundle $\mathcal{E} \to \mathcal{B}$ according to

$$\sigma_\tau = \phi_\tau(\sigma) , \quad \sigma = \sigma_0, \sigma_\tau \in \Gamma(\mathcal{E}) , \quad \tau \in [0,T]$$
$$\sigma_{\tau_1+\tau_2} = \phi_{\tau_1} \circ \phi_{\tau_2}(\sigma) , \quad \sigma_{\tau_1+\tau_2} \in \Gamma(\mathcal{E}) , \quad \tau_1, \tau_2, \tau_1 + \tau_2 \in [0,T]$$

for a certain $T \in \mathbb{R}^+$. We assume the existence of functions $v^\alpha \in C^\infty(J^n(\mathcal{E}))$ such that

$$v^\alpha \circ j^n(\sigma) = \partial_\tau \phi_\tau^\alpha(\sigma)|_{\tau=0} \tag{44}$$

is met. It is straightforward to show that $v = v^\alpha \partial_\alpha$ meets $v \in \Gamma\left(\pi_0^{n,*}(\mathcal{V}(\mathcal{E}))\right)$. In the case $n > 0$ the field v is not a tangent vector field and does not generate a flow, but it generates the semi group ϕ_τ by the set of partial differential equations

$$\dot{X}^i = 0 , \quad \dot{x}^\alpha = v , \tag{45}$$

provided suitable boundary conditions are added.

To prolong v to a field $j(v) = v + v_i^\alpha \partial_\alpha^i \in \Gamma\left(\pi_0^{n+1,*}(\mathcal{V}(J(\mathcal{E})))\right)$, we prolong $\phi_t(\sigma)$ to $j(\phi_t(\sigma))$. From the relation

$$j(\phi_\tau \circ \sigma)^*(\theta^\alpha) = 0 ,$$

we derive immediately

$$\partial_\tau \, j(\phi_\tau \circ \sigma)^*(\theta^\alpha)\big|_{\tau=0} = j(\sigma)^*(j(v)(\theta)) = 0 .$$

The Lie derivative of a contact form along $j(v)$ must be the sum of contact forms, since it lies in the kernel of $j(\sigma)^*$. After some calculations one derives the result $v_i^\alpha = d_i(v^\alpha)$. This result can easily be extended to higher order prolongations and one gets the prolongation formula

$$j^n(v) = v^\alpha \partial_i + \sum_{\#I=1}^{n} d_I(v^\alpha) \partial_\alpha^I , \quad d_I = (d_1)^{I(1)} \circ \cdots \circ (d_p)^{I(p)} \tag{46}$$

for $v \in \Gamma\left(\pi_0^{n,*}(\mathcal{V}(\mathcal{E}))\right)$.

Let us consider the functional $\mathfrak{F} : \Gamma(\mathcal{E}) \to \mathbb{R}$,

$$\mathfrak{F}(\sigma) = \int_\mathcal{D} f \circ j^m(\sigma) \, dX , \quad \sigma \in \Gamma(\mathcal{E}) , \quad f \in C^\infty(J^m(\mathcal{E})) . \tag{47}$$

The change of (47) along a semi flow ϕ_t, which meets (44), follows after a short calculation as

$$\partial_t \mathfrak{F}(\phi_t \circ \sigma)_{t=0} = \int_\mathcal{D} j^{m+n}(\sigma)^*(j^m(v)(f dX)) .$$

Therefore, we introduce the new functional $v(\mathfrak{F})(\sigma)$,

$$v(\mathfrak{F})(\sigma) = \int_\mathcal{D} j^{m+n}(\sigma)^*(j^m(v)(f dX)) , \tag{48}$$

which measures the change of $\mathfrak{F}$ into the direction of v at the point σ. It is worth mentioning, that (48) does not require that v is linked to a semi group by (45).

References

1. Choquet-Bruhat Y, Witt-Morette C (1982) Analysis, Manifolds and Physics. Elsevier, Amsterdam
2. Curtain R F, Zwart H J (1995) An Introduction to Infinite-Dimensional Linear System Theory. Springer Verlag
3. Ennsbrunner H (2006) Infinite Dimensional Euler-Lagrange and Port Hamiltonian Systems, PhD Thesis at the Johannes Kepler University, Linz, Austria
4. Ennsbrunner H, Schlacher K (2005) On the geometrical representation and interconnection of infinite dimensional port controlled hamiltonian systems, Proceedings of 44th IEEE Conference on Decision and Control
5. Giachetta G, Mangiarotti L, Sardanashvily G (1997) New Lagrangian and Hamiltonian Methods in Field Theory. World Scientific, Singapore
6. Hebey E (2000) Nonlinear Analysis on Manifolds: Sobolev Spaces and Inequalities. Courant Institute of Mathematical Sciences, New York
7. Isidori A (1995) Nonlinear Control Systems. Springer Verlag, London
8. Nowacki W (2006) Static and Dynamic Coupled Fields in Bodies with Piezoeffects or Polarization Gradient. Springer Verlag
9. Olver P J (1986) Applications of Lie Groups to Differential Equations. Springer Verlag, New York
10. Schlacher K (2006) Mathematical modelling for nonlinear control - a hamiltonian approach,Proceedings of 5th Vienna Symposium on Mathematical Modelling
11. van der Schaft A J (2000) L_2-Gain and Passivity Techniques in Nonlinear Control. Springer Verlag, New York
12. van der Schaft A J , Maschke B M (2002) Hamiltonian formulation of distributed-parameter systems with boundary energy flow, Journal of Geometry and Physics, 42:166–194
13. Zeidler E (1995) Applied Functional Analysis. Springer Verlag, New York

Observability and the Design of Fault Tolerant Estimation Using Structural Analysis

Marcel Staroswiecki

SATIE UMR CNRS 8029
Ecole Normale Supérieure de Cachan
4 avenue du Président Wilson
F-94235 Cachan
`marcel.staroswiecki@univ-lille1.fr`

Summary. This chapter presents a structural analysis approach for the design of fault tolerant estimation algorithms. The general fault tolerance problem setting is first given, and structural analysis is presented in the component based modeling frame. An original condition for structural observability is developed, which is constructive, since it allows to identify those Data Flow Diagrams by which unknown variables can be estimated, both in healthy and in faulty conditions. The link with two basic dependability concepts, namely critical faults and reliability is shown.

Keywords: Structural Analysis, Observability, Fault Tolerance.

1 Introduction

In increasingly complex systems, faults may lead to performance degradation, instability, loss of control. Fault tolerance is needed to preserve the ability of the system to achieve the objectives it has been assigned, or if this turns out to be impossible, to assign new (achievable) objectives and avoid catastrophic behaviors.

The design of Fault Tolerant Systems (FTS) is a recent research field [1], [14], [19], [29]. Most works use the quantitative system behavior model, for example, state and output equations in the time or in the symbolic domain, in order to design automatic procedures by which faults can be handled. However, integrating such approaches in large scale, complex, partially automated systems, where control and maintenance performances are dependent, needs the analysis to be carried out at the components / subsystems level. Structural analysis uses graph-theoretical based tools that allow this kind of analysis. It has been used for the decomposition of large scale systems [28], [13], [7], [8], the analysis of observability and controllability [16], [17], [12], [21], the design of control and diagnostic systems [23], [22], [5], [3], [24], [4], including sensor placement [2], [20].

This paper develops a structural analysis approach to the design of fault tolerant estimation. A clear setting of the Fault Tolerance (FT) design problem, based on behavior models, is first given in Section 2. In Section 3, the structural model of a system is presented at the component / subsystems level. Section 4 establishes an original condition for structural observability, which is applied in

C. Bonivento et al. (Eds.): Adv. in Control Theory and Applications, LNCIS 353, pp. 257–278, 2007.
springerlink.com   © Springer-Verlag Berlin Heidelberg 2007

Section 5 to the design of Fault Tolerant Estimation of observable variables. By establishing the link with the basic concept of critical faults, it is also shown that this approach allows a clear definition and evaluation of the achieved FT performance level.

2 Fault Tolerance

This section introduces the basic concepts that are needed to properly address the design of fault tolerant systems.

Given a set of nominal systems $S_N = \{\Sigma_n, n \in N\}$, a (disjoint) set of faulty systems $S_F = \{\Sigma_f,\ f \in F\}$, and a property P, we wish to design control and estimation algorithms such that P is true for every $\Sigma_n \in S_N$ (robustness) and for every $\Sigma_f \in S_F$ (fault tolerance - FT) [1]. In this paper, we are interested only in FT. The set S_F is the FT specification, i.e. the faults under which property P is wished to be invariant, while P is the design objective, e.g. stability, dynamic performance, etc. Obviously, $S_F \subseteq \mathcal{S}_\mathcal{F}$ where $\mathcal{S}_\mathcal{F}$ is a set of faults that are likely to occur (obtained via approaches like fault trees, fault modes and effects analysis). In the sequel, the notation $P(\Sigma, a)$ stands for "Property P is true for system Σ equipped with algorithm a".

2.1 Passive and Active Fault Tolerance

There are two ways to set the FT problem [1].

Passive fault tolerance (PFT), aims at designing an algorithm a that achieves property P in all cases:

$$\forall \Sigma \in S_N \cup S_F : P(\Sigma, a) \tag{1}$$

Note that this problem is identical to the robust control problem, where S_N has been replaced by $S_N \cup S_F$.

In active fault tolerance (AFT), property P is obtained (when possible) by designing an algorithm a for each *post-fault system*. Assume that fault $f \in F$ has occurred, the *post-fault system* may be chosen as the faulty system itself - Σ_f - or as the subsystem $\tilde{\Sigma}_f$ that is obtained by switching-off the faulty components. The first strategy: *adapt the algorithm to the faulty system*, is called fault accommodation (FA). It is defined by the statement

$$\forall \Sigma_f \in S_F : P(\Sigma_f, a(\Sigma_f)) \tag{2}$$

and it obviously needs the model Σ_f to be known in order to design $a(\Sigma_f)$. The second strategy: *adapt the algorithm to the healthy subsystem of the faulty system*, is named system reconfiguration (SR). It is defined by the statement

$$\forall \Sigma_f \in S_F : P(\tilde{\Sigma}_f, a\left(\tilde{\Sigma}_f\right)) \tag{3}$$

and since the models of the healthy components are known, it only needs the isolation of the faulty ones (to be switched-off). Note that SR may be impossible (e.g. there is no mechanism in the system to switch-off some of the faulty components). Therefore, mixed FA/SR strategies must often be used. In the sequel we are only interested in Active Fault Tolerance.

2.2 Recoverable Faults

A fault $\Sigma_f \in S_F$ is FA-recoverable (resp. SR-recoverable) if there exists a solution to the FA problem (resp. to the SR problem). Let $\mathcal{F}_r(P)$ be the set of all recoverable faults for property P, i.e.

$$\mathcal{F}_r(P) = \left\{ \Sigma_f \in \mathcal{S}_F : \exists a_f \text{ s.t. } P(\Sigma_f, a_f) \vee P(\tilde{\Sigma}_f, a_f) \right\}$$

The FT specification is met if and only if $S_F \subseteq \mathcal{F}_r(P)$.

2.3 Objective Reconfiguration

For non recoverable faults the original objective cannot be satisfied. Objective reconfiguration is then the only behaviour that makes sense from a control point of view. Objectives can be changed in two ways:

(1) by accepting performance degradation [15], i.e. by specifying a new property P^- which is weaker than P in the following sense

$$\forall a, \forall \Sigma \in S_N \cup S_F : \ P(\Sigma, a) \implies P^-(\Sigma, a)$$

(2) by abandoning the current mission and changing the users expectations, which results in an objective associated with a brand new set of properties P' [1].

The new objective P^* (which stands here either for P^- or for P') must be *consistent*, i.e. if fault Σ_f was not recoverable for P, it must be recoverable for P^* :

$$\forall \Sigma \notin \mathcal{F}_r(P), P^* \text{ consistent} \implies \Sigma \in \mathcal{F}_r(P^*).$$

A critical situation is reached when there is no consistent objective reconfiguration for some fault Σ_f.

3 Component Based Model and Structural Analysis

This section presents the component based modeling frame as the natural frame in which both fault tolerance and structural analysis can be clearly set.

3.1 Components, Systems and Subsystems

A system Σ is a set of interconnected (hardware and software) components. The normal behavior of each component $comp \in \Sigma$ is described by a pair $(C(comp), V(comp))$ where $V(comp)$ is a set of variables, whose values belong to some set $\mathbb{V}(comp)$ and $C(comp)$ is a set of dynamic and/or static constraints which apply to these variables. Interconnections imply that some variables are common to several components.

The behavior model of the system is the pair $(C(\Sigma), V(\Sigma))$, where

$$V(\Sigma) = \bigcup_{comp \in \Sigma} V(comp) \tag{4}$$

is the set of the system variables, while

$$\mathbb{V}(\Sigma) = \prod_{comp \in \Sigma} \mathbb{V}(comp) \tag{5}$$

is the set of their possible values, and

$$C(\Sigma) = \bigcup_{comp \in \Sigma} C(comp) \tag{6}$$

is the set of the system constraints. When necessary, parameters are taken into account, by including them as system variables.

The variables in $V(\Sigma)$ are inputs, outputs or internal variables, which can be decomposed into

$$V(\Sigma) = K(\Sigma) \cup X(\Sigma) \tag{7}$$

where $K(\Sigma)$ are known (computed or measured) and $X(\Sigma)$ are unknown. The set $K(\Sigma)$ can be expanded to $\bar{K}(\Sigma)$ by considering also, for each variable in $K(\Sigma)$, a number of its time derivatives (if continuous time is considered) or its shifted time values (if discrete time is considered). The size of the expansion need not be specified here.

A subsystem $\sigma \subset \Sigma$ is a subset of components along with their interconnections. Its variables $V(\sigma)$, set of values $\mathbb{V}(\sigma)$ and constraints $C(\sigma)$ are defined by (4), (5) and (6), replacing Σ by σ. More generally, subsystems are defined as subset of constraints along with the variables they are associated with. Interconnecting two subsystems whose behavior is (C_1, V_1) and (C_2, V_2) boils down to create the subsystem whose behavior is (C, V) where $C = C_1 \cup C_2$ and $V = V_1 \cup V_2$ (without repetitions of the shared variables).

In the sequel, since there is no ambiguity, the notations will be simplified into $C, V, \mathbb{V}, K, X$, and both the system and its behavior will be noted as (C, V). Deterministic behavior models are usually under the state-space form

$$\dot{x} = f(x, u, t) \tag{8}$$
$$y = g(x, u, t)$$

where x is the state, u is the control inputs, y is the sensors outputs. Then $C = \{f, g\}$, $V = \{x, u, y\}$, $K = \{u, y\}$, $X = \{x\}$. When static equations are included, internal variables are decomposed into static ones x_s and dynamic ones x_d:

$$\dot{x}_d = f(x_d, x_s, u, t) \tag{9}$$
$$0 = h(x_d, x_s, u, t)$$
$$y = g(x_d, x_s, u, t)$$

Then $C = \{f, g, h\}$, $V = \{x_d, x_s, u, y\}$, $K = \{u, y\}$, $X = \{x_d, x_s\}$. A canonical form is obtained by introducing a new set of variables and constraints

$$z = \delta\left(x_d\right) \triangleq \frac{dx_d}{dt} \tag{10}$$

and replacing $\dot{x}_d$ by z in (9). The system becomes

$$z = f(x_d, x_s, u, t) \tag{11}$$
$$0 = h(x_d, x_s, u, t)$$
$$y = g(x_d, x_s, u, t)$$
$$z = \delta\left(x_d\right)$$

Then $C = \{f, g, h, \delta\}$ where $\{f, g, h\}$ are static constraints and $\delta \triangleq \frac{d}{dt}$ is the set of dynamic constraints, $V = \{x_d, z, x_s, u, y\}$, $X = \{x_d, z, x_s\}$, $K = \{u, y\}$.

3.2   Faults

No matter whether a component *comp*, a subsystem σ or the whole system Σ is considered, denote by C and V its sets of constraints and variables and by H_0, $H_1 = \neg H_0$ the propositions "*comp*/σ/Σ is healthy" and " *comp*/σ/Σ is faulty".

Let $\hat{V} \in \mathbb{V}$ be a realization of the variables V, it is *consistent* with respect to the constraints C iff all the constraints are satisfied - this is noted $True\left[C, \hat{V}\right]$ - otherwise it is *inconsistent* (i.e. at least one constraint is *falsified* by $\hat{V}$) - this is noted $False\left[C, \hat{V}\right]$. Introducing $True\left[C, V\right]/False\left[C, V\right]$ instead of $True\left[C, \hat{V}\right]/False\left[C, \hat{V}\right]$ means that any realization $\hat{V}$ of the variables V is considered. Then one obviously has:

Definition. A healthy component / subsystem / system is defined by (12). A faulty component / subsystem / system is defined by (13):

$$H_0 \iff True\left[C, V\right] \tag{12}$$
$$H_1 \iff \exists c \in C : False\left[c, V\right] \tag{13}$$

A fault is therefore defined as the violation of one or several constraints in C. As there are $\left|2^C\right| - 1$ different ways to falsify the constraints C, it follows that any non empty subset of C is a theoretically possible fault mode.

Remark. At this stage, no model of the falsified constraints is necessary (for example additive or multiplicative). A model will only be needed when a fault accommodation algorithm will be designed.

3.3   Structure

The structure of a system (C, V) is an abstract representation of its behavior, where we are interested only in the existence of constraints, but not in their mathematical expression[1].

---

[1] Being independent on the mathematical expression of the constraints, structural analysis results are valid both for linear and for nonlinear systems.

Definition. The structure of the system (C, V) - or its structural graph - is a bi-partite graph $\mathcal{G}(C, V, E)$ with the two sets of nodes C and V and the set of edges $E \subset C \times V$ defined by $(c_i, v_j) \in E$ if and only if the variable v_j appears in the constraint c_i.

The bi-partite graph is not oriented: it only expresses that all the variables adjacent to a given node satisfy the (static or dynamic) constraint this node represents (whatever its form and parameters). System structures can be represented by their incidence matrix where the intersection of row i and column j is 1 if and only if $(c_i, v_j) \in E$ otherwise it is 0. Any chain between two nodes is an *alternated chain* because a node in C can only be adjacent to nodes in V and vice-versa.

Subsets of rows and columns of the incidence matrix define subgraphs of the bi-partite graph. Practical interpretations of some subgraphs are as follows:

- the set of rows associated with the constraints in $C\,(comp)$ represent the component *comp*, a subset of them represents a fault mode of component *comp*,
- several such sets are a subsystem formed by the interconnection of several components (by extension, a subsystem is a subset of rows),
- a subset of columns represents a subset of variables that may be known (controlled inputs, measured outputs), or unknown (non measured inputs, states), observable or not, controllable or not.

Remark. System structures can also be defined by directed graphs (digraphs) [5], [16], [17]. The set of nodes is then restricted to the set of variables, and edges become arcs that represent causal influences between variables. We use bi-partite graphs for their capability to integrate static as well as dynamic constraints.

It is usefull to define the two projections:

$$P_V : C \rightarrow V$$
$$c \mapsto P_V(c) = \{v \in V : (c, v) \in E\}$$

$$P_C : V \rightarrow C$$
$$v \mapsto P_C(v) = \{c \in C : (c, v) \in E\}$$

P_V associates with each constraint c the set of variables to which it applies, and P_C associates to each variable v the set of constraints to which it is submitted.

3.4   Matching

Let $\mathcal{G}(C, V, E)$ be the structure of the system (C, V).

- A *matching* M is a subset of disjoint edges of E (any two edges have no common node, neither in C nor in V). The set $\mathbb{M}$ of all matchings is a subset of the lattice 2^E, therefore it has maximal elements.

- A *maximal matching* M is such that: $\forall N \in 2^E$, $M \subset N$, N is not a matching (no edge can be added without violating the *no common node* property). The set $\mathbb{M}$ being partially ordered, there is in general more than one maximal matching. Let $\mathbb{M}^* \subseteq \mathbb{M}$ be the set of maximal matchings. Each matching can be associated with the subsets $\pi_C(M)$ and $\pi_Z(M)$ of its matched constraints and matched variables. From the definition, maximal matchings obviously satisfy

$$\forall M \in \mathbb{M}^* \quad \begin{cases} \pi_C(M) \subseteq C \\ \pi_Z(M) \subseteq V \end{cases}$$

from where it follows that

$$|M| \leq \min\{|C|, |V|\}.$$

Matchings for which the equality holds are called *complete*, namely a *complete matching on* C is such that $|M| = |C|$ while a *complete matching on* V is such that $|M| = |V|$.

3.5 Oriented Graph Associated with a Matching

Orientation. A matching on a structure graph defines an oriented graph as follows:

- if constraint c is matched, i.e. $\exists v \in V : (c, v) \in M$, the edge (c, v) is oriented from c to v (v is the output) and for any other variable $w \in P_V(c) \setminus \{v\}$ the edge (c, w) is oriented from w to c (the w are inputs),
- if constraint c is not matched, all edges (c, w), $w \in P_V(c)$ are oriented from w to c (inputs), the constraint has no output.

Paths. When oriented, alternated chains in $\mathcal{G}(C, V, E)$ produce alternated paths. The alternated chain $z_i - \delta_i - x_i$ associated with the ith dynamic constraint in (11) becomes a path in *integral causality*, if the matching contains (δ_i, x_i), and a path in *derivative causality* if the matching contains (δ_i, z_i).

Reachability. A variable v is *reachable* from a variable w iff there exists an alternated path from w to v. A variable v is reachable from a subset $\chi \subseteq V \setminus \{v\}$ iff there exists $w \in \chi$ such that v is reachable from w.

Loops. An alternated path from v to v is a *loop*. A loop that contains only static constraints is a *static loop*. Otherwise it is a *differential loop* (which means that it contains at least one path $x_i \to \delta_i \to z_i$, or $z_i \to \delta_i \to x_i$ for some index i, according to the fact that δ_i is in derivative or integral causality). An alternated path may include loops.

Example. Figure 1 shows a bi-partite graph where cercles are variables and bars are constraints. The matchings $n°1 : \{(c_1, x_1), (c_3, y)\}$, $n°2 : \{(c_2, x_2), (c_3, x_1)\}$, and $n°3 : \{(c_1, x_1), (c_2, u), (c_3, y)\}$ are respectively non-maximal, maximal, and complete w.r.t. the constraints.

Matching $n°3$ gives an oriented graph in which $x_2 \to c_1 \to x_1 \to c_3 \to y$ is an alternated path. Assume that constraint c_1 writes $c_1 : x_2 = \frac{d}{dt}x_1$, then it is matched in integral causality. Variable y is reachable from x_1 and x_2, but not from u. Under the above form of c_1 the complete matching $n°4 : \{(c_1, x_2), (c_2, x_1), (c_3, y)\}$ would give the differential loop $x_1 \to c_1 \to x_2 \to c_2 \to x_1$ with c_1 in derivative causality.

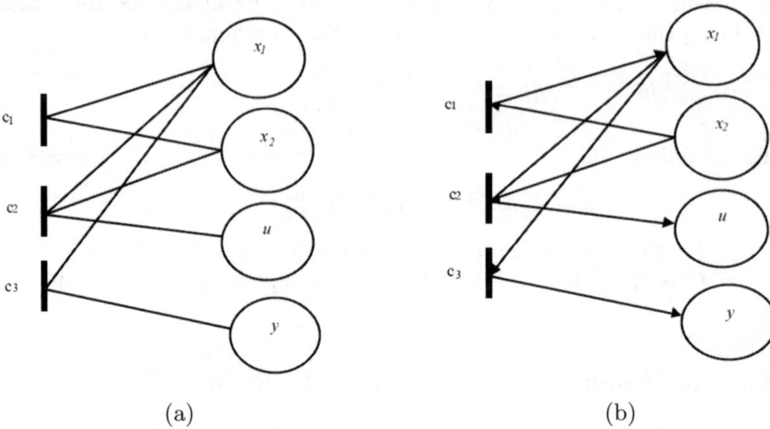

(a)   (b)

Fig. 1. Matching and oriented graph: (a) bi-partite graph; (b) oriented graph associated with matching n°3

4 Structural Observability

Given an unknown variable x, this section introduces specific subgraphs of the structural graph, namely x-rooted elementary graphs, and shows that they provide a constructive means to decide about the observability of x.

4.1 Elementary Graph

Let $\mathcal{G}(C, V, E)$ be a bi-partite graph and $\mathcal{G}_e(C_e, V_e, E_e)$ be a connected oriented sub-graph such that:

(i) each variable $v \in V_e$ satisfies $d_i(v) \leq 1$
(ii) each constraint $c \in C_e$ satisfies $d_e(c) = 1$

where $d_i(s)$ - resp. $d_e(s)$ - is the in-degree (resp. the out-degree) of node s.
Let us decompose the variables into:

- inputs: $\{v \in V_e : d_i(v) = 0, d_e(v) \geq 1\}$,
- internal variables: $\{v \in V_e : d_i(v) = 1, d_e(v) \geq 1\}$,
- outputs: $\{v \in V_e : d_i(v) = 1, d_e(v) = 0\}$.

$\mathcal{G}_e(C_e, V_e, E_e)$ is called elementary graph. It follows from (i) and (ii) that it is associated with a complete matching on the set of its internal and output variables. An elementary graph may be without any input (it is then called *autonomous*) and/or without any output.

Example. Figure 2 shows some elementary graphs: the simple path (a) has one input $\{y\}$, one internal variable $\{x_1\}$, one output $\{x_2\}$, and it is associated with the matching $\{(c_1, x_1), (c_2, x_2)\}$; the tree (b) has 2 inputs $\{u, y\}$, one internal variable $\{x_1\}$, one output $\{x_2\}$, and is associated with the matching $\{(c_1, x_1), (c_2, x_2)\}$; the loop (c) has 3 inputs $\{x_1, x_2, x_6\}$, 3 internal variables $\{x_3, x_4, x_5\}$, no output, and is associated with the matching $\{(c_3, x_3), (c_4, x_4), (c_5, x_5)\}$. Deleting the variables $\{x_1, x_2, x_6\}$ in graph (c) would create an autonomous graph. Superposing the 2 nodes x_1 and the 2 nodes x_2 of (b) and (c) would provide another elementary graph, with inputs $\{u, y, x_6\}$, internal variables $\{x_1, x_2, x_3, x_4, x_5\}$, no output, associated with the matching $\{(c_1, x_1), (c_2, x_2), (c_3, x_3), (c_4, x_4), (c_5, x_5)\}$.

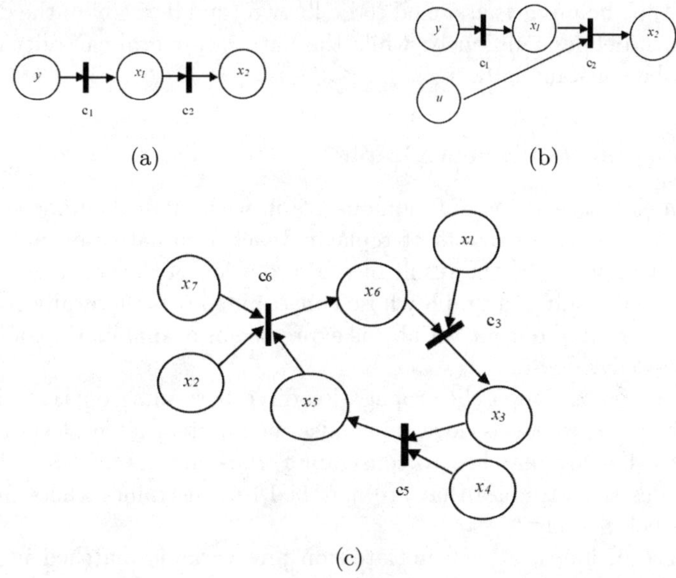

Fig. 2. Examples of elementary graphs: (a) simple path; (b) tree; (c) loop

4.2   Constraint Propagation

Resolution

Let $\mathcal{G}_e(\{c\}, P_V(c), E(c))$ be an elementary graph with the single constraint c and let $x \in P_V(c)$ be the matched variable, one has $E(c) = \{(c, x)\} \cup \{(w, c), w \in P_V(c) \setminus \{x\}\}$.

Constraint c writes $c(P_V(c)) = 0$, but it can also be represented by $x - \gamma(P_V(c) \setminus \{x\}) = 0$ where γ is a function that results from its resolution w.r.t. x. In other terms:

$$(c, x) \in E(c) : c(P_V(c)) = 0 \iff x = \gamma(P_V(c) \setminus \{x\}) \tag{14}$$

In the structural frame, this resolution is always possible. Indeed, if constraint c is static, one has generically

$$x \in P_V(c) \iff \frac{\partial c\left(P_V(c)\right)}{\partial x} \neq 0$$

and there exists a γ satisfying (14), at least locally (from the implicit functions theorem). If constraint c is dynamic, then from (11) it writes $z - \frac{d}{dt}x = 0$. The 2 possible matchings are:

(1) derivative causality: $(c, z) \in E(c) : x \to c \to z \iff z(t) = \frac{d}{dt}x(t)$,

(2) integral causality: $(c, x) \in E(c) : z \to c \to x \iff x(t) = x(0) + \int_0^t z(\tau)\, d\tau$.

Note that, if the input is assumed to be known (and derivable), the derivative causality path defines z uniquely, while the path in integral causality defines x only up to the constant $x(0)$.

Propagation and Equivalent Graph

Consider an elementary non autonomous graph with more than one constraint. Constraint propagation consists of replacing each internal or output variable, everywhere it appears, by the result of the resolution of the constraint where it is matched. The result is a graph whose constraints are structurally equivalent, where each internal or output variable is expressed as a combination of functions of the inputs only.

For graphs without loops, the propagation directly results from their decomposition into hierarchical levels (level 0 includes the input variables, level 1 includes the variables that are matched to constraints whose inputs are all at level 0, .. level i includes the variables that are matched to constraints whose inputs are all at levels below i, etc.)

In the case of loops, a path initiated on any variable matched in the loop terminates on the variable itself, therefore the propagation provides a constraint which links this variable only to the inputs of the loop, and can be generically solved.

Example. Let us illustrate constraint propagation and equivalent graphs on the three examples of Figure 2, the equivalent graphs are respectively shown on Figures 3, 4 and 5.

Simple path (a): $c_1(x_1, y) = 0$ gives $x_1 = \gamma_1(y)$. The second constraint $c_2(x_1, x_2) = 0$ gives $x_2 = \gamma_2(x_1)$ and by replacing x_1 we get $x_2 = \gamma_2(\gamma_1(y)) \triangleq \Gamma_2(y)$.

Tree (b): According to the same process, one successively gets $c_1(x_1, y) = 0 \to x_1 = \gamma_1(y)$, then $c_2(x_1, x_2, u) = 0 \to x_2 = \gamma_2(x_1, u)$ and finally $x_2 = \gamma_2(\gamma_1(y), u) \triangleq \Gamma_2(y, u)$.

Loop (c): Starting for example with x_3 one has $c_3(x_1, x_3, x_5) = 0 \to x_3 = \gamma_3(x_1, x_5)$. Then $c_4(x_3, x_4) = 0 \to x_4 = \gamma_4(\gamma_3(x_1, x_5))$ and $c_5(x_2, x_4, x_5, x_6) = 0 \to x_5 = \gamma_5(x_2, \gamma_4(\gamma_3(x_1, x_5)), x_6)$ from which it follows that $x_5 = \Gamma_5(x_1, x_2, x_6)$. The same process, repeated from the starting nodes x_5 and x_4, gives respectively $x_4 = \gamma_4(\gamma_3(x_1, \gamma_5(x_2, x_6, x_4))) \triangleq \Gamma_4(x_1, x_2, x_6)$ and $x_3 = \gamma_3(x_1, \gamma_5(x_2, x_6, \gamma_4(x_3))) \triangleq \Gamma_3(x_1, x_2, x_6)$.

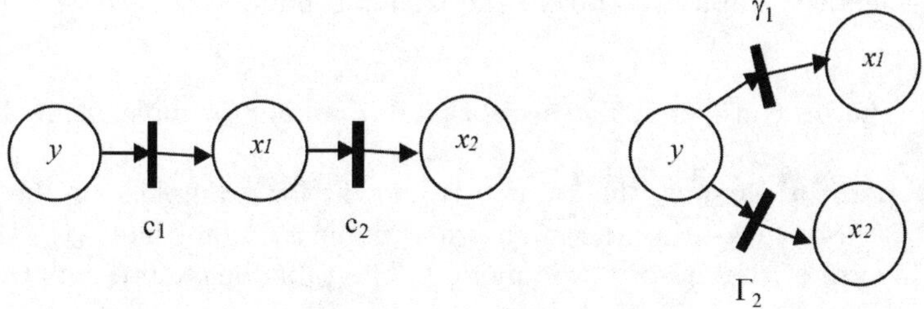

Fig. 3. Simple path (a) and its equivalent graph

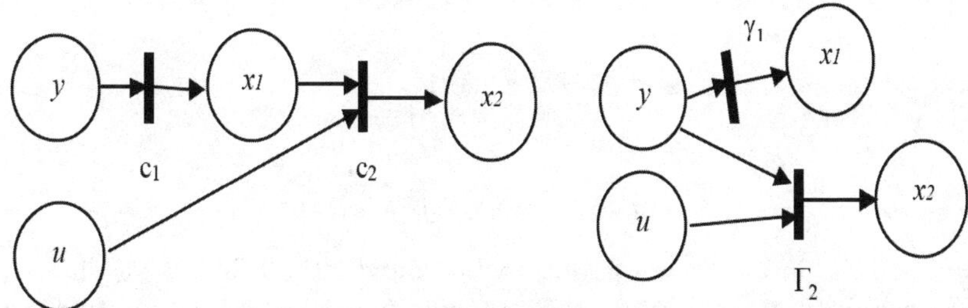

Fig. 4. Tree (b) and its equivalent graph

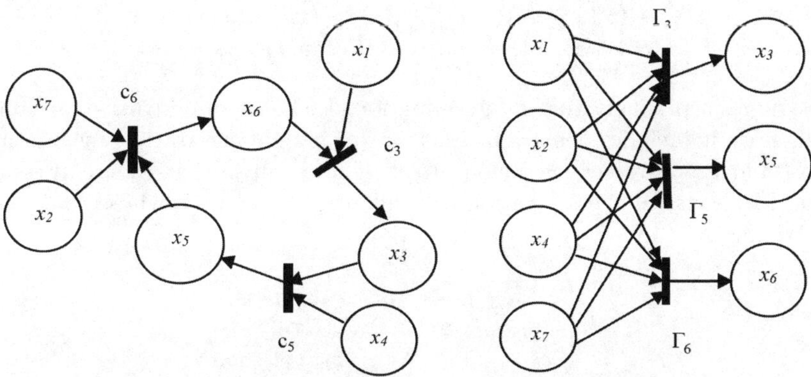

Fig. 5. Loop (c) and its equivalent graph

Static and Dynamic Constraints

In an elementary graph, a given path will cross a number of static and dynamic constraints. Constraint propagation between two nodes produces a resultant constraint whose dynamic order is the number of dynamic constraints crossed between the two nodes. Therefore, it may happen that, even when the inputs of the elementary graph are known, the terminal node is defined only up to a constant.

Example. Let us consider again cases (a) and (c) of our example, using different causalities.

Simple path (a): assume the input y is known (for example, y is the output signal of a sensor measuring the unknown variable x_1; note that $\gamma_1(y)$ is therefore the inverse of the sensor static model). The table below presents the three possible cases associated with constraint c_2 that links x_1 and x_2, namely "s" (static constraint), or "d" (dynamic under derivative causality), or "i" (dynamic under integral causality). As already noticed in 4.2, the function $\Gamma_2(y)$ that results from constraint propagation defines x_2 only up to a constant in the last case.

$$c_2 : s \; x_2 = \gamma_2(\gamma_1(y))$$

$$c_2 : d \; x_2 = \left[\frac{\partial}{\partial y}\gamma_1(y)\right]\dot{y}$$

$$c_2 : i \; x_2 = x_2(0) + \int_0^t \gamma_1(y)d\tau$$

Loop (c): the inputs x_1, x_2 and x_6 being known, the table below shows that when c_4 is a dynamic constraint, the matching under integral causality gives the result only up to a constant, as previously.

$$c_4 : i \begin{cases} x_3 = \gamma_3(x_1, \gamma_5(x_2, x_6, x_3(0) + \int_0^t x_3 d\tau)) \\ x_4 = x_4(0) + \int_0^t \gamma_3(x_1, \gamma_5(x_2, x_6, x_4)d\tau \\ x_5 = \gamma_5(x_2, x_3(0) + \int_0^t \gamma_3(x_1, x_5)d\tau, x_6) \end{cases}$$

Maybe more surprisingly, this indetermination also holds for derivative causality. Indeed, since it includes constraint c_4, loop (c) is differential, that means that both x_3 and its derivative x_4 belong to it. The result is that each expression produced by constraint propagation is a differential equation, whose resolution will introduce an (unknown) initial condition.

$$c_4 : d \begin{cases} x_3 = \gamma_3(x_1, \gamma_5(x_2, x_6, \frac{d}{dt}x_3)) \\ x_4 = \frac{d}{dt}(\gamma_3(x_1, \gamma_5(x_2, x_6, x_4)) \\ x_5 = \gamma_5(x_2, \frac{d}{dt}(\gamma_3(x_1, x_5)), x_6) \end{cases}$$

4.3 Elementary Graph on x

Let $x \in X \subset V$. This variable is observable if and only if its value can be estimated as a function of the extended known variables $\bar{K}$ ([6], [10]). Let $O(x)$ be the property "x is structurally observable", we now show that its analysis follows simply from the analysis of specific elementary graphs.

Definition. Let $x \in X \subset V$. An *elementary graph on* x, $\mathcal{G}_e(x)$ - or *x-rooted elementary graph (x-REG)* - is a maximal elementary graph such that x is its only output or it is an internal variable (if the graph has no output).

An x-REG is noted $\mathcal{G}_e(x) = (C_x, V_x, E_x)$. Let V_x^i be the internal variables. It follows from the definition that the known variables $V_x \cap K$ can only be inputs. Note that the set of inputs may be empty (the graph is autonomous), and that some input variables may be unknown. Remind that $\mathcal{G}_e(x)$ defines a complete matching on the set of its constraints and on the set of variables $V_x^i \cup \{x\}$ (if there is an output) or V_x^i (if there is no output). Finally, it is clear that a given unknown variable x can possibly be the root of several elementary graphs.

Example. Figure 6 gives two examples of x_1-REG, the first one (a) with an output, and the second one (b) without output.

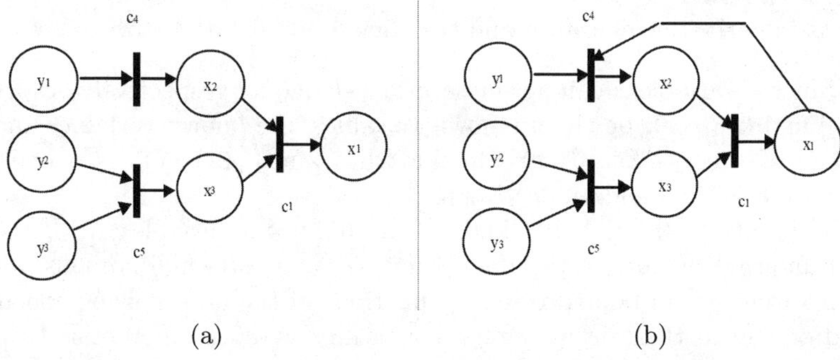

Fig. 6. Two examples of elementary graphs on x_1: (a) x_1 is an output; (b) x_1 is internal

Obtaining x-Rooted Elementary Graphs

The following algorithm builds an x-REG. It manipulates 4 sets: the roots of the current iteration $R_{current}$, the roots for the next iteration R_{next}, the labelled variables VL and the labelled constraints CL.

Algorithm. Initialise $R_{current} = \{x\}$, $R_{next} = \varnothing, VL = \varnothing$ et $CL = \varnothing$.
    Start If $R_{current} \neq \varnothing$
        1 select one variable $v \in R_{current}$
        2 update $VL = VL \cup \{v\}$ and $R_{current} = R_{current} \setminus \{v\}$
        3 If $P_C(v) \cap CL = \varnothing$ go to Start
        If $P_C(v) \cap CL \neq \varnothing$
          a) select a non labelled constraint $c \in P_C(v) \cap CL$
          b) update $CL = CL \cup \{c\}$ and $R_{next} = R_{next} \cup P_V(c)$
          c) match v to c
          d) go to Start
        If $R_{current} = \varnothing$
        4 If $R_{next} \neq \varnothing$
          a) initialise the next ieration with $R_{current} = R_{next}$
          b) go to Start
        If $R_{next} = \varnothing$ an x-REG has been obtained.

This algorithm can obviously be repeated to construct all the x-REG associated with any $x \in X$, by selecting each time a different constraint at step 3.a) (following a depth-first search, for example).

Observability and Elementary Graphs

The interest of x-REG is motivated by the following theorem:

Theorem. x is structurally observable if there is at least one non autonomous x-REG such that

(1) all its inputs belong to K,
(2) it is in derivative causality and contains no differential loop.

Proof. Since a variable x can admit several x-REG, let respectively $X\left[\mathcal{G}_e(x)\right]$, $K\left[\mathcal{G}_e(x)\right]$ and $C\left[\mathcal{G}_e(x)\right]$ be the unknown variables, the known variables and the constraints of the x-REG $\mathcal{G}_e(x)$. Remind that any $\mathcal{G}_e(x)$ defines a complete matching on $C\left[\mathcal{G}_e(x)\right]$ and on $X\left[\mathcal{G}_e(x)\right]$.

All the inputs of $\mathcal{G}_e(x)$ being known, the unknown variables $X\left[\mathcal{G}_e(x)\right]$ are therefore internal or output variables. Since $\mathcal{G}_e(x)$ is non autonomous, all the unknown variables can be expressed as functions of the inputs using constraint propagation. Since there is no integral causality and no differential loop, no initial condition is needed for constraint resolution.

An x-REG that satisfies conditions (1) - (2) of the above theorem is said to be *valid*. Note that if $\mathcal{G}_e(x)$ is a valid x-REG, then all the variables $X\left[\mathcal{G}_e(x)\right]$ are observable. A valid x-REG is a data flow diagramm (DFD) [9], [18] associated with one possible way to compute x as a function of the known variables. Call it a *version* of the estimation service of x (this is the link with the fonctional models of the system [11]). Note also that the order of derivation that is necessary in $\bar{K}\left[\mathcal{G}_e(x)\right]$ for the estimation of x automatically follows from the number of dynamic constraints that are crossed along the alternated paths of $\mathcal{G}_e(x)$. Finally, translating the DFD associated with $\mathcal{G}_e(x)$ into an algorithm is possible if and only if the models of all the constraints in $C\left[\mathcal{G}_e(x)\right]$ are known.

5 Fault Tolerant Estimation

5.1 Versions of the Estimation Algorithm

Let $x \in X$ and let $\mathcal{G}(x) = \left\{\mathcal{G}_e^1(x), \mathcal{G}_e^2(x), ...\right\}$ be the set of the valid x-REG. There are $|\mathcal{G}(x)|$ different DFDs that allow the computation of x from K, i.e. $|\mathcal{G}(x)|$ different versions of the estimation algorithm. Note that these versions are *minimal* in the sense that the results provided by different DFDs could always be merged (by any fusion procedure e.g. taking the means). This also implies minimality in the sense that it is necessary and sufficient to know the models of the constraints in $C\left[\mathcal{G}_e^i(x)\right]$ to implement version $n°i$ of the algorithm.

Example. The sequel will be illustrated by the system whose structure appears on Figure 7. The unknown variables are noted x, the known ones are noted y. We are interested in the observability of x_1.

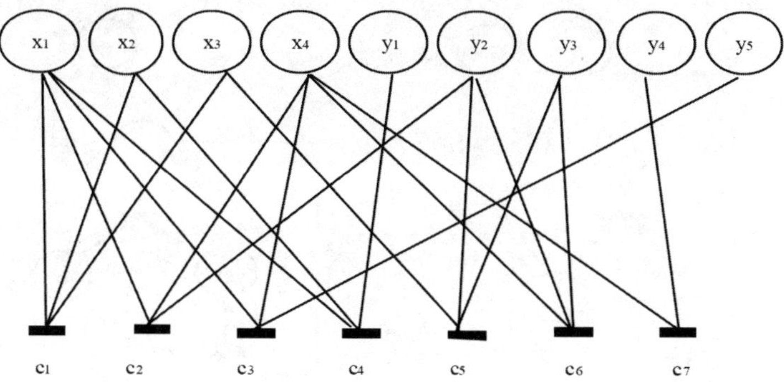

Fig. 7. Structure of the example

The elementary graphs on x_1 are drawn on Figure 8, we assume they are all valid. There are 4 versions of the estimation algorithm for x_1, namely:

$$v1 : x_1 - c_1(c_4(x_1, y_2), c_5(y_2, y_3)) = 0 \tag{15}$$
$$v2 : x_1 - c_2(c_6(y_2, y_3), y_2) = 0$$
$$v3 : x_1 - c_2(c_7(y_4), y_2) = 0$$
$$v4 : x_1 - c_3(c_2(x_1, y_2), y_5) = 0$$

Remark. Versions v2 and v3 give x_1 explicitly because they are provided by two x_1-*REG* without loop, while versions v1 and v4 give x_1 implicitely. By construction, all versions imply only static operations and derivations. This is of course a well known problem in practical implementations, where inputs might be very noisy. A practical implementation could rely on a *realization* of the input/output relation that is given by the DFD by means of a filter. The present analysis provides all the possibilities allowed by the system structure to estimate an unknown variable. The problem of their realization is not addressed here.

5.2 Fault Recovery

Assume that the version associated with the ith elementary graph $\mathcal{G}_e^i(x)$ is the one currently in use at time t, and that fault mode $n° f$ of component *comp* occurs, thus falsifying the subset of constraints $C_f(comp)$. This fault would obviously

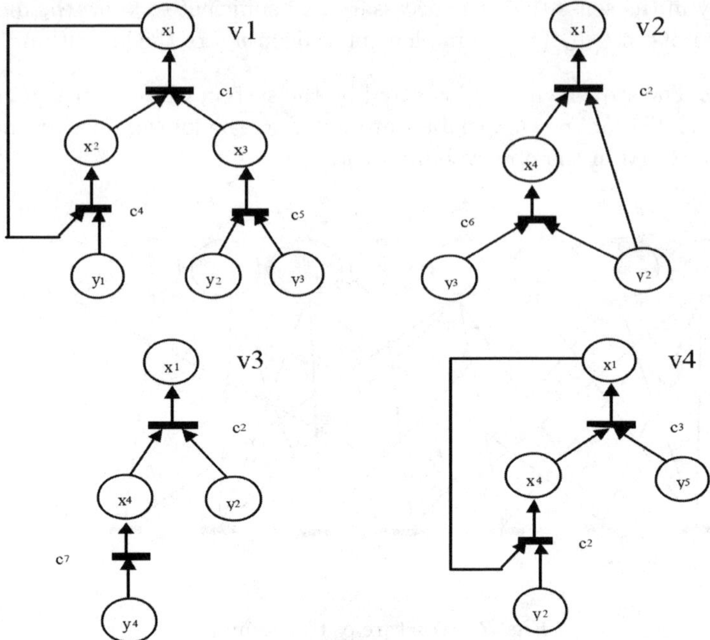

Fig. 8. The four valid elementary graphs on x_1

have no incidence on property $O(x)$ if $C\left[\mathcal{G}_e^i(x)\right] \cap C_f(comp) = \varnothing$, and the estimation functionality would still be available, under its version $\mathcal{G}_e^i(x)$. On the contrary, if $C\left[\mathcal{G}_e^i(x)\right] \cap C_f(comp) \neq \varnothing$ the conclusions will be different according to whether the constraints $C_f(comp)$ - i.e. the model of the faulty system - are known or not.

Fault accommodation. If the model of the faulty system is identified, the variable $x \in X$ is still structuraly observable, and its estimation is still possible by all the versions of the estimation algorithm [2]. Indeed, it is enough to replace the nominal models of the constraints in $C\left[\mathcal{G}_e^i(x)\right] \cap C_f(comp)$ by their model under fault f to adapt the DFD $\mathcal{G}_e^i(x)$ to the new situation.

System reconfiguration. If the model of the faulty system is not available, the DFD $\mathcal{G}_e^i(x)$ can obviously not be used any longer. Fault tolerance of property $O(x)$ can only result from the existence of another valid x-REG for exemple $\mathcal{G}_e^j(x)$, that does not contain any faulty constraint. Replacing the algorithm

---

[2] The choice of the version to be used under given circumstances is not developed here. One can easily imagine that a pre-order is defined on the set of versions, according to their performances in specific operation conditions [11].

associated with the DFD $\mathcal{G}_e^i(x)$ by the one associated with $\mathcal{G}_e^j(x)$ is the reconfiguration strategy.

Example. Assume that x_1 is currently estimated by version v3 of the algorithm, and that a fault occurs that changes constraint c_7 into the new $\tilde{c}_7$ (for example, c_7 was $y_4 = x_4$ and it becomes $y_4 = x_4 + bias$).

1) First note that if the fault would have concerned neither c_2 nor c_7, it would have had no impact on the estimation of x_1 using v3.

2) Whatever the new expression $\tilde{c}_7$, as soon as it is identified by the diagnostic algorithm[3], the accommodated estimation

$$x_1 - c_2\left(\tilde{c}_7\left(y_4\right), y_2\right) = 0$$

can be used to recover from the fault [4].

3) if the new expression $\tilde{c}_7$ is unknown, it is still possible to estimate x_1 by using one of the versions v1, v2 or v4, that rely on subsets of constraints that do not include c_7.

5.3   Critical Faults

Since structural properties are invariant under the fault accommodation strategy, we are interested in the sequel only in system reconfiguration (SR). At time t, the system components can be decomposed into two classes, the healthy ones (index 0) and the faulty ones (index 1). The resulting constraints decomposition is

$$C = C_0(t) \cup C_1(t)$$

where only the constraints in $C_0(t)$ are usable by DFDs for the estimation of the unknown variables.

Let $x \in X$ and $\mathcal{G}(x) = \left\{\mathcal{G}_e^1(x), \mathcal{G}_e^2(x), ...\right\}$ the set of all x-REG. The necessary and sufficient condition for version $\mathcal{G}_e^i(x)$ of the estimation service of x to be available at time t is that:

$$C\left[\mathcal{G}_e^i(x)\right] \subseteq C_0(t).$$

As a result, any fault situation $C_1(t)$ such that:

$$\forall \mathcal{G}_e^i(x) \in \mathcal{G}_e(x) : C\left[\mathcal{G}_e^i(x)\right] \cap C_1(t) \neq \varnothing \tag{16}$$

is not SR-recoverable.

Definition. A critical fault for property $O(x)$ is a minimal subset of constraints $\bar{C}(x) \subseteq C$ such that subsystem $C\backslash\bar{C}(x)$ has property $\neg O(x)$.

---

[3] Structural identifiability conditions are not developed here: the possibility to identify the new model of the faulty constraint c_7 is assumed.

[4] A transient period is obviously to be considered between the time at which the fault occurs and the time at which it is accommodated. This aspect is not addressed here (see [30], [27] for example, for proposals of control strategies during this transient).

Remark. From (16) critical faults are *minimal hitting sets* [5] of the set $\mathcal{G}(x)$. Once $\mathcal{G}(x)$ has been determined (by the algorithm in 4.3 for example) the critical faults are easily computed by associating with $\mathcal{G}(x)$ the logical expression:

$$O(x) = \left[\bigvee_{\mathcal{G}_e^i(x) \in \mathcal{G}(x)} \left(\bigwedge_{c \in C[\mathcal{G}_e^i(x)]} H_0(c) \right) \right] \tag{17}$$

which reads "x is structuraly observable if and only if the logical expression is true" (with $H_0(c)$ true iff constraint c is not falsified). It is easily seen that there may exist several critical faults for $O(x)$. Let $\bar{C}_{cr}(x) = \{\bar{C}_{cr}^1(x), \bar{C}_{cr}^2(x), ...\}$ be the set of all critical faults.

Example. From the 4 versions of the estimation algorithms of x_1 given by (15), $O(x_1)$ is described by the Boolean expression:

$$O(x_1) = c_2.c_3 + c_2.c_6 + c_2.c_7 + c_1.c_4.c_5$$

where $c_i.c_j$ (resp. $c_i + c_j$) are condensed notations for $H_0(c_i) \wedge H_0(c_j)$ (resp. $H_0(c_i) \vee H_0(c_j)$). The set of critical faults is easily found to be:

$$C_{cr}(x) = \{\{c_1, c_2\}, \{c_2, c_4\}, \{c_2, c_5\}, \{c_1, c_3, c_6, c_7\}, \{c_3, c_4, c_6, c_7\}, \{c_3, c_5, c_6, c_7\}\}. \tag{18}$$

5.4   Evaluating the Fault Tolerance Level

Property $O(x)$ is fault tolerant as long as

$$\exists \bar{C}_{cr}^i(x) \in \bar{C}_{cr}(x) : \bar{C}_{cr}^i(x) \not\subseteq C_1(t)$$

Since $\bar{C}_{cr}(x)$ is determined off-line, one can easily check on-line, when components fail and are switched-off by the reconfiguration procedure, if the set $C_1(t)$ satisfies this condition. Even better, it is possible to evaluate the "residual fault tolerance" of $O(x)$, at time t, as follows.

Redundancy degrees. The cardinal of each element of $\bar{C}_{cr}^k(x) \in \bar{C}_{cr}(x)$, $k = 1, 2, ...$ is the number of constraints whose violation is needed to falsify $O(x)$. An evaluation of the residual fault tolerance level at time t, is therefore given by the number of components [6] whose switching-off would lead to a set $C_1(t)$ such that:

$$\exists \bar{C}_{cr}^i(x) \in \bar{C}_{cr}(x) : \bar{C}_{cr}^i(x) \subseteq C_1(t).$$

---

[5] A minimal hitting set of $\mathcal{G}(x) = \{\mathcal{G}_e^1(x), \mathcal{G}_e^2(x), ...\}$ is a set that has a non empty intersection with every $\mathcal{G}_e^i(x), i = 1, 2, ...$

[6] A fault mode is the falsification of a subset of constraints. The link between subsets of constraints and subsets of components is straightforward: the reconfiguration strategy, by switching-off a component, deletes all the constraints that are associated with it from the set of constraints that are usable by the estimation algorithm (note that some non-faulty constraints are deleted too).

The smallest number of such components is called "strong redundancy degree", the largest one is the "weak redundancy degree" [26], [25].

Reliability. Strong and weak redundancy degrees evaluate the residual fault tolerance by the number of components (respectively in the "pessimistic" and the "optimistic" case) whose fault falsifies the observability property. However, more elaborate indexes can be used, since faults are events whose probability can be evaluated, namely through reliability data, if available. Thus, fault tolerance of property $O(x)$ can be evaluated by [26]:

- $R_x(C_0(t), t, T)$, the probability for $O(x)$ to remain true on the time interval $[t, T[$, under initial non-falsified constraints $C_0(t)$ at time t,
- $MTOF_x(C_0(t), t)$, the mean time to property $\neg O(x)$ to become true, under initial condition $C_0(t), t$.

Example. Assume that the 7 constraints of the example model a system of 3 interconnected components $C(comp1) = \{c_1, c_2\}$, $C(comp2) = \{c_3, c_4, c_5\}$, and $C(comp3) = \{c_6, c_7\}$. Figure 9 shows for each configuration of in-service components (identified on the first line of the box), the set of available versions of the estimation algorithm of x_1 (on the second line of the box). Top-down arrows show configuration changes that result of switching-off components after faults, while bottom-up arrows show configuration changes that result from switching components on again (e.g. after repair).

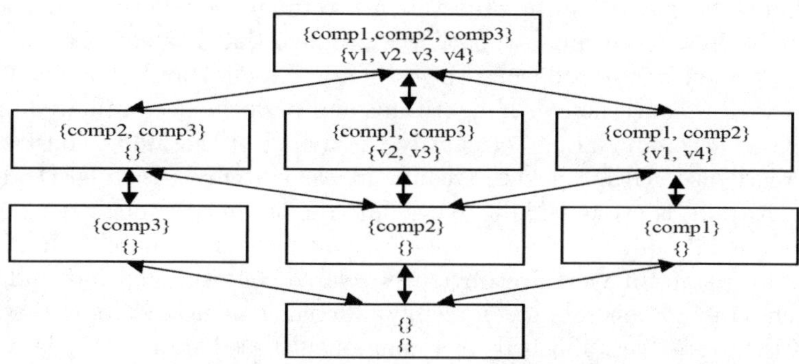

Fig. 9. Possible configurations and available estimation versions

In this system, the strong redundancy degree is 1, since the single fault of *comp1* is enough to falsify $O(x_1)$; the weak redundancy degree is 2 since there are cases where two components must be faulty for the property to be lost. On another hand, if all three components are assumed to be healthy at time 0 and their reliabilities are known, the reliability of the estimation function of x_1 is computed by:

$$R_{x_1}(\{comp1, comp2, comp3\}, 0, T) = ...$$
$$... = P_{[0,T]}(comp1, comp2, comp3) + P_{[0,T]}(comp1, comp3) + P_{[0,T]}(comp1, comp2)$$

where $P_{[0,T]}(a, b, c)$ is the probability for the 3 components a, b, c, that are heathy at time 0, to remain healthy during the whole time interval $[0, T]$. Classical computations also give the mean time to non-observability

$$MTOF_{x_1}(\{comp1, comp2, comp3\}, 0).$$

6   Conclusion

The general problem of fault tolerance, and the contribution of structural analysis to the design of fault tolerant systems have been presented in this paper, in a component-based modeling frame. Through the specific application to the design of estimation algorithms, it has been shown that strutural analysis provides a simple and constructive approach to identify the observable part of a system, as well as the possible reconfigurations of the estimation scheme when faults occur. The approach can be extended to other structural properties like e.g. controlability, monitorability, etc.

More generally, fault tolerance can be considered from at least two points of view. The functional point of view is qualitative: when a fault occurs, is the system still able to accomplish its mission ? (here estimate a given unknown variable x); if it is, how to proceed ? (use a DFD defined by a valid x-REG); if not, what to do ? (change the objective, i.e. estimate some other variable than x, or try to do without any estimation, a problem that has not been addressed in this paper since it goes beyond the estimation application). The quantitative point of view is concerned with algorithms and performance evaluation: how to design fault accommodation ? (identify the model of the faulty constraints in the currently used DFD); how to reconfigure the system ? (use a DFD where no faulty constraint is present); how to evaluate the obtained fault tolerance level ? (analyze critical faults, compute redundancy degrees and reliability indexes).

The two points of view are obviously related, but the analysis and design approaches rely in general on completely different models and tools. The structure of the system, established in a component-based framework, is a design model that allows to unify the architectural (components and interconnections), the behavioral (variables and constraints), and the functional (existence of a property) points of view, and to design fault tolerance strategies that are clear and integrate well established dependability concepts (critical faults, reliability, etc.). Note that this is not the case in classical approaches based on state space or input/output models, where the components are "rubbed out". The presented approach also provides a frame for architecture design, by identifying the constraints (i.e. the components) that must be added to a given architecture in order to obtain a given property, or to improve the fault tolerance level of an existing property [26]. An obvious by-product is the definition of maintenance strategies that allow to obtain a specified fault tolerance level [25].

References

1. Blanke M, Kinnaert M, Lunze J, Staroswiecki M (2003) Diagnosis and Fault Tolerant Control, Springer-Verlag, Berlin-Heidelberg, Germany
2. Carpentier T, Litwak R, Cassar J -Ph (1997) Criteria for the evaluation of FDI systems, Application to sensors location, In: Proceedings of IFAC Safeprocess'97, 1083-1088
3. Commault C, Dion J M, Sename O, Motyeian R (2000) Observer-based fault detection and isolation of structured systems, In: Proceedings of 39th IEEE Conference on Decision and Control, Sydney, Australia
4. Declerck P, Staroswiecki M (1991) Characterization of the Canonical Components of a Structural Graph for Fault Detection in Large Scale Industrial Plants, European Control Conference, Grenoble, France
5. Dion J M, Commault C, van der Woude J (2003) Generic properties and control of linear structured systems: a survey, Automatica, (39)7:1125-1144
6. Diop S (2002) From the Geometry to the Algebra of Nonlinear Observability, In: Contemporary Trends in Nonlinear Geometric Control Theory and its Applications, editors Anzaldo-Meneses A, Monroy-Pérez F, Bonnard B, Gauthier J P, World Scientific Publishing, Singapore
7. Dulmage A L, Mendelsohn N S (1958) Covering of bi-partite graphs, Canadian Journal of Mathematics, 10:517-534
8. Dulmage A L, Mendelsohn N S (1959) A structure theory of bi-partite graphs of finite exterior dimension, Transaction of the Royal Society of Canada, Section III, 53:1-13
9. Gane C, Sarson T (1978) Structured Systems Analysis: Tools and Techniques, Englewood Cliffs, Prentice-Hall
10. Gauthier J P, Kupka I A K (2001) Deterministic Observation Theory and Applications, Cambridge University Press, Cambridge, UK
11. Gehin A L, Staroswiecki M (1999) A Formal Approach to Reconfigurability Analysis, Application to the three Tank Benchmark, European Control Conference '99, Karlsruhe, Germany
12. Golver K, Silverman L M (1976) Characterization of structural controllability, IEEE Transaction on Automatic Control, 4: 534-537
13. Harary F (1962) A graph theoretic approach to matrix inversion by partitioning, Numerische Mathematik, 4:128-135
14. Isermann R (2006) Fault-Diagnosis Systems, Springer, Berlin, Germany
15. Jiang J Zhang Y M (2006) Accepting performance degradation in fault-tolerant control system design, IEEE Transaction on Control Systems Technology, 14(2):284—292
16. Lin C T (1974) Structural Controllability, IEEE Transaction on Automatic Control, 3:201-208
17. Lin C T (1977) System structure and minimal structure controllability, IEEE Transaction on Automatic Control, 5:855-862
18. Liu X, Liu J, Eker J, Lee E A (2003) Heterogeneous Modeling and Design of Control Systems, Software-Enabled Control: Information Technology for Dynamical Systems, editors Tariq Samad and Gary Balas, Wiley-IEEE Press
19. Mahmoud M, Jiang J, Zhang Y M (2003) Active Fault Tolerant Control Systems: Stochastic Analysis and Synthesis, LNCIS-287, Springer, Berlin, Germany
20. Meyer M, Le Lann J M, Koehret B, Enjalbert M (1994) Optimal selection of sensor location on a complex plant using a graph oriented approach, Computers and Chemical Engineering, 18:S535-S540

21. Murota K (1987) Systems analysis by graphs and matroïds. Structural solvability and controllability, Springer Verlag
22. Reinschke K J (1988) Multivariable control : a graph theoretic approach, Springer-Verlag
23. Schizas C, Evans F J (1981) A graph theoretic approach to multivariables control system design, Automatica, 17-2:371-377
24. Staroswiecki M, Cassar J P, Declerck P (2000) A structural framework for the design of FDI in large scale industrial plants, Issues of Fault Diagnosis for Dynamic Systems, editors Patton R, Frank P, Clark R, Springer Verlag, 2000
25. Staroswiecki M (2003) On Fault Tolerant Estimation in Sensor Networks, In: Proceedings of European Control Conference, Cambridge, UK
26. Staroswiecki M, Hoblos G, Aitouche A (2004) Sensor Network Design for Fault Tolerant Estimation, International Journal of Adaptive Control and Signal Processing, 18: 55 − 72
27. Staroswiecki M (2004) Progressive accommodation of actuator faults in the linear quadratic control problem, In: Proceedings of 43rd IEEE Conference on Decision and Control, Paradise Island, The Bahamas, pages 5234-5241
28. Steward D V (1962) On an approach to techniques for the analysis of the structure of large systems of equations, SIAM Review, 4: 321-342
29. Zhang Y M, Jiang J (2003) Bibliographical review on reconfigurable fault-tolerant control systems, In: Proceedings of IFAC Symposium on SAFEPROCESS'03, pages 265—276, Washington, D.C., USA
30. Zhang X, Parisini T and Polycarpou M M, (2004), Adaptive Fault-Tolerant Control of Nonlinear Uncertain Systems: An Information-based Diagnostic Approach, IEEE Transaction on Automatic Control, 49(8): 1259-1274

Robust Hybrid Control Systems: An Overview of Some Recent Results

Andrew R. Teel

Department of Electrical and Computer Engineering
University of California
Santa Barbara, CA 93106-9560
teel@ece.ucsb.edu

Summary. This paper gives an overview of a framework for analyzing hybrid dynamical systems. The emphasis is on modeling assumptions that guarantee robustness. These conditions lead to a general invariance principle and to results on the existence of smooth Lyapunov functions (converse theorems) for hybrid systems. In turn, the stability analysis tools motivate novel hybrid control algorithms for nonlinear systems.

Keywords: Hybrid Systems, Robustness, Stability Theory, Hybrid Control Systems.

1 Hybrid Systems

The capabilities of nonlinear feedback control can be enhanced by using dynamic controllers with states that can make jumps. Such controllers are called "hybrid" controllers and yield "hybrid control systems". Until recently, an open question has been whether hybrid controllers enjoy the same generic robustness properties as their "non-hybrid" counterparts. This paper will review certain recent advances in the development of hybrid control systems with a focus on robustness. The scope is narrow, focusing on results in which the author has participated.

To begin, the notion of "hybrid" is clarified. A continuous-time arc is an absolutely continuous function defined on the nonnegative real numbers. An illustration is depicted in Figure 1(a). A discrete-time arc is a function defined on the nonnegative integers, as in Figure 1(b). A "hybrid-time arc" (or simply, "hybrid arc") is, intuitively, a combination of a continuous-time arc and a discrete-time arc, like in Figure 2. A hybrid arc is defined on a hybrid time domain, which is a set of points of the form (t, j) where t is a nonnegative real number and j is a nonnegative integer. A hybrid time domain E has the property that, for each $(T, J) \in E$ there exists times $\{t_i\}_{i=0}^{J+1}$ satisfying $0 = t_0 \leq t_1 \leq \cdots t_{J+1} < \infty$ such that

$$E \bigcap ([0, T] \times \{0, \ldots, J\}) = \bigcup_{j=0}^{J} ([t_j, t_{j+1}] \times \{j\}) .$$

A hybrid arc is a function, defined on a hybrid time domain, that is absolutely continuous with respect to continuous time t for each discrete time index j.

C. Bonivento et al. (Eds.): Adv. in Control Theory and Applications, LNCIS 353, pp. 279–302, 2007.
springerlink.com

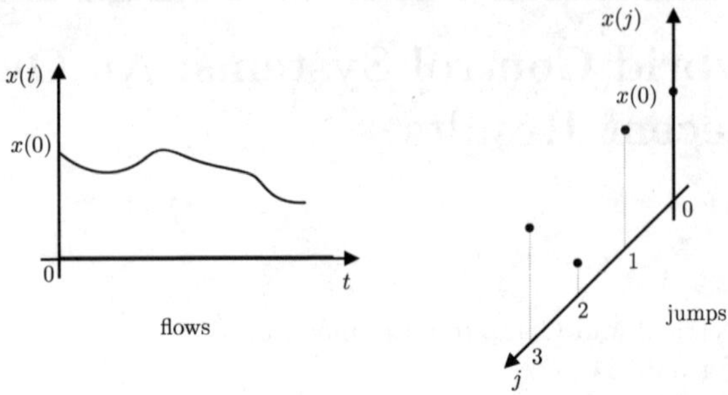

(a) A continuous-time arc.   (b) A discrete-time arc.

Fig. 1. Arcs

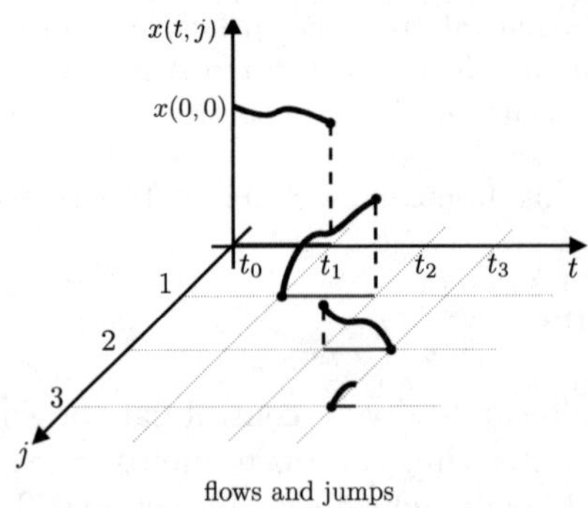

Fig. 2. Hybrid arc

The hybrid time domain/arc characterization of functions that make jumps is very convenient for bookkeeping purposes. For example, it easily allows for the possibility of multiple jumps at the same continuous time instant. It is also natural from a pedagogical point of view. However, these are not its biggest selling points. The biggest selling point for hybrid arcs is that they provide a natural way of characterizing when a sequence of arcs has a subsequence converging to a hybrid arc. Indeed, the notion of set and graphical convergence is well-suited for such studies. For more details, the reader should consult [12], and also the fundamental reference [23]. Hybrid time domains were also considered in [7] and [10]. They appear implicitly in [20].

To make the idea of graphical convergence more concrete, consider the sequence of hybrid arcs depicted in Figure 3. They are solutions of a hybrid "bouncing ball" model (see Example 3) showing the position of the ball when dropped for successively lower heights, each time with zero velocity. The sequence of graphs created by these hybrid arcs converges to a graph of a hybrid arc with hybrid time domain given by $\{0\} \times \{\text{nonnegative integers}\}$ where the value of

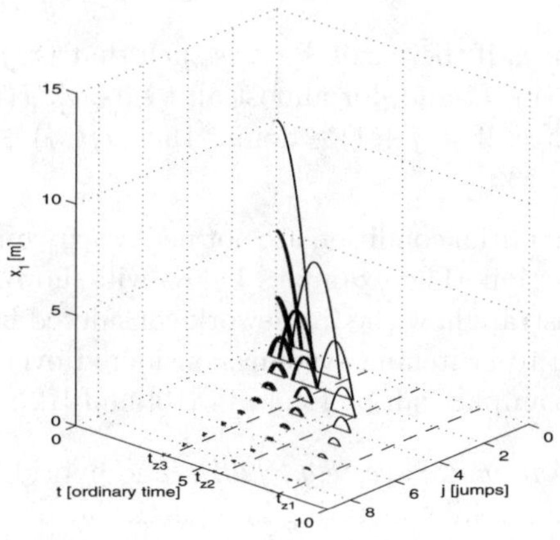

Fig. 3. Solutions to the bouncing ball system

the arc is zero everywhere on its domain. If this hybrid arc is a solution then the hybrid system is said to have a "compactness" property. This attribute for the solutions of hybrid systems is critical for robustness properties. It is the hybrid generalization of a property that automatically holds for continuous differential equations and difference equations, where nominal robustness of asymptotic stability is guaranteed. (For more details, see [9, §7 and §8].)

Solutions of hybrid systems are hybrid arcs that are generated in the following way: Let C and D be subsets of $\mathbb{R}^n$ and let f, respectively g, be mappings from C, respectively D, to $\mathbb{R}^n$. The hybrid system $\mathcal{H} := (f, g, C, D)$ can be written in the form

$$\begin{aligned}
\dot{x} &= f(x) & x &\in C \\
x^+ &= g(x) & x &\in D \ .
\end{aligned} \tag{1}$$

The map f is called the "flow map", the map g is called the "jump map", the set C is called the "flow set", and the set D is called the "jump set". The state x may contain variables taking values in a discrete set (logic variables), timers, etc. Consistent with such a situation is the possibility that $C \cup D$ is a strict subset of $\mathbb{R}^n$. For simplicity, assume that f and g are continuous functions. At times it is useful to allow these functions to be set-valued mappings, which will denote by F and G, in which case F and G should have a closed graph and be locally bounded, and F should have convex values. (For more information, see [12].) In this case, we will write

$$\begin{aligned}
\dot{x} &\in F(x) & x &\in C \\
x^+ &\in G(x) & x &\in D \ .
\end{aligned} \tag{2}$$

A solution to the hybrid system (2) starting at a point $x_\circ \in C \cup D$ is a hybrid arc x with the following properties:

1. $x(0,0) = x_{\circ}$;
2. given $(s,j) \in \mathrm{dom}\ x$, if there exists $\tau > s$ such that $(\tau, j) \in \mathrm{dom}\ x$, then, for all $t \in [s, \tau]$, $x(t,j) \in C$ and, for almost all $t \in [s, \tau]$, $\dot{x}(t,j) \in F(x(t,j))$;
3. given $(t,j) \in \mathrm{dom}\ x$, if $(t, j+1) \in \mathrm{dom}\ x$ then $x(t,j) \in D$ and $x(t, j+1) \in G(x(t,j))$.

Solutions from a given initial condition are not necessarily unique, even if the flow map is a smooth function. The examples below will show how non-uniqueness can arise and will illustrate how the framework considered here can be related to hybrid automata and to switching systems considered over a class of switching signals. Additional examples can be found in [10] and [12].

Example 1 (Hybrid Automata; see, e.g., [20]). The hybrid automaton depicted in Figure 4 is such that, while operating in a particular discrete mode q, a jump is enabled when any of the "guard" conditions $d_i(q, z) \geq 0$ is satisfied, in which case the mode can transition to mode i with the state being reset to the value $g_{q,i}(z)$. Otherwise, flows are enabled and satisfy the equation $\dot{z} = f_q(z)$. The overall hybrid system can be written as

$$\dot{z} \;=\; f_q(z) \qquad\qquad z \in C_q := \bigcap_i C_{q,i}$$

$$\begin{bmatrix} z \\ q \end{bmatrix}^+ \in \left\{ \begin{bmatrix} g_{q,i}(z) \\ i \end{bmatrix}, \; z \in D_{q,i} \right\} \quad z \in D_q := \bigcup_i D_{q,i}\ .$$

where $C_{q,i} := \{z : d_{q,i}(z) \leq 0\}$ and $D_{q,i} := \{z : d_{q,i}(z) \geq 0\}$. Thus, the overall state is $x := [z^T q]^T$, the flow map is

$$f(x) := \begin{bmatrix} f_q(z) \\ 0 \end{bmatrix}$$

the flow set is $C := \{(z,q) : z \in C_q\}$, the jump set is $D := \{(z,q) : z \in D_q\}$, and the jump map is the set-valued mapping given above. The solutions of the hybrid system are not necessarily unique, since more than one guard condition may be satisfied at the same time, and because it may be possible to "graze" the guard condition $d_{q,i}(z) \geq 0$, meaning that it may be possible to flow from the condition $d_{q,i}(z) = 0$ by keeping $d_{q,i}(z) \leq 0$. ∎

Example 2 (Average dwell-time switching). Consider the switching system $\dot{z} = f_q(z)$ where q makes jumps in a discrete set Q and its hybrid time domain is required to satisfy the constraint

$$j - i \leq \delta(t - s) + N \qquad \forall (t,j), (s,i) \in \mathrm{dom}\ q\ , \; t + j > s + i$$

where $\delta \geq 0$ and N is a positive integer. This constraint on jumps was introduced in [16] (though not expressed in terms of hybrid time domains) and was called an "average dwell-time condition". Moreover, if q jumps then the state z is allowed to jump according to the rule $z^+ = g_q(z)$. It has been shown in [4] that this class

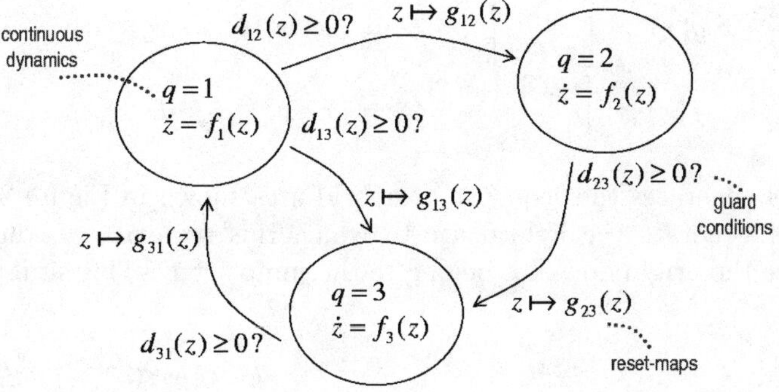

Fig. 4. A hybrid automaton

of switching signals and the resulting dynamics is captured completely by the hybrid system

$$\left.\begin{array}{l} \dot{z} = f_q(z) \\ \dot{q} = 0 \\ \dot{\tau} \in \left\{ \begin{array}{ll} \delta & \tau \in [0, N) \\ [0, \delta] & \tau = N \end{array} \right. \end{array}\right\} \quad \tau \in [0, N]$$

$$\left.\begin{array}{l} z^+ = g_q(z) \\ q^+ \in Q \\ \tau^+ = \tau - 1 \end{array}\right\} \quad \tau \in [1, N] \, .$$

Clearly, the solutions of this hybrid system are not unique since the flow and jump sets overlap and q^+ is set valued. A similar hybrid system can be written down to cover the case of "reverse average dwell-time" switching introduced in [15]. For details, see [4]. ∎

2 Compactness and Generalized Solutions

2.1 Compactness [12]

A natural question to ask is: What conditions on the system data (f, g, C, D) guarantee that each sequence of (locally eventually bounded[1]) solutions has a subsequence converging to a solution?

Example 3 (Bouncing Ball). To see what can go wrong, consider a possible model for a bouncing ball, as depicted in Figure 5. Taking $x_1 := h$ and $x_2 := \dot{h}$, this system has the data

---

[1] See [12, p. 579] for a definition of this phrase. Roughly speaking, by restricting the domains of the solutions to a given compact set, the solutions should be uniformly bounded for all sufficiently large index numbers.

$$g(x) := \begin{bmatrix} 0 \\ -\gamma x_2 \end{bmatrix} , \quad D := \{x : x_1 = 0 , \ x_2 < 0\}$$

$$f(x) := \begin{bmatrix} x_2 \\ -g \end{bmatrix} , \quad C := \{x : x_1 \geq 0\} \setminus D .$$

This model generates the sequence of hybrid arcs shown in Figure 3. However, it does not generate the hybrid arc to which this sequence of solutions converges since the origin does not belong to the jump set D. This situation can be

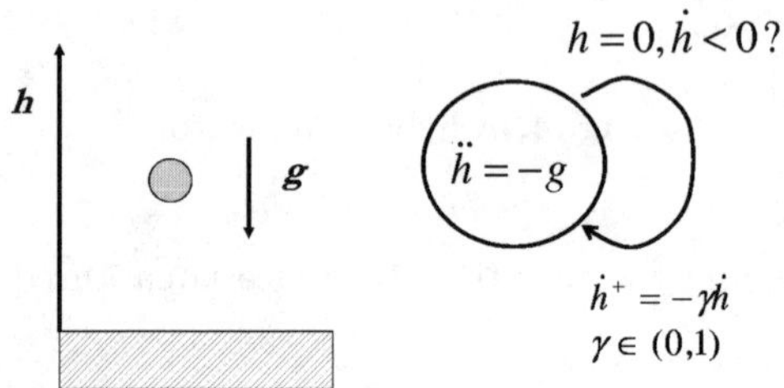

Fig. 5. Diagram for the bouncing ball system

remedied by including the origin in the jump set D. This amounts to replacing the jump set D by its closure. One can also replace the flow set C by its closure, although this has no effect on the solutions.   ∎

It turns out that *whenever the flow set and jump set are closed, the solutions of the corresponding hybrid system enjoy a useful compactness property: every locally eventually bounded sequence of solutions has a subsequence converging to a solution.* For more details, see [12, Theorem 4.4].

2.2   Generalized Solutions [26]

If a hybrid system is given with a flow or jump set that is not closed, it may be worthwhile to redefine the system data, taking the new flow and jump sets to be the closure of the original sets (and extending the definition of f and g to the closure of the original flow and jump set if necessary; for simplicity, assume that f and g were already defined on the closures). As with the bouncing ball system, using the closures may introduce new solutions. The set of solutions one gets when using the closures will be called "generalized solutions" of the original hybrid system. The generalized solutions, i.e., the solutions to

$$\begin{aligned} \dot{x} &= f(x) \qquad x \in \overline{C} \\ x^+ &= g(x) \qquad x \in \overline{D} \end{aligned} \tag{3}$$

enjoy the sequential compactness property, as indicated at the end of the previous subsection. Moreover, generalized solutions have a natural interpretation in terms of hybrid control systems with measurement noise. They are equivalent to the limiting solutions one gets to the hybrid system

$$\begin{aligned}
\dot{x} &= f(x+e) & x+e &\in C \\
x^+ &= g(x+e) & x+e &\in D
\end{aligned} \tag{4}$$

when letting the noise signal e converge to zero. For more specifics, see [26] where this result was first reported, as a generalization of an analogous result for discontinuous continuous-time systems in [14, 13, 8]. This equivalence suggests that the behavior of generalized solutions will be a good indicator of robustness. If the generalized solutions of a hybrid control system are not well behaved then it should be expected that arbitrarily small measurement noise can produce ill-behaved solutions in the closed-loop system. This will be illustrated next.

2.3   Robustness [10, 12]

Example 4 (Non-robust zero-cross detection). Let $0 < \varepsilon \ll 1$ and consider the hybrid system with the following data:

$$g(x) := \begin{bmatrix} 0 & 0 \\ -\varepsilon & 0 \end{bmatrix} x \qquad D := \{x : x_2 = 0\}$$

$$f(x) = \begin{bmatrix} 1 & 1 \\ -1 & 1 \end{bmatrix} x \qquad C := \mathbb{R}^2 \backslash D \ .$$

As depicted in Figure 6, solutions initially spiral away from the origin in the clockwise direction. When a solution reaches the x_1-axis, it jumps to the x_2-axis, with a significant reduction in magnitude. The net effect is to cause the solution to tend toward the origin. In fact, the origin is global exponentially stable. (One may also observe that each initial condition $x_i := (1, -1/i)$ produces a solution that flows around to a point on the negative x_1 axis, whereas the limiting initial condition $x_\infty := (1, 0)$ produces a unique solution that jumps to $(0, -\varepsilon)$. So, not every sequence of locally eventually bounded solutions has a subsequence converging to a solution.)

Now consider the behavior of generalized solutions. Since the closure of the flow set is $\mathbb{R}^2$, there now exist solutions that never jump, and thus spiral out to infinity. See Figure 7. Thus, when considering generalized solutions, the origin is unstable. This is consistent with the fact that detecting whether the state is on the x_1-axis is not a robust operation as modeled in the given hybrid system. Indeed, it is easy to inject arbitrarily small measurement noise into the system so that $x_2 + e$ is never equal to zero.   ∎

Example 5 (Robust zero-cross detection). When simulating the previous example in a software package like matlab/simulink, one would most likely simulate the system with "zero-cross detection" activated and would only see the behavior

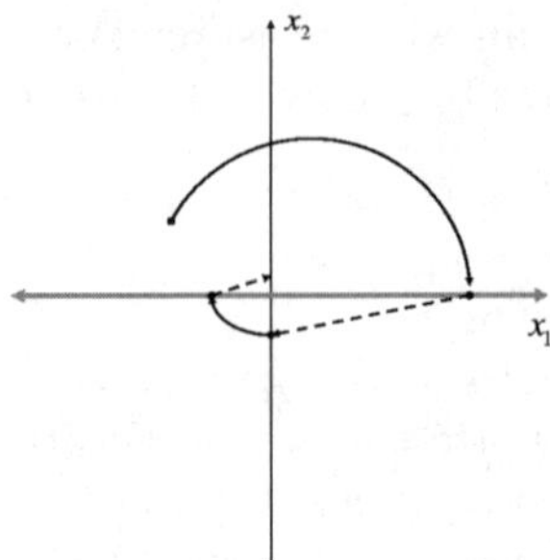

Fig. 6. Unstable spiral with stabilizing jumps

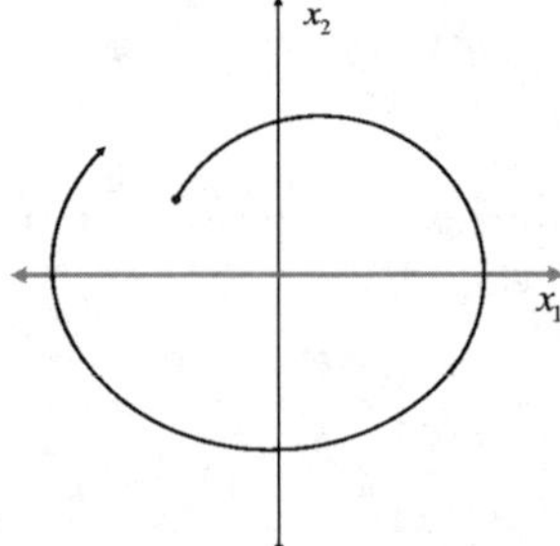

Fig. 7. Unstable spiral without stabilizing jumps

depicted in Figure 6. Implementing the hybrid system (4) with zero-cross detection corresponds to simulating a hybrid system that is an extension of the system (4) with an extra state x_3 that is used to keep track of which side of the x_2-axis the solution has been on recently. Flows are only allowed when x_3 and x_2 have the same sign. The extra state x_3 remains constant during flows and is updated at jumps according to the rule $x_3^+ \in \mathrm{SGN}(x_2^+)$, where SGN is one when its argument is positive, negative one when its argument is negative, and equal to the set $\{-1, 1\}$ when its argument is zero. (Note that the two-point set $\{-1, 1\}$ is forward invariant for the state x_3). The flow condition is modified to the condition $C := \{x : x_2 x_3 \geq 0\}$. The jump set is left unchanged. These flow and jump sets are already closed, so the generalized solutions match the regular solutions. Moreover, it is not possible for a solution to flow through the x_1-axis, as this would correspond to x_2 changing sign without x_3 changing sign.   ∎

Example 4 illustrates that generalized solutions can lead to non-uniqueness: generalized solutions starting on the x_1-axis can either jump or flow. One may be compelled to ask: "Isn't non-uniqueness the ultimate in non-robustness?" A reasonable answer is: "Yes and no. Non-uniqueness will make it impossible to predict exactly which course is going to be taken, but it may still be possible to accurately (robustly) predict the final outcome." To explain this statement, consider the problem of stabilizing a point on a circle using a hybrid controller.

Example 6 (Global stabilization of a point on a circle). The control system is taken to be

$$\dot{x} = \begin{bmatrix} 0 & -1 \\ 1 & 0 \end{bmatrix} xu \qquad x \in C_\circ := \left\{ x : x_1^2 + x_2^2 = 1 \right\}$$

where $u \in [-1, 1]$. The goal is to globally asymptotically stabilize the point $(x_1, x_2) = (0, -1)$ as indicated in Figure 8. Consider the hybrid controller with logic state $q \in \{-1, 1\}$ where flows and jumps are enabled in the sets

$$C := \left\{ (x, q) : x \in C_\circ, |x_2| \geq -qx_1 \right\} , \qquad D := \overline{C_\circ \backslash C} .$$

The feedback law is chosen to be a continuous function that moves the solution toward the desired point on the circle, as shown in Figure 8. In the region where $-x_2 \geq |x_1|$, where flowing in either mode is possible, the control can be independent of q: the flow is clockwise for $x_1 > 0$ and counterclockwise for $x_1 < 0$. In the region where $x_2 \geq |x_1|$, which is also where flowing in either mode is possible, the flow with $q = 1$ is in the clockwise direction whereas the flow with $q = -1$ is in the counterclockwise direction. Solutions are not unique for the initial condition (corresponding to θ_2 in Figure 8) satisfying $x_2 = x_1$, $q = -1$ and the initial condition (corresponding to θ_1 in Figure 8) satisfying $x_2 = -x_1$, $q = 1$. From these points it is possible to have no jumps while flowing to the desired point in a direction consistent with the initial value for q. It is also possible to have one jump immediately and then to flow to the desired point in a direction consistent with the negative of the initial value for q. In the presence of measurement noise, one cannot be sure which of these two solutions would occur. Nevertheless, one can be sure of the final outcome: each solution will approach the desired point on the circle. It is typical in hybrid control systems that make decisions (in this case, whether to approach the point on the circle by going clockwise or counterclockwise) that there will be points at which generalized

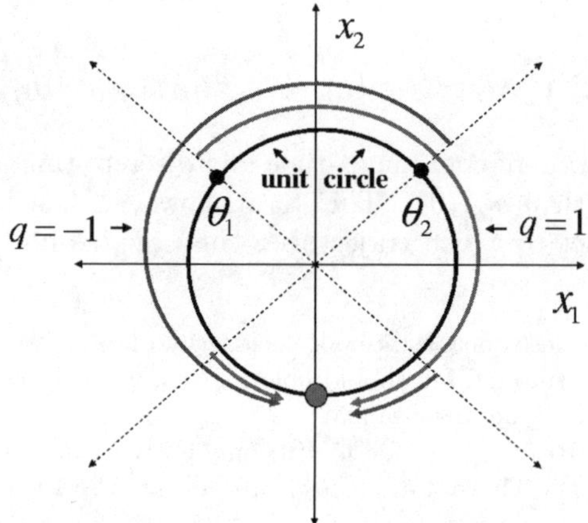

Fig. 8. Robustly stabilizing a point on a circle

solutions are not unique, so that the particular solution with measurement noise will be ambiguous, and yet certainly a decision will be made.   ∎

3   Value Added: Invariance Principles, Converse Lyapunov Theorems, and Robust Stability

The price paid for considering generalized solutions is that extra solutions are introduced. It has been argued that these extra solutions are quite reasonable from a control point of view so that, in a sense, this price paid is low. Conversely, the advantages gained from considering generalized solutions are tremendous. This section will emphasize the availability of general invariance principles, converse Lyapunov theorems, and generic robustness for asymptotically stable compact sets.

3.1   Invariance Principles [25]

Certain invariance principles for hybrid systems have appeared in [20] and [6]. Both of these results require, among other things, unique solutions which is not generic for hybrid control systems. In [25], general invariance principles were established that do not require uniqueness. The work in [25] includes several invariance results, some involving integrals of functions, like for continuous-time systems in [2] or [24], and some involving nonincreasing energy functions, like in the work of LaSalle [18] or [19]. One such result will be described here.

Suppose one can find a continuously differentiable function $V : \mathbb{R}^n \to \mathbb{R}$ such that

$$
\begin{aligned}
u_c(x) &:= \langle \nabla V(x), f(x) \rangle \leq 0 & \forall x \in \overline{C} \\
u_d(x) &:= V(g(x)) - V(x) \leq 0 & \forall x \in \overline{D} .
\end{aligned}
\tag{5}
$$

Let $x(\cdot, \cdot)$ be a bounded solution with an unbounded hybrid time domain. Then there exists a value r in the range of V such that x tends to the largest weakly[2] invariant set inside the set[3]

$$
M_r := V^{-1}(r) \cap \left(u_c^{-1}(0) \cup \left(u_d^{-1}(0) \cap g(u_d^{-1}(0)) \right) \right) .
\tag{6}
$$

The naive combination of continuous-time and discrete-time results would omit the intersection with $g(u_d^{-1}(0))$. This term, however, can be quite useful for zeroing in on the set to which trajectories converge, as illustrated in the next example.

---

[2] This term is used since solutions are not necessarily unique. "Weakly" indicates that the property this term modifies should hold for at least one solution rather than for all solutions. For more details, see [25].

[3] The notation $u_d^{-1}(0)$ (similarly for $u_c^{-1}(0)$ and $V^{-1}(r)$) means the set of points x satisfying $u_d(x) = 0$. The notation $g(u_d^{-1}(0))$ means the set of points $g(y)$ where $y \in u_d^{-1}(0)$.

Example 7 (Illustrating the invariance principle). Consider the hybrid system with data

$$f(x) = g(x) := \begin{bmatrix} x_2 \\ -x_1 \end{bmatrix} \ , \ C := \{x : x_2 \geq 0\} \ , \ D := \overline{\mathbb{R}^2 \backslash C} \ .$$

Note that $\overline{C} = C$ and $\overline{D} = D$, i.e., the flow set and jump set are closed. Solutions of this system with initial conditions $x_{1_o} \leq 0$, $x_{2_o} = 0$ are not unique, since both flowing and jumping are possible from such points. Solutions that start in a given circle remain in that circle for all time. The asymptotic behavior of solutions can be established as follows. Consider the function $V(x) := x^T x$. The function $u_c(x)$ defined in (5) is zero for all $x \in C$ and it can be taken to be $-\infty$ for $x \notin C$. Thus $u_c^{-1}(0) = C$. The function $u_d(x)$ defined in (5) is zero for all $x \in D$ and it can be taken to be $-\infty$ for $x \notin D$. Thus $u_d^{-1}(0) = D$. Moreover, $g(u_d^{-1}(0)) = g(D) = \{x : x_1 \leq 0\}$. Then, $u_d^{-1}(0) \cap g(u_d^{-1}(0))$ is equal to the lower left quadrant. The invariance principle then states that the solutions starting in a given circle converge to the largest weakly invariant set in the circle intersected with top half plane and the lower left quadrant. However, points in the strict lower left quadrant can be excluded because they cannot be part of a weakly invariant set: to reach these points requires starting in the lower right quadrant. One concludes that the trajectories converge to the circle intersected with the top half plane and the negative x_2-axis. In fact, this statement is not conservative. This is exactly the ω-limit set of each trajectory starting on the given circle. ■

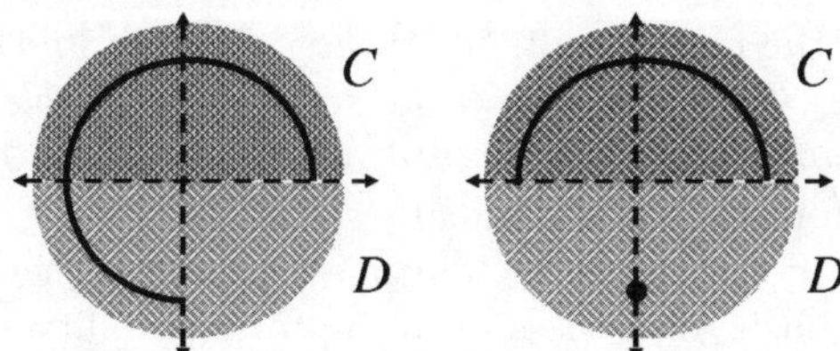

Fig. 9. The three-quarters circle on the left represents the set M_r in (6) for Example 7. On the right, the semicircle plus the point at $(0, -1)$ represent the largest weakly invariant set contained in M_r.

3.2 Converse Lyapunov Theorems [5, 3, 4]

Some early results on the existence of nonsmooth Lyapunov functions for hybrid systems appeared in [32]. The first results on the existence of *smooth* Lyapunov functions, which are intimately linked to robustness, appeared in [5]. These results required open basins of attraction, but this requirement has since been relaxed in [3]. The simplified discussion here is borrowed from this later work.

Let $\mathcal{O}$ be an open subset of the state space containing a given compact set $\mathcal{A}$. and let $\omega : \mathcal{O} \to \mathbb{R}_{\geq 0}$ be a continuous function that is zero for all $x \in \mathcal{A}$, is positive otherwise, and that grows unbounded as its argument grows unbounded or approaches the boundary of $\mathcal{O}$. Such a function is called a proper indicator for the compact set $\mathcal{A}$ on the open set $\mathcal{O}$. An example of such a function is the norm function defined on $\mathbb{R}^n$ which is a proper indicator for the origin. More generally, the distance to a compact set $\mathcal{A}$ is a proper indicator for the set $\mathcal{A}$ on $\mathbb{R}^n$.

Given an open set $\mathcal{O}$, a proper indicator ω and hybrid data (f, g, C, D), a function $V : \mathcal{O} \to \mathbb{R}_{\geq 0}$ is said to be a smooth Lyapunov function for $(f, g, C, D, \omega, \mathcal{O})$ if it is smooth and there exist functions α_1, α_2 belonging to class-$\mathcal{K}_\infty$ such that

$$
\begin{aligned}
\alpha_1(\omega(x)) \leq V(x) \leq \alpha_2(\omega(x)) &\qquad \forall x \in \mathcal{O} \\
\langle \nabla V(x), f(x) \rangle \leq -V(x) &\qquad \forall x \in \overline{C} \cap \mathcal{O} \\
V(g(x)) \leq e^{-1} V(x) &\qquad \forall x \in \overline{D} \cap \mathcal{O} \,.
\end{aligned}
\tag{7}
$$

Supposing that such a function exists, it is easy to verify that every (generalized) solution to the hybrid system (f, g, C, D) starting in $\mathcal{O} \cap (\overline{C} \cup \overline{D})$ satisfies

$$
\omega(x(t,j)) \leq \alpha_1^{-1}\left(e^{-t-j}\alpha_2(\omega(x(0,0)))\right) \qquad \forall (t,j) \in \mathrm{dom}\ x \,.
$$

In particular,[4]

- (pre-stability of $\mathcal{A}$) for each $\varepsilon > 0$ there exists $\delta > 0$ such that $x(0,0) \in \mathcal{A} + \delta\mathbb{B}$ implies, for every generalized solution, that $x(t,j) \in \mathcal{A} + \varepsilon\mathbb{B}$ for all $(t,j) \in \mathrm{dom}\ x$, and
- (pre-attractivity of $\mathcal{A}$ on $\mathcal{O}$) every generalized solution starting in $\mathcal{O} \cap (\overline{C} \cup \overline{D})$ is bounded and if its time domain is unbounded then it converges to $\mathcal{A}$.

According to one of the main results in [4] *there exists a smooth Lyapunov function for $(f, g, C, D, \omega, \mathcal{O})$ if and only if the set $\mathcal{A}$ is pre-stable and pre-attractive on $\mathcal{O}$ (as defined above) and $\mathcal{O}$ is forward invariant* (i.e., $x(0,0) \in \mathcal{O} \cap (\overline{C} \cup \overline{D})$ implies $x(t,j) \in \mathcal{O}$ for all $(t,j) \in \mathrm{dom}\ x$).

One of the primary interests in converse Lyapunov theorems is that they can be used to establish robustness of asymptotic stability to various types of perturbations. Some such perturbations are recalled in the next subsection.

3.3 Robust Stability [12, 5, 3, 4]

General Observations [4]

If the function V satisfies the first inequality in (7) then, for each pair (ℓ_1, ℓ_2) with $0 < \ell_1 < \ell_2 < \infty$, the set $\{x \in \mathcal{O} : \ell_1 \leq V(x) \leq \ell_2\}$ is a compact subset of

---

[4] The prefix "pre-" is used here to distinguish from the case where it is assumed that all solutions are complete. The distinction is important, since systems that might seem to be unstable may actually be pre-stable in the sense defined below. For example, a planar linear flow map with system matrix having complex eigenvalues with positive real part is unstable as a continuous-time system but is stable as part of a hybrid system with a flow set that excludes a cone emanating from the origin and with an empty jump set. For more details, see [4].

$\mathcal{O} \backslash \mathcal{A}$. If V also satisfies the remaining inequalities in (7), perhaps with f and g replaced by a set-valued mappings F and G satisfying appropriate assumptions, then, given $0 < \ell_1 < \ell_2 < \infty$, there exists $\rho > 0$ such that

$$\max_{f \in F_\rho(x)} \langle \nabla V(x), f \rangle \leq -\tfrac{1}{2} V(x) \qquad \forall x \in C_\rho \cap \{x \in \mathcal{O} : \ell_1 \leq V(x) \leq \ell_2\}$$

$$\max_{g \in G_\rho(x)} V(g) \leq e^{-1/2} V(x) \qquad \forall x \in D_\rho \cap \{x \in \mathcal{O} : \ell_1 \leq V(x) \leq \ell_2\} \quad (8)$$

$$\max_{g \in G_\rho(x)} V(g) \leq \ell_1 \qquad \forall x \in D_\rho \cap \{x \in \mathcal{O} : V(x) \leq \ell_1\} \ .$$

where

$$F_\rho(x) := \overline{\mathrm{co}} F((x + \rho \overline{\mathbb{B}}) \cap \overline{C}) + \rho \overline{\mathbb{B}}, \ \forall x \in \mathcal{O},$$
$$G_\rho(x) := \{v \in \mathcal{O} : v \in g + \rho \overline{\mathbb{B}}, \ g \in G((x + \rho \overline{\mathbb{B}}) \cap \overline{D})\}, \ \forall x \in \mathcal{O},$$
$$C_\rho := \{x \in \mathcal{O} : (x + \rho \overline{\mathbb{B}}) \cap \overline{C} \neq \emptyset\},$$
$$D_\rho := \{x \in \mathcal{O} : (x + \rho \overline{\mathbb{B}}) \cap \overline{D} \neq \emptyset\}$$

and $\overline{\mathrm{co}}$ denotes the closed convex hull. It follows from (8) that, for the system

$$\mathcal{H}_\rho := \begin{cases} \dot{x} \in F_\rho(x) & x \in C_\rho \\ x^+ \in G_\rho(x) & x \in D_\rho \ , \end{cases} \qquad (9)$$

the compact set $\{x \in \mathcal{O} : V(x) \leq \ell_1\}$ is pre-stable and pre-attractive from the forward invariant open set $\{x \in \mathcal{O} : V(x) < \ell_2\}$. It also follows from (7) that the set $\{x \in \mathcal{O} : V(x) \leq \ell_1\}$ converges to $\mathcal{A}$ as $\ell_1 \to 0$ and the set $\{x \in \mathcal{O} : V(x) < \ell_2\}$ converges to $\mathcal{O}$ as $\ell_2 \to \infty$. Using terminology from parameterized differential equations, the set $\mathcal{A}$ is said to be *semiglobally* (with respect to $\mathcal{O}$) *practically pre-asymptotically stable in the parameter ρ*. These ideas are now applied to robustness with respect to slowly-varying, weakly-jumping parameters, to temporal regularization, and to "average dwell-time" perturbations.

3.4   Slowly-Varying, Weakly-Jumping Parameters

Consider a parameterized hybrid system

$$\mathcal{H} := \begin{cases} \dot{\xi} \ = f(\xi, p) & (\xi, p) \in C \\ \xi^+ = g(\xi, p) & (\xi, p) \in D \ , \end{cases} \qquad (10)$$

where the state is taken to be (ξ, p) and C and D are assumed to be closed. Suppose this system has the compact set $\mathcal{A}$ pre-stable and pre-attractive on $\mathbb{R}^n$. Since $\mathcal{A}$ is pre-attractive and p does not change along solutions, the parameter vector p is restricted to a compact set. Using converse Lyapunov theorems, one can establish robustness to slow variations in the parameter p and also small jumps in the parameter p, even those that are not synchronized with jumps of the hybrid system. So, for example, the original system may not have any jumps (a constrained differential equation) and yet the hybrid systems framework permits showing robustness to small jumps in the parameter.

In order to allow jumps in the parameter p that are not synchronized with the jumps of the system, add a timer state τ that limits how often these extra jumps can be inserted. The parameter $\tau^* > 0$ is used to determine this separation. It can be arbitrarily small. Then, based on the value of $\tau^* > 0$ and the data of the system, a perturbation level $\delta > 0$ will be determined for the allowed size of the jumps and variations in the parameter. To see how this works, let $\tau^* > 0$ and $\delta \geq 0$, and consider the hybrid system $\mathcal{H}_{\delta,\tau^*} := (F, G, \widetilde{C}, \widetilde{D})$ with the state $x := (\xi, p, \tau)$ where

$$\widetilde{C} := C \times \mathbb{R}_{\geq 0} , \quad \widetilde{D} := (D \times \mathbb{R}_{\geq 0}) \cup ((C \cup D) \times [\tau^*, \infty))$$

$$F(x) := \begin{bmatrix} f(\xi, p) \\ \delta\overline{\mathbb{B}} \\ 1 - \tau + \tau^* \end{bmatrix} , \quad G(x) := \begin{cases} G_1(x) & \text{for } x \in D \times [0, \tau^*) \\ G_1(x) \bigcup G_2(x) & \text{for } x \in D \times [\tau^*, \infty) \\ G_2(x) & \text{for } x \in (C \backslash D) \times [\tau^*, \infty) , \end{cases}$$

where

$$G_1(x) = \begin{bmatrix} g(\xi, p) \\ p + \delta\overline{\mathbb{B}} \\ \tau \end{bmatrix} , \quad G_2(x) = \begin{bmatrix} \xi \\ p + \delta\overline{\mathbb{B}} \\ 0 \end{bmatrix} .$$

When $\delta = 0$, the parameter p is constant along solutions and all of the solutions of (10) are enabled as the (ξ, p) component of the solution. The new enabled solutions are those containing "jumps" via $G_2(\cdot)$, but these jumps are separated by a flow with at least $\ln\left(\frac{1+\tau^*}{\tau^*}\right)$ seconds, since that is the amount of time required for $\dot{\tau} = 1 - \tau + \tau^*$ to increase from 0 to τ^*. So $\mathcal{H}_{0,\tau^*}$ has the set $\widetilde{\mathcal{A}} := \mathcal{A} \times [0, 1 + \tau^*]$ pre-stable and pre-attractive on $\mathbb{R}^{n+1}$. This system then admits a smooth Lyapunov function from which robustness of asymptotic stability for for $\delta > 0$ sufficiently small can be deduced. This corresponds to the situation where the parameter p is allowed to change slowly during flows, it is allowed to make small jumps when the hybrid system would be jumping anyway, and it is also allowed to make additional jumps when the timer τ reaches or exceeds the value τ^*.

3.5   Temporal Regularization

A hybrid arc is said to be Zeno if its hybrid time domain is bounded in the ordinary time direction. Zeno behavior is frequently encountered in hybrid or switched control systems. To eliminate Zeno behavior in applications, temporal regularization (i.e. to force the interval between jumps to be at least some amount of time) is an effective recipe. This subsection shows how to recover the result on semiglobal practical robustness under temporal regularization, reported in [12, Example 6.8], via converse Lyapunov theorems.

Suppose one is given a hybrid system $\mathcal{H} := (f, g, C, D)$ where C and D are closed and suppose that the compact set $\mathcal{A}$ is pre-stable and pre-attractive on $\mathbb{R}^n$. Now let $\delta \geq 0$ and consider a related system $\mathcal{H}_\delta := (\widetilde{f}, \widetilde{g}, C_\delta, D_\delta)$ with the state $\widetilde{x} := (x, \tau)$ where

$$\widetilde{f}(\widetilde{x}) := \begin{bmatrix} f(x) \\ 1 - \tau \end{bmatrix} , \qquad C_\delta := (C \times \mathbb{R}_{\geq 0}) \cup (\mathbb{R}^n \times [0, \delta]) ,$$

$$\widetilde{g}(\widetilde{x}) := \begin{bmatrix} g(x) \\ 0 \end{bmatrix} , \qquad D_\delta := D \times [\delta, \infty) .$$

When $\delta = 0$, flowing is possible only if $x \in C$, since $\dot{\tau} = 1 - \tau$ and the flow set for τ when $x \notin C$ is the point $\tau = 0$. Thus the x component of the solution with $\delta = 0$ matches the solution of $\mathcal{H}$ and the τ component converges to the interval $[0, 1]$. So the system $\mathcal{H}_0$ has the compact set $\widetilde{\mathcal{A}} := \mathcal{A} \times [0, 1]$ pre-stable and pre-attractive from $\mathbb{R}^{n+1}$.

When $\delta > 0$, in each hybrid time domain of each solution, each time interval is at least δ seconds long, since $\dot{\tau} \leq 1$ for all $\tau \in [0, \delta]$. In particular, Zeno solutions, if there were some, have been eliminated. Nevertheless, using converse Lyapunov theorems, it follows that the set $\widetilde{\mathcal{A}}$ is semiglobally practically pre-stable and pre-attractive with respect to δ.

3.6 Average Dwell Time

Consider the differential equation $\dot{x} = f(x)$ where the state x may contain logical modes that remain constant. Suppose that the compact set $\mathcal{A}$ is stable and attractive from $\mathbb{R}^n$, and that there is interest in injecting jumps, on occasion, through a jump inclusion $x^+ \in G(x)$ while maintaining (semiglobal practical) pre-stability and pre-attractivity. In order to achieve this goal, suppose

$$G(\mathcal{A} \cap D) \subset \mathcal{A} \tag{11}$$

and that the jumps satisfy an average dwell-time condition, as discussed in Example 2. Thus, augment the system as in Example 2 (where (z, q) of Example 2 is associated with x here.) In the special case where $\delta = 0$, at most N jumps are allowed in the time domain of a solution. It follows from (11) that these jumps do not destroy global asymptotic stability of $\mathcal{A}$ for the x component of the solution. For the composite system, we have that the set $\widetilde{\mathcal{A}} := \mathcal{A} \times [0, N]$ is pre-stable and pre-attractive from $\mathbb{R}^{n+1}$. It follows, via converse Lyapunov theorems, that this set is semiglobally practically pre-asymptotically stable in the parameter $\delta > 0$ that quantifies the average dwell-time condition.

4 Feedback Control Applications

4.1 Hybrid Stabilization Implies Input-to-State Stabilization [3, 4]

In the pioneering paper [29] it was shown, for continuous-time control systems, that smooth stabilization implies smooth input-to-state stabilization with respect to input additive disturbances. The proof relied on converse Lyapunov theorems for continuous-time systems. As shown in [4, 3], the result generalizes to hybrid control systems via the converse Lyapunov theorem mentioned earlier. In particular, if one can find a hybrid controller, with the type of regularity used throughout this paper, to achieve asymptotic stability (for example, see Sections 4.2 and 4.3) then input-to-state stability with respect to input additive disturbances can also be achieved. (The reader may also wish to compare this idea with the ideas in [21] where input-to-state stabilization is established using discontinuous, continuous-time feedback and nonsmooth Lyapunov functions.)

Here consider the special case where the hybrid controller is a logic-based controller where the logic variable takes values in a finite set. Consider the hybrid control system

$$\mathcal{H} := \begin{cases} \dot{\xi} = f_q(\xi) + \eta_q(\xi)(u_q + v_q d) & \xi \in C_q , \ q \in Q \\ \begin{bmatrix} \xi \\ q \end{bmatrix}^+ \in G_q(\xi) & \xi \in D_q , \ q \in Q \end{cases} \tag{12}$$

where Q is a finite index set, for each $q \in Q$, f_q, $\eta_q : C_q \to \mathbb{R}^n$ are continuous functions, C_q and D_q are closed and G_q has a closed graph and is locally bounded. The signal u_q is the control, and d is the disturbance, while v_q is vector that is independent of the state, input, and disturbance. Suppose $\mathcal{H}$ is stabilizable by logic-based continuous feedback; that is, for the case where $d = 0$, there exist continuous functions k_q defined on C_q such that, with $u_q := k_q(\xi)$, the nonempty and compact set $\mathcal{A} = \bigcup_{q \in Q} \mathcal{A}_q \times \{q\}$ is pre-stable and globally pre-attractive. Converse Lyapunov theorems can then be used to establish the existence of a logic-based continuous feedback that renders the closed-loop system input-to-state stable with respect to d. The feedback has the form

$$u_q = \kappa_{q,\varepsilon}(\xi) := k_q(\xi) - \varepsilon\eta_q^T(\xi)\nabla V_q(\xi)$$

where $\varepsilon > 0$ and $V_q(\xi)$ is a smooth Lyapunov function that follows from the assumed asymptotic stability when $d \equiv 0$. There exist class-$\mathcal{K}_\infty$ functions α_1 and α_2 such that, with this feedback control, the following estimate holds:

$$|\xi(t,j)|_{\mathcal{A}_{q(t,j)}} \leq$$
$$\max\left\{ \alpha_1^{-1}(2\exp(-t-j)\alpha_2(|\xi(0,0)|_{\mathcal{A}q(0,0)})), \alpha_1^{-1}\left(\tfrac{\max_{q \in Q} |v_q|^2}{2\varepsilon}||d||_\infty^2 \right) \right\}$$

where $||d||_\infty := \sup_{(s,i)\in\text{dom } d} |d(s,i)|$.

4.2   Patchy Smooth Control Lyapunov Functions [11]

While control design by means of a continuously differentiable control-Lyapunov function (clf) is well established for input-affine nonlinear control systems, it is well known that not every controllable input-affine nonlinear control system admits a continuously differentiable clf. A prominent example where no such clf exists is the so-called "Brockett" or "nonholonomic" integrator, which is described in the example below. While this system does not admit a continuously differentiable control Lyapunov function, it has been established recently that it admits a smooth "patchy" control-Lyapunov function.

The patchy clf concept, which was introduced in [11], is inspired not only by the classical clf idea but also by the approach to feedback stabilization based on patchy vector fields proposed in [1]. The patchy clf idea was conceived to overcome a limitation of discontinuous feedbacks like those coming from patchy feedbacks, which is a lack of robustness to measurement noise. In [11] it has been shown that every asymptotically controllable nonlinear system admits a smooth patchy clf if one allows for the possibility that the number of patches may need to be infinite. Moreover, it has been shown how to construct a robustly stabilizing hybrid feedback from a patchy clf. Here the idea when the number of patches is finite is outlined and then specialized to the nonholonomic integrator.

In general, a global patchy smooth control-Lyapunov function for the origin for the control system $\dot{x} = f(x, u)$ in the case of a finite number of patches is a collection of functions V_q and sets Ω_q and Ω'_q where $q \in Q := \{1, \ldots, m\}$, such that

1. for each $q \in Q$, Ω_q and Ω'_q are open and
   a) $\mathcal{O} := \mathbb{R}^n \setminus \{0\} = \bigcup_{q \in Q} \Omega_q = \bigcup_{q \in Q} \Omega'_q$
   b) for each $q \in Q$, the outward unit normal to $\partial\Omega_q$ is continuous on $\left(\partial\Omega_q \setminus \bigcup_{r>q} \Omega'_r\right) \cap \mathcal{O}$,
   c) for each $q \in Q$, $\overline{\Omega'_q} \cap \mathcal{O} \subset \Omega_q$;
2. for each $q \in Q$, V_q is a smooth function defined on a neighborhood (relative to $\mathcal{O}$) of $\overline{\Omega_q}$.
3. there exist a continuous positive definite function α and class-$\mathcal{K}_\infty$ functions $\overline{\gamma}$ and $\underline{\gamma}$ such that

   a) $\underline{\gamma}(|x|) \leq V_q(x) \leq \overline{\gamma}(|x|) \qquad \forall q \in Q , \ x \in \overline{\left(\Omega_q \setminus \bigcup_{r>q} \Omega'_r\right)} \cap \mathcal{O}$;
   b) for each $q \in Q$ and $x \in \Omega_q \setminus \bigcup_{r>q} \Omega'_r$ there exists $u_{x,q}$ such that

   $$\langle \nabla V_q(x), f(x, u_{x,q}) \rangle \leq -\alpha(x) ;$$

   c) for each $q \in Q$ and $x \in \left(\partial\Omega_q \setminus \bigcup_{r>q} \Omega'_r\right) \cap \mathcal{O}$ there exists $u_{x,q}$ such that

   $$\langle \nabla V_q(x), f(x, u_{x,q}) \rangle \leq -\alpha(x)$$
   $$\langle n_q(x), f(x, u_{x,q}) \rangle \leq -\alpha(x)$$

   where $x \mapsto n_q(x)$ denotes the outward unit normal to $\partial\Omega_q$.

From this patchy clf one can construct a robust hybrid feedback stabilizer, at least when the set $\{u : v \cdot f(x, u) \leq c\}$ is convex for each real number c and each real vector v, with the following data

$$u_q := \kappa_q(x) , \quad C_q = \overline{\left(\Omega_q \setminus \bigcup_{r>q} \Omega'_r\right)} \cap \mathcal{O}$$

where κ_q is defined on C_q, continuous and such that

$$\begin{aligned}
\langle \nabla V_q(x), f(x, \kappa_q(x)) \rangle &\leq -0.5\alpha(x) \qquad \forall x \in C_q \\
\langle n_q(x), f(x, \kappa_q(x)) \rangle &\leq -0.5\alpha(x) \qquad \forall x \in \left(\partial\Omega_q \setminus \bigcup_{r>q} \Omega'_r\right) \cap \mathcal{O} .
\end{aligned} \qquad (13)$$

The jump set is given by

$$D_q = (\mathcal{O}\backslash\Omega_q) \cup \left(\bigcup_{r>q} \overline{\Omega_r'} \cap \mathcal{O}\right)$$

and the jump map is

$$G_q(x) = \begin{cases} \{r \in Q : x \in \overline{\Omega_r'} \cap \mathcal{O} , \ r > q\} & x \in \left(\bigcup_{r>q} \overline{\Omega_r'} \cap \mathcal{O}\right) \cap \Omega_q \\ \{r \in Q : x \in \overline{\Omega_r'} \cap \mathcal{O}\} & x \in \mathcal{O}\backslash\Omega_q \ . \end{cases}$$

With this control, the index increases at each jump except possibly the first one. Thus, the number of jumps is finite (possibly zero), and the state converges to the origin, which is also stable.

Example 8 (Nonholonomic integrator). A patchy smooth control-Lyapunov function for the nonholonomic integrator system

$$\left.\begin{array}{l} \dot{x}_a = u_a \\ \dot{x}_b = u_b \\ \dot{x}_c = x_a u_b - x_b u_a \end{array}\right\} =: B(x)u$$

will be demonstrated. Only two patches are needed, as indicated in Figure 10. Pick ρ so that $\rho > 1$ and $\rho + 0.5\rho^{1/2} < 2$. Then define

$$\begin{array}{ll} V_1(x) := \left(\rho^{1/2} + 0.5\right)|x_c|^{1/2} - x_a \qquad & V_2(x) := 0.5x^T x \\ \Omega_1 \ := \mathbb{R}^3\backslash\{0\} & \Omega_2 \ := \{x : x_a^2 + x_b^2 > |x_c|\} \\ \Omega_1' := \mathbb{R}^3\backslash\{0\} & \Omega_2' := \{x : x_a^2 + x_b^2 > \rho|x_c|\} \ . \end{array}$$

Note that V_1 is smooth on $\Omega_1\backslash\Omega_2'$ and, with $u_1(x) = \begin{bmatrix} 1 \ 0 \end{bmatrix}^T$, for all $x \in \Omega_1\backslash\Omega_2'$, it follows that

$$\begin{aligned} \langle \nabla V_1(x), B(x)u_1 \rangle &= -0.5\left(\rho^{1/2} + 0.5\right)|x_c|^{-1/2}x_b - 1 \\ &\leq 0.5\left(\rho^{1/2} + 0.5\right)\rho^{1/2} - 1 \\ &< 0 \ . \end{aligned}$$

In addition, V_2 is smooth and, with

$$u_2(x) = -\begin{bmatrix} x_a \\ x_b \end{bmatrix} + \frac{3x_c}{x_a^2 + x_b^2}\begin{bmatrix} x_b \\ -x_a \end{bmatrix}$$

it follows that (cf. (13))

$$\begin{array}{ll} \langle \nabla V_2(x), B(x)u_2(x) \rangle = -x_a^2 - x_b^2 - 3x_c^2 \qquad & x \in \overline{\Omega_2} \cap \mathcal{O} \\ \langle n_2(x), B(x)u_2(x) \rangle < 0 & x \in \partial\Omega_2 \cap \mathcal{O} \end{array}$$

where $x \mapsto n_2(x)$ denotes the outward unit normal to $\partial\Omega_2$. The resulting robustly stabilizing hybrid controller is $u := u_q(x)$, $C_1 := \{x : x_a^2 + x_b^2 \leq \rho|x_c|\}$, $C_2 := \{x : x_a^2 + x_b^2 \geq |x_c|\}$, $D_1 = \overline{\mathbb{R}^3\backslash C_1}$, $D_2 : \overline{\mathbb{R}^3\backslash C_2}$, and $g_q = \text{toggle}(q)$. This is similar to the controller proposed in [17]; see also [10]. ∎

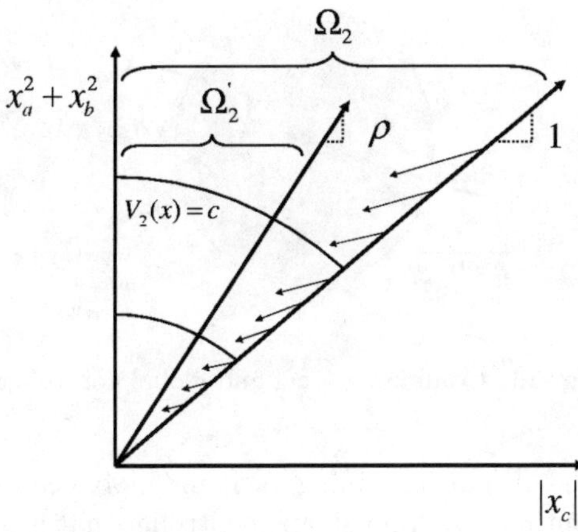

Fig. 10. Smooth patchy clf for nonholonomic integrator

4.3 "Throw-and-Catch" Control

Hysteresis Switching Between Global and Local Stabilizers

In the paper [22], it was shown how to combine local and global state feedbacks to achieve global stabilization and local performance. The idea, which exploits hysteresis switching, is quite simple. Two continuous functions, κ_{global} and κ_{local}, are given where the feedback $u = \kappa_{global}(x)$ renders the origin of the control system $\dot{x} = f(x, u)$ globally asymptotically stable whereas the feedback $u = \kappa_{local}(x)$ renders the origin of the control system locally asymptotically stable with basin of attraction containing the open set $\mathcal{O}$, which contains the origin. Then one takes C_{local} to be a compact subset of the $\mathcal{O}$ that contains the origin in its interior and one takes D_{global} to be a compact subset of C_{local}, again containing the origin in its interior and such that, when using the controller κ_{local}, trajectories starting in D_{global} never reach the boundary of C_{local}. See Figure 11. Finally, the hybrid control that achieves global asymptotic stabilization while using the controller κ_2 for small signals is the following: Define $D_{local} := \overline{\mathbb{R}^n \backslash C_{local}}$ and $C_{global} := \overline{\mathbb{R}^n \backslash D_{global}}$ and

$$u := \kappa_q(x) \qquad C := \{(x, q) : x \in C_q\}$$
$$g(q, x) := \text{toggle}(q) \qquad D := \{(x, q) : x \in D_q\} \ .$$

Hysteresis Switching Between a Global Controller, Local Stabilizers at Different Locations, and Open-Loop Controls [28]

In the problem of uniting local and global controllers, one can view the global controller as a type of "bootstrap" controller that is guaranteed to bring the system to a region where another controller can control the system adequately.

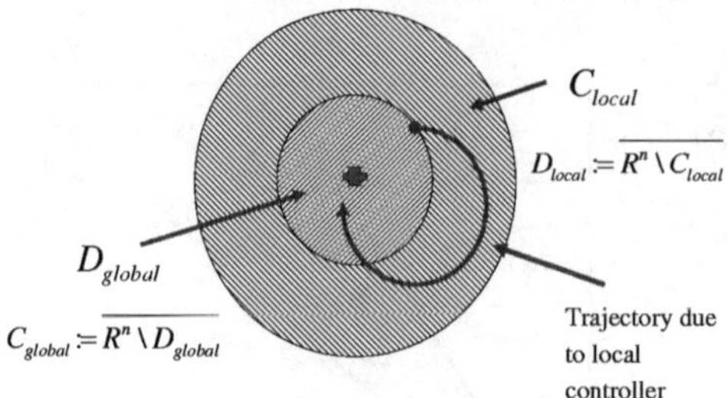

Fig. 11. Combining local and global controllers

An extension of the idea of combining local and global controllers is to assume the existence of a continuous, "bootstrap" controller that is guaranteed to bring the system, in finite time, into a neighborhood of a set of points, not just a neighborhood of the desired final destination (the controller doesn't need to be able to keep the state in this neighborhood); moreover, these sets of points form chains that terminate at the desired final destination and along which controls are known to steer (or "throw") from one point in the chain to the next point in the chain. Moreover, in order to minimize the propagation error along a chain, a local stabilizer is known for each point, except perhaps those points at the start of a chain. These can be used to "catch" each throw. To make this idea clearer, consider the problem of globally asymptotically stabilizing the upright position for the pendubot system.

Example 9 (Pendubot). The "pendubot", introduced in [30], corresponds to a two-link arm where a control torque τ can be applied at the end of the first link (shoulder) while the second link is free to rotate at the other end of the first link (elbow). One can think of the state of the pendubot system as evolving on $(S_1 \times \mathbb{R})^2$ where S_1 denotes the unit circle in the plane, as depicted in Figure 12. The goal here is to stabilize the position corresponding to both links oriented upward, with zero velocity.

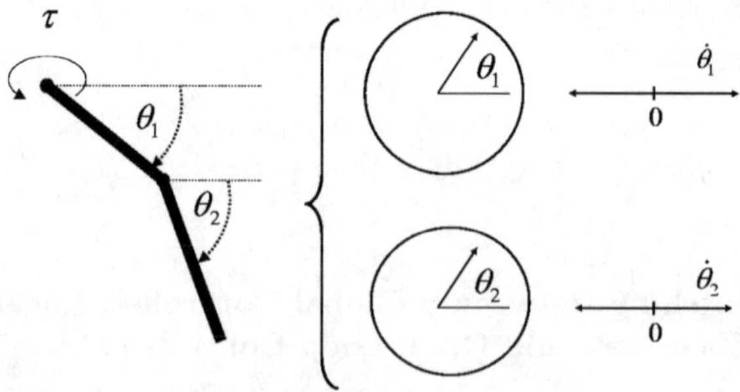

Fig. 12. Pendubot: diagram (left) and state space (right)

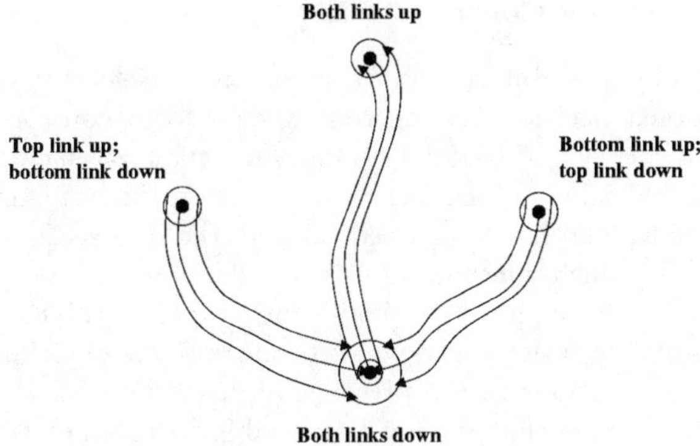

Fig. 13. Throw-and-catch control for the pendubot

The pendubot has four open-loop equilibria, corresponding to both links oriented in the up position, both links oriented in the down position, and two additional equilibria where one link is up and the other is down. The "bootstrap" controller for this system is of the "$-L_gV$" type where V is the kinetic plus potential energy of the system. This feedback causes every initial condition to eventually reach a neighborhood of one of the four open-loop equilibrium in finite time, but not necessarily stay there. For example, only the straight down equilibrium point is stable under "$-LgV$" control. This bootstrap controller thus brings the state of the system close to a point from which signals are known to steer to the desired final destination by following a "chain" of points. See Figure 13. Indeed, it is not difficult to find control signals to steer the system from either of the "mixed" equilibrium to the down equilibrium, and to find a third control signal to steer the system from the equilibrium with both links down to the equilibrium with both links up. Moreover, one can construct local asymptotic stabilizers for the point corresponding to both links oriented downward and for the point corresponding to both links oriented upward. The latter is useful for when the state of the system gets close to the desired final destination. The former is useful for "catching" the throw from a mixed equilibrium point to the down equilibrium point, helping to ensure that the subsequent throw to the up position is successful. See Figure 13 for a graphical depiction. These simple ingredients can be used to build a hybrid feedback controller that globally asymptotically stabilizes the up equilibrium for the pendubot. The hybrid control involves logic states to keep track of when the bootstrap controller should be used, when a control signal to "throw" from near one point to another point in a chain can be used, and when a particular local stabilizer should be used to "catch" a throw. It also involves a timer that is used to keep track of when a control signal that throws the state from one point to another should have completed its task. If this timer runs out before a catch is executed, the controller defaults to the "bootstrap" controller. The details of the control construction are beyond the scope of this paper, but can be found in [28]. ■

4.4 Decision-Making Control [31, 27]

The problem of global stabilization of a point on a circle provides a canonical example of a robust decision-making requirement for a control system. Other similar examples correspond to determining, for a planar vehicle, whether to go to the left or to the right of a moving obstacle, or on the occurrence of a traffic light turning yellow, whether to proceed through the intersection or stop before the intersection. The introduction of hysteresis decision making through hybrid feedback control has been shown to significantly increase robustness to noise in [31]. Another similar situation corresponds to the problem of picking between two possible planar destinations in the presence of drift. This situation is depicted in Figure 14 and reveals that it may be possible to produce trajectories that correspond to several changes in the target destination over time, even in the absence of disturbances, but with a guarantee that a decision is eventually made and convergence is achieved.

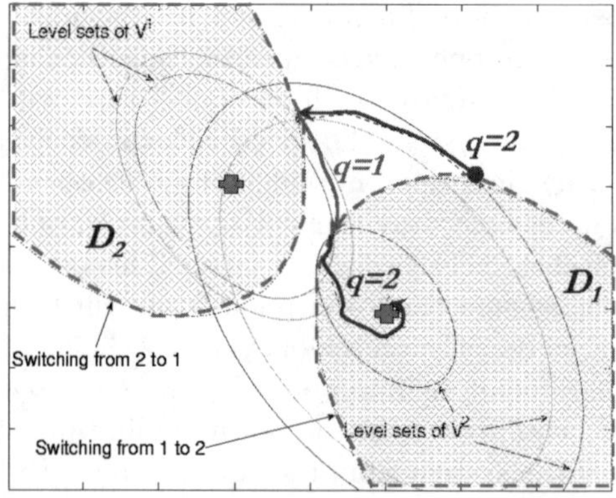

Fig. 14. Robust stabilization of a disconnected set of points

The behavior in Figure 14 is in contrast to the situation that could occur when using a discontinuous feedback. In that case, the boundaries of D_1 and D_2 would coalesce into a single decision line that would necessarily contain a Filippov equilibrium point, i.e., a point at which arbitrarily small measurement noise could keep the state of the system. For more details, see [27].

5 Conclusion

When it comes to stabilization with some guaranteed level of robustness, there are some problems that cannot be solved with ordinary (possibly discontinuous) feedback but can be solved with hybrid feedback. In order to grow familiar with such control methodologies, it is important to gain a firm understanding of the behavior of hybrid systems and the basic convergence and stability properties that are available for them. In this particular paper, some of the basic results have been given along with references to where more details can be found in the literature.

Acknowledgments

The results summarized in this paper are due to the hard work of my students, Ricardo Sanfelice, Chaohong Cai, Emre Tuna, and Michael Messina, and my colleagues Rafal Goebel and Christophe Prieur. Much of this work has been supported in part by the Air Force Office of Scientific Research under grant numbers F49620-03-1-0203 and F9550-06-1-0134, Army Research Office under grant DAAD19-03-1-0144, and the National Science Foundation under Grant numbers CCR-0311084 and ECS-0324679.

References

1. Ancona F, Bressan A (1999) Patchy vector fields and asymptotic stabilization, ESAIM: Control, Optimisation and Calculus of Variations, 4:445–471
2. Byrnes C I, Martin C F (1995) An integral-invariance principle for nonlinear systems, IEEE Transactions on Automatic Control, 40:983–994
3. Cai C, Teel A R, Goebel R (2007) Results on existence of smooth Lyapunov functions for asymptotically stable hybrid systems with nonopen basin of attraction, submitted to the 2007 American Control Conference
4. Cai C, Teel A R, Goebel R (2006) Smooth Lyapunov functions for hybrid systems Part I: Existence is equivalent to robustness, Part II: (Pre-)asymptotically stable compact sets, submitted
5. Cai C, Teel A R, Goebel R (2005) Converse Lyapunov theorems and robust asymptotic stability for hybrid systems, Proceedings of 24th American Control Conference, 12–17
6. Chellaboina V, Bhat S P, Haddad W H (2003) An invariance principle for nonlinear hybrid and impulsive dynamical systems, Nonlinear Analysis, 53:527–550
7. Collins P (2004) A trajectory-space approach to hybrid systems, Proceedings of 16th International Symposium on Mathematical Theory of Networks and Systems
8. Coron J M, Rosier L (1994) A relation between continuous time-varying and discontinuous feedback stabilization, Journal of Mathematical Systems Estimation and Control , 4:67–84
9. Filippov A F (1988) Differential Equations with Discontinuous Righthand Sides. Kluwer
10. Goebel R, Hespanha J, Teel A R, Cai C, Sanfelice R G (2004) Hybrid systems: generalized solutions and robust stability, Proceedings of 6th IFAC Symposium on Nonlinear Control Systems
11. Goebel R, Prieur C, Teel A R (2006) smooth patchy control Lyapunov functions, Proceedings of 45th IEEE Conference on Decision and Control
12. Goebel R, Teel A R (2006) Solutions to hybrid inclusions via set and graphical convergence with stability theory applications, Automatica, 42:573–587
13. Hájek O (1979) Discontinuous differential equations I, Journal Differential Equations, 32:149–170

14. Hermes H (1967) Discontinuous vector fields and feedback control, in Differential Equations and Dynamical Systems, Academic Press, 155–165
15. Hespanha J P, Liberzon D, Teel A R (2005) On input-to-state stability of impulsive systems, Proceedings of 44th IEEE Conference on Decision and Control, 3992–3997
16. Hespanha J P, Morse A S (1999) Stability of switched systems with average dwell-time, Proceedings of 38th IEEE Conference on Decision and Control, 2655–2660
17. Hespanha J P, Morse A S (1999) Stabilization of nonholonomic integrators via logic-based switching, Automatica, 35:385–393
18. LaSalle J P (1967) An invariance principle in the theory of stability, in Differential equations and dynamical systems, Academic Press, New York
19. LaSalle J P (1976) The stability of dynamical systems, Regional Conference Series in Applied Mathematics, SIAM
20. Lygeros J, Johansson K H, Simić S N, Zhang J, Sastry S S (2003) Dynamical properties of hybrid automata, IEEE Transactions on Automatic Control, 48:2–17
21. Malisoff M, Rifford L, Sontag E D (2004) Global asymptotic controllability implies input-to-state stabilization, SIAM Journal on Control and Optimization, 42:2221–2238
22. Prieur C (2001) Uniting local and global controllers with robustness to vanishing noise, Mathematics Control Signals Systems, 14:143–172
23. Rockafellar R T, Wets R J B (1998) Variational Analysis. Springer Verlag
24. Ryan E P (1998) An integral invariance principle for differential inclusions with applications in adaptive control, SIAM Journal on Control and Optimization, 36:960–980
25. Sanfelice R G, Goebel R, Teel A R (2005) Results on convergence in hybrid systems via detectability and an invariance principle, Proceedings of 2005 American Control Conference, 551–556
26. Sanfelice R G, Goebel R, Teel A R (2006) A feedback control motivation for generalized solutions to hybrid systems, In J. P. Hespanha and A. Tiwari, editors, Hybrid Systems: Computation and Control: 9th International Workshop, volume LNCS 3927, 522–536
27. Sanfelice R G, Messina M J, Tuna S E, Teel A R (2006) Robust hybrid controllers for continuous-time systems with applications to obstacle avoidance and regulation to disconnected set of points, Proceedings of 2006 American Control Conference, 3352–3357
28. Sanfelice R G, Teel A R (2007) A "throw-and-catch" hybrid control strategy for robust stabilization of nonlinear systems, submitted to the 2007 American Control Conference.
29. Sontag E (1989) Smooth stabilization implies coprime factorization, IEEE Transactions on Automatic Control, 34:435–443
30. Spong M W, Block D J (1995) The pendubot: A mechatronic system for control research and education, Proceedings of 34th Conference on Decision and Control, 555–556
31. Tuna S E, Sanfelice R G, Messina M J, Teel A R (2005) Hybrid MPC: Open-minded but not easily swayed, in L. Biegler R. Findeisen, F. Allgower, editor, Preprints of the International Workshop on Assessment and Future Directions of Nonlinear Model Predictive Control, Freudenstadt-Lauterbad, Germany, 169–180
32. Ye H, Michel A N, Hou L (1998) Stability theory for hybrid dynamical systems. IEEE Transactions on Automatic Control, 43:461–474

List of Contributors

Frank Allgöwer
Institute for Systems Theory and
Automatic Control,
University of Stuttgart,
Germany
allgower@ist.uni-stuttgart.de

Tamer Başar
University of Illinois at Urbana-
Champaign, Urbana,
USA
tbasar@control.csl.uiuc.edu

Manuel Béjar
Universidad Pablo de Olavide,
Spain
mbejdom@upo.es

Claudio Bonivento
Center for Research on Complex
Automated Systems, DEIS,
Università di Bologna,
Italy
cbonivento@deis.unibo.it

Marco C. Campi
Dipartimento di Elettronica per
l'Automazione,
Università di Brescia,
Italy
campi@ing.unibs.it

Ruggero Carli
Department of Information Engineer-
ing,
Università di Padova,
Italy
calirug@dei.unipd.it

Federico Cuesta
Universidad de Sevilla,
Spain
fede@cartuja.us.es

Agostino De Santis
Dipartimento di Informatica e
Sistemistica,
Università di Napoli Federico II,
Italy
agodesa@unina.it

Christian Ebenbauer
Laboratory for Information and
Decision Systems,
Massachusetts Institute of Technology,
USA
ebenbauer@mit.edu

Simone Garatti
Dipartimento di Elettronica e
Informatica,
Politecnico di Milano,
Italy
sgaratti@elet.polimi.it

Luca Gentili
Center for Research on Complex
Automated Systems, DEIS,
Università di Bologna,
Italy
lgentili@deis.unibo.it

Alessandro Giua
Dipartimento di Ingegneria Elettrica
ed Elettronica,
Università di Cagliari,
Italy
giua@diee.unica.it

Orhan C. Imer
General Electric Global Research
Center,
Niskayuna, NY,
USA
imer@research.ge.com

Alberto Isidori
Dipartimento di Informatica e
Sistemistica,
Sapienza - Università di Roma,
Italy,
isidori@ese.wustl.edu

Arthur J. Krener
University of California,
CA, USA,
Naval Postgraduate School,
CA, USA
ajkrener@ucdavis.edu

Stéphane Lafortune
Department of Electrical Engineering
and Computer Science,
University of Michigan, Ann Arbor,
Michigan,
USA
stephane@eecs.umich.edu

Vincenzo Lippiello
Dipartimento di Informatica e
Sistemistica,
Università di Napoli Federico II,
Italy
vincenzo.lippiello@unina.it

Lorenzo Marconi
Center for Research on Complex
Automated Systems, DEIS,
Università di Bologna,
Italy
lmarconi@deis.unibo.it

Anibal Ollero
Universidad de Sevilla,
Spain
aollero@cartuja.us.es

Andrea Paoli
Center for Research on Complex
Automated Systems, DEIS,
Università di Bologna,
Italy
apaoli@deis.unibo.it

Maria Prandini
Dipartimento di Elettronica e
Informatica,
Politecnico di Milano,
Italy
prandini@elet.polimi.it

Carlo Rossi
Center for Research on Complex
Automated Systems, DEIS,
Università di Bologna,
Italy
crossi@deis.unibo.it

Kurt Schlacher
Institute of Automatic Control and
Control Systems Technology,
Johannes Kepler University, Linz,
Austria
kurt.schlacher@jku.at

Carla Seatzu
Dipartimento di Ingegneria Elettrica
ed Elettronica,
Università di Cagliari,
Italy
seatzu@diee.unica.it

Bruno Siciliano
Dipartimento di Informatica e
Sistemistica,
Università di Napoli Federico II,
Italy
siciliano@unina.it

Marcel Staroswiecki
SATIE UMR,
Ecole Normale Supérieure de Cachan,
France
marcel.staroswiecki@
univ-lille1.fr

Andrew R. Teel
Department of Electrical and
Computer Engineering,
University of California,
USA
teel@ece.ucsb.edu

Andrea Tilli
Center for Research on Complex
Automated Systems, DEIS,
Università di Bologna,
Italy
atilli@deis.unibo.it

Manuel Toniato
Center for Research on Complex
Automated Systems, DEIS,
Università di Bologna,
Italy
mtoniato@deis.unibo.it

Luigi Villani
Dipartimento di Informatica e
Sistemistica,
Università di Napoli Federico II,
Italy
lvillani@unina.it

Sandro Zampieri
Department of Information Engineer-
ing,
Università di Padova,
Italy
zampi@dei.unipd.it

Printing: Mercedes-Druck, Berlin
Binding: Stein+Lehmann, Berlin

Lecture Notes in Control and Information Sciences

Edited by M. Thoma, M. Morari

Further volumes of this series can be found on our homepage:
springer.com

Vol. 353: Bonivento C.; Isidori A.; Marconi L.;
Rossi C. (Eds.)
Advances in Control Theory and Applications
305 p. 2007 [978-3-540-70700-4]

Vol. 352: Chiasson, J.; Loiseau, J.J. (Eds.)
Applications of Time Delay Systems
358 p. 2007 [978-3-540-49555-0]

Vol. 351: Lin, C.; Wang, Q.-G.; Lee, T.H., He, Y.
LMI Approach to Analysis and Control of
Takagi-Sugeno Fuzzy Systems with Time Delay
204 p. 2007 [978-3-540-49552-9]

Vol. 350: Bandyopadhyay, B.; Manjunath, T.C.;
Umapathy, M.
Modeling, Control and Implementation of Smart
Structures 250 p. 2007 [978-3-540-48393-9]

Vol. 349: Rogers, E.T.A.; Galkowski, K.;
Owens, D.H.
Control Systems Theory
and Applications for Linear
Repetitive Processes 482 p. 2007 [978-3-540-
42663-9]

Vol. 347: Assawinchaichote, W.; Nguang, K.S.;
Shi P.
Fuzzy Control and Filter Design
for Uncertain Fuzzy Systems
188 p. 2006 [978-3-540-37011-6]

Vol. 346: Tarbouriech, S.; Garcia, G.; Glattfelder,
A.H. (Eds.)
Advanced Strategies in Control Systems
with Input and Output Constraints
480 p. 2006 [978-3-540-37009-3]

Vol. 345: Huang, D.-S.; Li, K.; Irwin, G.W. (Eds.)
Intelligent Computing in Signal Processing
and Pattern Recognition
1179 p. 2006 [978-3-540-37257-8]

Vol. 344: Huang, D.-S.; Li, K.; Irwin, G.W. (Eds.)
Intelligent Control and Automation
1121 p. 2006 [978-3-540-37255-4]

Vol. 341: Commault, C.; Marchand, N. (Eds.)
Positive Systems
448 p. 2006 [978-3-540-34771-2]

Vol. 340: Diehl, M.; Mombaur, K. (Eds.)
Fast Motions in Biomechanics and Robotics
500 p. 2006 [978-3-540-36118-3]

Vol. 339: Alamir, M.
Stabilization of Nonlinear Systems Using
Receding-horizon Control Schemes
325 p. 2006 [978-1-84628-470-0]

Vol. 338: Tokarzewski, J.
Finite Zeros in Discrete Time Control Systems
325 p. 2006 [978-3-540-33464-4]

Vol. 337: Blom, H.; Lygeros, J. (Eds.)
Stochastic Hybrid Systems
395 p. 2006 [978-3-540-33466-8]

Vol. 336: Pettersen, K.Y.; Gravdahl, J.T.;
Nijmeijer, H. (Eds.)
Group Coordination and Cooperative Control
310 p. 2006 [978-3-540-33468-2]

Vol. 335: Kozłowski, K. (Ed.)
Robot Motion and Control
424 p. 2006 [978-1-84628-404-5]

Vol. 334: Edwards, C.; Fossas Colet, E.;
Fridman, L. (Eds.)
Advances in Variable Structure and Sliding Mode
Control
504 p. 2006 [978-3-540-32800-1]

Vol. 333: Banavar, R.N.; Sankaranarayanan, V.
Switched Finite Time Control of a Class of
Underactuated Systems
99 p. 2006 [978-3-540-32799-8]

Vol. 332: Xu, S.; Lam, J.
Robust Control and Filtering of Singular Systems
234 p. 2006 [978-3-540-32797-4]

Vol. 331: Antsaklis, P.J.; Tabuada, P. (Eds.)
Networked Embedded Sensing and Control
367 p. 2006 [978-3-540-32794-3]

Vol. 330: Koumoutsakos, P.; Mezic, I. (Eds.)
Control of Fluid Flow
200 p. 2006 [978-3-540-25140-8]

Vol. 329: Francis, B.A.; Smith, M.C.; Willems,
J.C. (Eds.)
Control of Uncertain Systems: Modelling,
Approximation, and Design
429 p. 2006 [978-3-540-31754-8]

Vol. 328: Loría, A.; Lamnabhi-Lagarrigue, F.;
Panteley, E. (Eds.)
Advanced Topics in Control Systems Theory
305 p. 2006 [978-1-84628-313-0]

Vol. 327: Fournier, J.-D.; Grimm, J.; Leblond, J.;
Partington, J.R. (Eds.)
Harmonic Analysis and Rational Approximation
301 p. 2006 [978-3-540-30922-2]

Vol. 326: Wang, H.-S.; Yung, C.-F.; Chang, F.-R.
H_∞ Control for Nonlinear Descriptor Systems
164 p. 2006 [978-1-84628-289-8]

Vol. 325: Amato, F.
Robust Control of Linear Systems Subject to Uncertain
Time-Varying Parameters
180 p. 2006 [978-3-540-23950-5]

Vol. 324: Christofides, P.; El-Farra, N.
Control of Nonlinear and Hybrid Process Systems
446 p. 2005 [978-3-540-28456-7]

Vol. 323: Bandyopadhyay, B.; Janardhanan, S.
Discrete-time Sliding Mode Control
147 p. 2005 [978-3-540-28140-5]

Vol. 322: Meurer, T.; Graichen, K.; Gilles, E.D. (Eds.)
Control and Observer Design for Nonlinear Finite and Infinite Dimensional Systems
422 p. 2005 [978-3-540-27938-9]

Vol. 321: Dayawansa, W.P.; Lindquist, A.; Zhou, Y. (Eds.)
New Directions and Applications in Control Theory
400 p. 2005 [978-3-540-23953-6]

Vol. 320: Steffen, T.
Control Reconfiguration of Dynamical Systems
290 p. 2005 [978-3-540-25730-1]

Vol. 319: Hofbaur, M.W.
Hybrid Estimation of Complex Systems
148 p. 2005 [978-3-540-25727-1]

Vol. 318: Gershon, E.; Shaked, U.; Yaesh, I.
H_∞ Control and Estimation of State-multiplicative Linear Systems
256 p. 2005 [978-1-85233-997-5]

Vol. 317: Ma, C.; Wonham, M.
Nonblocking Supervisory Control of State Tree Structures
208 p. 2005 [978-3-540-25069-2]

Vol. 316: Patel, R.V.; Shadpey, F.
Control of Redundant Robot Manipulators
224 p. 2005 [978-3-540-25071-5]

Vol. 315: Herbordt, W.
Sound Capture for Human/Machine Interfaces:
Practical Aspects of Microphone Array Signal Processing
286 p. 2005 [978-3-540-23954-3]

Vol. 314: Gil', M.I.
Explicit Stability Conditions for Continuous Systems
193 p. 2005 [978-3-540-23984-0]

Vol. 313: Li, Z.; Soh, Y.; Wen, C.
Switched and Impulsive Systems
277 p. 2005 [978-3-540-23952-9]

Vol. 312: Henrion, D.; Garulli, A. (Eds.)
Positive Polynomials in Control
313 p. 2005 [978-3-540-23948-2]

Vol. 311: Lamnabhi-Lagarrigue, F.; Loría, A.; Panteley, E. (Eds.)
Advanced Topics in Control Systems Theory
294 p. 2005 [978-1-85233-923-4]

Vol. 310: Janczak, A.
Identification of Nonlinear Systems Using Neural Networks and Polynomial Models
197 p. 2005 [978-3-540-23185-1]

Vol. 309: Kumar, V.; Leonard, N.; Morse, A.S. (Eds.)
Cooperative Control
301 p. 2005 [978-3-540-22861-5]

Vol. 308: Tarbouriech, S.; Abdallah, C.T.; Chiasson, J. (Eds.)
Advances in Communication Control Networks
358 p. 2005 [978-3-540-22819-6]

Vol. 307: Kwon, S.J.; Chung, W.K.
Perturbation Compensator based Robust Tracking Control and State Estimation of Mechanical Systems
158 p. 2004 [978-3-540-22077-0]

Vol. 306: Bien, Z.Z.; Stefanov, D. (Eds.)
Advances in Rehabilitation
472 p. 2004 [978-3-540-21986-6]

Vol. 305: Nebylov, A.
Ensuring Control Accuracy
256 p. 2004 [978-3-540-21876-0]

Vol. 304: Margaris, N.I.
Theory of the Non-linear Analog Phase Locked Loop
303 p. 2004 [978-3-540-21339-0]

Vol. 303: Mahmoud, M.S.
Resilient Control of Uncertain Dynamical Systems
278 p. 2004 [978-3-540-21351-2]

Vol. 302: Filatov, N.M.; Unbehauen, H.
Adaptive Dual Control: Theory and Applications
237 p. 2004 [978-3-540-21373-4]

Vol. 301: de Queiroz, M.; Malisoff, M.; Wolenski, P. (Eds.)
Optimal Control, Stabilization and Nonsmooth Analysis
373 p. 2004 [978-3-540-21330-7]

Vol. 300: Nakamura, M.; Goto, S.; Kyura, N.; Zhang, T.
Mechatronic Servo System Control
Problems in Industries and their Theoretical Solutions
212 p. 2004 [978-3-540-21096-2]

Vol. 299: Tarn, T.-J.; Chen, S.-B.; Zhou, C. (Eds.)
Robotic Welding, Intelligence and Automation
214 p. 2004 [978-3-540-20804-4]

Vol. 298: Choi, Y.; Chung, W.K.
PID Trajectory Tracking Control for Mechanical Systems
127 p. 2004 [978-3-540-20567-8]